The Conservation of Plant Biodiversity takes an evolutionary perspective to the [illegible] of plant biodiversity, stressing the need to explore both curr[illegible]m issues. It highlights three conservation strategies: com[illegible]ed in nature reserves; individual species conserved in nature [illegible] botanic gardens; and domesticated plants, preserved mainly in germplasm collections. Because conservation biology has such a broad scope and relevance to the community at large, it is faced with several controversial issues such as:

- The dichotomy between the preservation of individual species versus a broader focus on the ecosystem
- The relative importance to give to endangered species
- The design and management of reserves and the value of corridors
- The drive for increasing agricultural productivity through plant improvement versus the drive to maintain traditional peasant varieties in cultivation.

Here, these and other issues are examined, by emphasizing and exploring the underlying scientific principles. The conservation of whole communities emerges as the paramount strategy for maintaining the evolutionary potential of plant life.

This book will be required reading for all those interested in conservation and the environment.

THE CONSERVATION OF PLANT BIODIVERSITY

THE CONSERVATION OF PLANT BIODIVERSITY

OTTO H. FRANKEL
ANTHONY H. D. BROWN
and
JEREMY J. BURDON

Published by the Press Syndicate of the University of Cambridge
The Pitt Building, Trumpington Street, Cambridge CB2 1RP
40 West 20th Street, New York, NY 10011-4211, USA
10 Stamford Road, Oakleigh, Melbourne 3166, Australia

First published 1995

Printed in Great Britain at the University Press, Cambridge

A catalogue record for this book is available from the British Library

Library of Congress cataloguing in publication data

Frankel, O. H. (Otto Herzberg), 1900–
The conservation of plant biodiversity/Otto H. Frankel, Anthony
H. D. Brown, and Jeremy J. Burdon.
p. cm.
Includes bibliographical references (p.) and index.
1. Plant conservation. 2. Germplasm resources, Plant.
3. Biological diversity conservation. I. Brown, A. H. D.
II. Burdon, J. J. (Jeremy James) III. Title.
QK86.A1F73 1995
639.9'9–dc20 95-6492 CIP

ISBN 0 521 46165 0 hardback
ISBN 0 521 46731 4 paperback

SE

Contents

Preface

This book is the successor to *Conservation and evolution* by O. H. Frankel and M. E. Soulé, published in 1981. That book helped to generate an understanding of and concern for biological diversity as an essential condition for the continuing existence of life on earth.

Since then biodiversity has become a world issue, recognized by politicians and publicists, by non-government organizations and by concerned people. It became linked to the problems of the environment, and the Convention on Biological Diversity forms a major achievement of the United Nations Conference on the Environment and Development at Rio de Janeiro in 1992.

Biodiversity started as a scientific issue, relating genetic diversity to the long-term survival and continuing evolution of plants and animals. It has become a widely acknowledged human preoccupation, touching, and being recognized by all concerned with global welfare and the human environment. Its relevance extends far into the future.

In all the writing that biological diversity and its precarious state have generated, a zoological bias is often apparent. Unlike its predecessor, the focus of this book is deliberately restricted to plants, though animals take their place in the chapters on the biology and conservation of communities. We thus explore and give emphasis to components of ecosystems which have a secondary role in the mostly zoologically-orientated literature. Plants of economic value, including the crops that sustain our own species, receive prominence, as do endangered species whose continued existence is in jeopardy.

The book is the result of joint planning, and all chapters bear the marks of consultation, critical exposure and constructive comment. However,

Chapters 1, 3, 4 and 7 are mainly the work of O.H.F., Chapters 2, 5 and 6 that of A.H.D.B. and Chapters 8 to 10 that of J.J.B.

We acknowledge with gratitude our many colleagues who have provided much help and encouragement. We are particularly grateful to Lloyd Evans, for his continuing support and help; to Curt Brubaker, William Hartley, Peter Kanowski, Gurdev Khush, Don Marshall, Eric Roberts, Lindsey Withers, Doug Yen and Andrew Young for their assistance and critical comments; to Mona Brown for compiling the list of references; and to our respective families.

September 1994

Otto H. Frankel
Division of Plant Industry, CSIRO, Canberra

Anthony H. D. Brown and Jeremy J. Burdon
Centre for Plant Biodiversity Research,
Division of Plant Industry, CSIRO, Canberra

1

The biological system of conservation

1.1 DIVERSITY PERCEIVED

The all-pervading diversity of living beings will have been part of the human consciousness since our ancestors learned to distinguish between food and foe among the animals and plants around them. Observation and experience of diversity led to the recognition of discrete entities, arrayed by Linnaeus in a hierarchy of species, genera, families and beyond. It also resulted in the recognition of diversity within species, and in the selection of superior strains of animals and plants, going back to antiquity, but greatly developed since the eighteenth century.

Darwin recognized the dynamics of diversity, and Johannsen discovered the boundaries and the interactions between heredity and environment, confirming the integrity of genetic diversity irrespective of environmental diversity. Explorers, from Christopher Columbus to Sir Joseph Banks, opened new horizons of the world's diversity. For gain or for knowledge, plant collectors throughout the nineteenth century scoured the earth for new plants to record, to use, or to enjoy. Early this century, Vavilov, the great scientist-explorer, identified the geographical centres of diversity where most of the present-day crop species originated and diversified. These discoveries of genetic reservoirs, coinciding with the rapid development of genetics and its application in plant breeding, stimulated widespread interest in the study of diversity in crop species.

1.2 DIVERSITY COMING UNDER THREAT

While these discoveries opened up global perspectives of diversity, observations on a local or regional scale revealed that the diversity within natural communities was coming under threat. In Britain, where industrialization had started earlier than in other European countries, by the end of the nineteenth century vegetation changes were beginning to be recognized. Vegetation surveys which had been recorded since the end of the eighteenth century provided evidence of losses in species representation. These were attributed to land use changes, smoke pollution, the picking of flowers and collecting of specimens, and the draining of the moors (Moss 1900). Surveys by naturalist societies contributed information on the effect of human activities on species composition and extinction in many parts of England. Salisbury (1924), in a study of the Hertfordshire flora, found that 45 species had become less common, and 33 extinct. He attributed this decline to the increasing human pressure resulting from population growth over 122 years from 144 to 400 per square mile (56–154 per km^2).

The concern for nature conservation was slow to grow, although the Society for the Promotion of Nature Reserves (SPNR) was established as early as 1913. The British Ecological Society did not become actively involved in nature conservation until 1942, when it began to participate, with SPNR and other concerned organizations, in discussions on a national park system for Britain (Sheail 1982). The eventual outcome was the establishment of the Nature Conservancy in 1949, 'to improve scientific advice on the conservation and control of the natural flora and fauna of Great Britain'. This resulted in the establishment of nature reserves selected and managed on ecological principles. It also provided facilities and support for ecological research. Ecologists came to play a leading role in nature conservation, in Britain and elsewhere, by providing a scientific framework for a rapidly expanding movement.

The development of the conservation movement was by no means confined to Britain. Indeed, some aspects of progress were more rapid elsewhere. In North America the world's first national park was established in 1874 (Yellowstone) following the declaration of the Yosemite State Park in 1864 (Haines 1974). While the initial emphasis in these parks was on public access and recreation rather than conservation, they represent the genesis of a movement that soon grew to embody protection of plants and animals for their own sake.

1.3 CONSERVATION: THE ECOLOGICAL PERSPECTIVE

The intervention of ecologists in nature conservation initiated what twenty years later was to be called conservation biology (Soulé & Wilcox 1980). It brought about a drastic change of emphasis, from species representation within communities, to communities in their environments. Survival and extinction of species remained a major concern, but were recognized as part of the functioning of the community as a whole. These ideas on the ecology of communities and their constituent species were basic to the International Biological Programme (IBP) which was initiated in 1964 and terminated in 1974. Ecological principles pervaded many aspects of this multi-disciplinary endeavour, and in particular the section Terrestrial Conservation (CT) led by E. M. Nicholson, the head of the Nature Conservancy in Great Britain.

The CT section set out to provide a scientific basis for the conservation of the world's ecosystems, seeing that IUCN, the International Union for the Conservation of Nature and Natural Resources, was mainly active in management and politics of nature conservation. The objective was to establish a check-list of the range of ecosystems world-wide and to assess the extent to which main types and their significant variants were adequately protected in national parks or similar reserves. Substantial results were obtained from individual countries, but financial stringency inhibited world-wide coverage (Worthington 1975).

Principles for ecosystem conservation were further developed in the UNESCO-sponsored international programme 'Man and the Biosphere' (MAB) which succeeded IBP. It initiated a network of biosphere reserves, representative of biomes and ecosystems of the world. These were to be surveyed and inventoried, to serve as benchmarks for long-term studies of the dynamics of species and their associations. The emphasis was on the conservation of ecosystems: species should be conserved 'as parts of ecosystems rather than receiving species-by-species attention' (MAB Project 8, unpublished).

1.4 CONSERVATION: THE EVOLUTIONARY PERSPECTIVE

Whether expressed or not, conservation has a time dimension: the period for which a particular project, e.g. a nature reserve, is expected to remain operative, subject to prevailing natural and social conditions at any future time. Conservation planning inevitably contains a time factor, with implications for site and size, design and management. This concept has been called the *time scale of concern* (Frankel 1974). It can extend from a year or a generation to an indefinite future (Table 1.1).

Small nature reserves whose primary function of preserving a community of plants and animals has been suborned into other objectives such as recreation, tourism or education are unlikely to maintain those communities in anything like their natural state. Such reserves are an important part of an

Table 1.1. *The time scale of concern*

	Period	Operator	Objective	Time scale
Wildlife	To 8000 BC	Hunter-gatherer	Next meal	1 day
Domesticated plants	To 1850 AD	'Primitive' or 'traditional' peasant farmer	Next crop	1 year
	From 1850	Plant breeder	Next variety	10 years
	From 1900	Crop evolutionist	To broaden the genetic base	100 years
Wildlife	Today	Genetical conservationist	Dynamic wildlife conservation	10,000+ years
		Politician	Current public interest	Next election

Source: From Frankel (1974), with permission.

overall conservation strategy but have a limited useful life. At the other end of the time scale of concern are larger nature reserves which are dedicated to securing stable and lasting habitats for communities and the species they contain. These reserves should be protected from all other uses and managed with a philosophy guided by a 'for ever' connotation. Where needed, buffer zones that accommodate the legitimate needs of traditional inhabitants may have to be established to ensure this outcome. It is the longer time scale and reserves of larger size that we contend should receive top priority in conservation.

Under conditions of long-term conservation, communities and their component species are inevitably exposed to changes in the environment – physical, biological, technological, economic or social. Drastic impacts may affect the species composition. More commonly, changes result in adaptive responses moderated by the available genetic diversity of the organisms represented in a community. The integrated genetic diversity within an ecosystem constitutes its *evolutionary potential*. Ecosystem, species and genetic diversity within species can thus be recognized as the elements of biological conservation.

Let us now consider the evolutionary potential in the three conservation systems which are elaborated in the chapters that follow. They are, in reverse order: first, *communities* conserved in nature reserves or cognate natural habitats (Chapters 9 and 10); second, *species* preserved *in situ* in nature reserves (Chapter 6) or *ex situ* in botanic gardens (Chapter 7). They include species used in forestry, as pasture or drug plants, in research or in plant breeding, as well as species that are threatened or endangered in their habitats. The third category is the *domesticates*, preserved mainly, but by no means solely, *ex situ* (Chapter 4).

Populations and species in nature reserves are in a dynamic state, responsive to environmental change (including the inevitable human impact), i.e. they are subject to natural selection, in place and in time. In botanic gardens we face, as a rule, marginal intraspecies genetic diversity, retained with a minimum of recombination; there is little scope for evolutionary change except through domestication in horticulture. The genetic resources of domesticates that are preserved in genetic resources centres are maintained 'frozen', which in many cases is literally true (Chapter 4). Their evolutionary potential is enormous, but it needs to be realized through recombination, genetic engineering, and selection.

These comparisons illuminate the very nature and purpose of conservation. Genetic resources of domesticates are *preserved*, not for their own sake, but because of their immediate or potential usefulness to humans, be it in breeding or in some other form of research or development. The reason for nature *conservation*, as we see it, is diametrically different. Its essence is for some forms of life to remain in existence in their natural state, to continue to evolve as have their ancestors throughout evolutionary time. The uses and benefits that we derive from nature reserves will be more fully demonstrated in Chapters 6 and 9. Here we stress their unique role in being dedicated and committed to safeguard and maintain the continuity of self-regulating communities with their infinitely complex adaptive balance, which no man-made system could attempt to recover, and which may not exist in such complexity anywhere else in the universe. Species conservation is complementary to ecosystem conservation in filling gaps, in protecting taxa of particular interest or importance, and in highlighting the preservation of rare and endangered species.

1.5 THE INTEGRATION: BIODIVERSITY

The term 'biodiversity' signifies the integration of ecology and genetics in conservation theory. It was introduced by W. G. Rosen (quoted by E. O. Wilson 1988) in the mid-1980s. It represents diversity at all levels of biological organization – the community, the species, the organism, and the gene (Figure 1.1). It forms the link between the evolutionary past, through the current era of attrition and depletion, to future survival, adaptation and continuing evolution, or decline and extinction.

Diversity is indeed 'the essence of life' (Frankel 1970). It is essential for

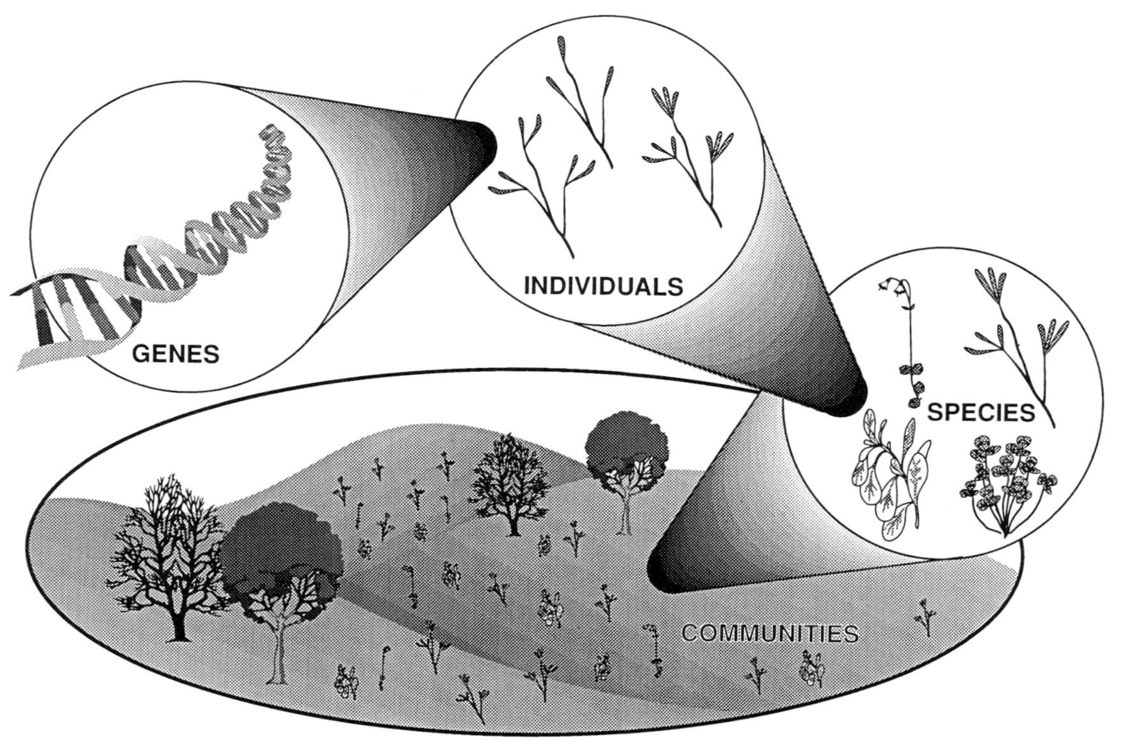

Figure 1.1. Diversity from the gene to the community.

survival, in time and in place, and for adaptation to specific environments, in the global context. It has a leading role in competition, symbiosis and parasitism, the impact of climate, the absorption of nutrients and the effect of deficiencies.

From all this its importance in long-term conservation is abundantly evident, hence biodiversity is the subject of four chapters that are basic to the biological systems of conservation. Chapter 2 is an overview of patterns of diversity in wild plants, of measurement and sampling methods, and of the use of wild gene resources in plant improvement. Chapter 3 contains corresponding information for domesticates; the diversity of pathways for domestication, the diversity of genes promoting or safeguarding productivity, and the genetic information built into cultivars of diverse ages and regions. These chapters introduce biodiversity at the level of the *gene* and the *individual*, which in Chapter 5 is extended to the level of the *population*.

By comparison, biodiversity of *natural communities* is at the highest level of complexity with their diversity of species, individuals and genes in dynamic interaction, enhanced by natural selection in diverse environments. The structure and function of natural communities is the subject of Chapter 8.

The *species* has a central place in conservation science. It embodies the array of diversity from gene to population and provides a measure for the diversity of communities. Species are readily identified and classified; and they can be *counted*. About 1.4 million species have been identified (including 250,000 higher plants) out of a total number of species estimated between five and 30 million (Wilson 1988). Chapters 6 and 8 deal with the diversity of species.

Two categories of species call for special attention because of their interest. The first are the domesticates – the species which directly or indirectly sustain humankind; their wild relatives used in breeding; other wild species used in agriculture, horticulture, forestry or other industries, and species of interest to a wide range of research areas. Chapters 2 and 3 contain a brief account of the diversity and conservation of this large and important category.

The second category causing widespread concern consists of species that are known to be, or likely to become, threatened or endangered in their habitats. Their preservation has become an issue of great significance in conservation literature, planning and politics and in public consciousness. This is caused by the drastic increase in the rate of extinction of plant species – referred to in section 7.1 – and the much higher extinction rates to be expected from the current rate of destruction of tropical forest communities.

For the defence of endangered species the two main avenues of conservation are actively pursued – *in situ* in national parks and other reserves, *ex situ* mainly in botanic gardens. The former is the subject of section 6.2.5, the latter of Chapter 7.

These two categories, species used by humans and species endangered (largely) by humans, have one common feature: they depend on human support for continued existence. In other respects they differ. The 'useful' species are protected because of their perceived usefulness – extending from the species that sustain the ever growing numbers of our own species, to the 'under-utilized' and 'new' crops (see section 3.2.7) and even what we have called 'boutique' crops, which enrich the diet of the well-to-do. On the other hand, rare and endangered species have no such overall protective shield like 'usefulness to humans', though some or all may be regarded as useful, or even essential, components of ecosystems which between them maintain living beings in their natural state of existence.

1.6 THE CONSERVATION OF BIODIVERSITY

As previously noted, the *conservation of species* has two main streams: *in situ*, i.e. within the confines of communities in their present and, hopefully, their future state of existence; and *ex situ*, i.e. subject to continuing active support and management by humans. Both methods are applied to preserving both wild and domesticated species, a choice, if available, depending on the breeding system, the life cycle, etc., and on the purpose and time scale of preservation. For example, landraces of crops are preserved in cultivation which facilitates adaptation to environmental change (see section 4.2), and in seed storage, for long-term security (section 4.4.2); endangered species are maintained in botanic gardens (section 7.3.2) and in nature reserves (section 6.2.5), and their seeds may be preserved in seed storage (section 4.4.2).

It is at the species level that biological conservation is most readily understood and operated upon, whatever the identity of the target, from the large mammals to rare and inconspicuous desert plants. Outside the circle of plants and animals exposed to human use or interest – economic, cultural, scientific, popular – there is a reservoir of organisms with as yet unknown and unappreciated uses for mankind. We may call it the 'genetic reserve'. Can this be safeguarded, and how?

As we shall see in Chapter 10, this question extends beyond concerns for human needs. It involves the survival of diversity as such, present and future, with a time scale of concern without limit. Chapters 8 to 10 explore the conditions for survival and adaptation in natural communities.

While it is at the species level that human needs, interests and expectations are directly identified and safeguarded, diversity itself, the essence of life, can only be preserved in the natural communities that we and our descendants safeguard and, we hope, cherish.

2

The genetic diversity of wild plants

2.1 INTRODUCTION – WILD PLANT GENES AS A RESOURCE

At its elemental level, biological diversity consists of differences between individuals and species in the presence of particular DNA sequences, or their location in the genome. Its building blocks thus include: diversity encoded by specific genes that some organisms possess but others lack; differences in sequences that regulate gene expression; differences in other non-coding sequences; diversity arising from differing copies of homologous or related DNA sequences (e.g. allelic variation); and diversity due to translocation of a sequence from one chromosomal site to another (e.g. position effect).

Genetic variation present in natural populations of wild plant species is the theme of this chapter; Chapter 3 deals with variation in domesticated species. Here, our leading question is: why should the genes contained in natural populations be the focus of conservation interest and be considered a 'resource'? Two reasons are pre-eminent. First, this genetic variation is a resource for the species' own survival and its future evolution. Second, a small fraction of the genes are a potential resource for improving the productivity of other populations or species.

In both these areas, wild genes have become prominent today. On the one hand, natural ecosystems are subject to increasing pressure from habitat fragmentation and accelerated changes in the environment, placing new demands on the genetic repertoire of plant species. On the other hand, modern biotechnology and genetic engineering provide new tools to extract useful genes from wild species and, unconstrained by mating barriers, to transfer them to related or unrelated species.

This chapter begins with an overview of the general patterns of genetic diversity in populations of seed plant species and the major forces that act upon them. Such patterns furnish indirect evidence that a portion of the variation is the key to the long-term future of populations. We then turn to the improvement of crop plants by means of wild genes, particularly their genes for disease and pest resistance. Such uses of wild genes underline the need to conserve wild species as resources for the plants that sustain humankind. The collection and use of wild genetic resources are sampling processes, which call for optimal sampling strategies. Strategies that improve efficiency in the choice of individuals or populations are crucial to better knowledge, conservation and use of genetic diversity of wild plants. The final section of the chapter addresses these sampling issues.

2.2 LEVELS AND PATTERNS OF DIVERSITY

A major conclusion to emerge from the last two decades of population genetic research in plants is that species differ greatly in their levels and patterns of genetic variation (Hamrick & Godt 1989). Some species, such as the crop *Zea mays*, the weed *Echium plantagineum*, the wild tomato relative *Solanum pennelli* and the forest tree *Pinus contorta*, are replete with large amounts of genetic diversity. In contrast, others such as the widespread *Pinus resinosa*, or the highly localized herb *Clarkia franciscana* are examples of genetically depauperate species (Karron 1991). The great variety in genetic composition and population structure is a fact, basic to planning the conservation of plant biodiversity. One immediate implication is that the optimal strategies for sampling individuals and populations, for the purposes of conserving or using their genes, will differ among species.

2.2.1 Measurement of genetic variation

How is diversity measured at the gene level? It is useful to recall the two related yet distinct notions of diversity, basic to the study of genetic variation. One is the *richness* of any population or sample from it, i.e. the total number of different genotypes present. The other is the equality or *evenness* in frequency of different types in the population or sample. Under the first notion, a population with three flower colours is more diverse than one with two colours, regardless of the frequencies of the colours in the populations. Under the second notion, a population with two equally frequent flower colours is more diverse than another population with three colours if the colours occur in very dissimilar or uneven frequencies (such as 0.90, 0.05 and 0.05).

In the case of alleles at a single genetic locus, the first notion corresponds to *allelic richness*, and the measure is the total number (A) of distinct alleles at that locus in the population or sample. The second notion is evenness, or lack of variation among the frequencies of alleles, and the common measure is the coefficient of *gene diversity* (Nei 1973). This is the probability (h) that two gametes randomly chosen from the population or sample will differ at a locus. The measure is largely, but not solely, dependent on evenness in allelic frequencies. If p_i denotes the frequency of the ith allele at a locus $\{i=1, \ldots, A\}$ and $var\ (p_i)$ the variance in these allele frequencies, then

$$\begin{aligned} h &= 1 - \sum p_i^2 \\ &= 1 - var\ (p_i) - 1/A \end{aligned}$$

Thus the gene diversity statistic (h) is high when the number of alleles (A) is large, but even more so when there is equality or low variance in their frequency.

The basic data that the two measures (A, h) summarize are the frequencies of the various genotypes present in the sample. As well as the two kinds of diversity outlined above, three measures of structure in the variation are often used in population genetic studies. These are (i) the observed level of heterozygosity (H), which measures the arrangement of alleles into genotypes; and Wright's *fixation index* (F), which measures the deviation of genotypic frequencies from panmictic expectations (which Weir (1990) calls the Hardy–Weinberg disequilibrium); (ii) the degree of *linkage disequilibrium* (D), which measures the arrangement of alleles at several linked loci, as the deviation from random assortment of alleles in gametes or zygotes; and (iii) the degree of population divergence (Wright's F_{ST} or Nei's G_{ST}), which measures the arrangement of alleles in populations, using variation in the frequency of each specific allele among the different populations. The G_{ST} statistic is a multiple allele extension of Wright's F_{ST}. The reader is referred to Weir's (1990) recent book for a thorough treatment of the procedures for estimating all these parameters from genotypic data.

Recent decades have seen a substantial growth in knowledge of the genetic diversity present in plant populations. Studies at the protein level, mostly using isozymes, have greatly augmented the classic studies of ecotypic divergence. These are now being extended directly at the DNA level. Also interest in the analysis of variation in natural host–pathogen systems has recently increased (Burdon 1987). What is really remarkable about all these results is the great range in levels and patterns exhibited by plant species.

2.2.2 Allozyme evidence

Hamrick and Godt (1989) summarized allozyme data from over 450 taxa of plants. The genetic statistics they computed from the published isozyme data of each study were: the percentage of loci that were polymorphic (P), the allelic richness (A) averaged over all loci including both monomorphic and polymorphic loci, and the gene diversity (h). Overall, plant species are polymorphic at about half their isozyme loci ($P=50\%$) with an average of two alleles per locus and a mean gene diversity (h) of 0.15. Within the average population, about one-third of the loci are polymorphic, with 1.5 alleles and a gene diversity of 0.11. The proportion of the total gene diversity in excess of that found within the average population, an index of population divergence, is about 20%, as measured by the G_{ST} statistic ($G_{ST}=1-h_P/h_S$. The subscripts P and S refer to within population and within species, respectively.)

These overall averages subsume a wide range of species and conditions. Hamrick & Godt (1989) classified each plant species according to eight attributes – taxonomic status, life form, geographic range, latitudinal distribution, breeding system, seed dispersal and successional status. Table 2.1 is an extract of their analysis. Of the several species attributes, *geographic range* had the greatest effect on variation at the species level. The average gene diversity (h) of species with a widespread distribution was 0.20, almost double that of narrow endemics. The same trend was apparent for variation within the average population, but at this level the *breeding system* was more influential than geographic range. Populations of self-pollinated species were about half as polymorphic as were populations of partially or completely outbreeding species. The breeding system was also the primary attribute determining the degree of population divergence. Populations of inbreeding species are much more divergent than those of outbreeders.

Genetic variation at the species level showed the same trend as variation at the population level for the following categories (which are not included in Table 2.1). For *taxonomic status*, gymnosperm species were more variable than monocotyledonous species which in turn were more variable than dicots. For *latitudinal distribution*, boreal species were more variable than temperate or tropical species. For *successional stage*, species from late stages were more variable than mid- or early successional species. In this case, the trend for levels of population divergence (G_{ST}) ran counter to that for the overall species diversity (h_S) with populations of early successional (weedy, colonizing) species being particularly divergent.

Such studies help to discern major trends in genetic diversity and the overall effect of important factors like breeding system. However, the range of values among different species within life-history categories, and among populations within species, is also remarkable. Some individual species depart substantially from the overall mean of their groups. For example,

Table 2.1. *Allozyme variation at the level of species and populations for higher plant species*

		Within species (*S*)			Within populations (*P*)			
Category	*N*	*P*	*A*	*h*	*P*	*A*	*h*	G_{ST}
Breeding system								
Selfing	113	42	1.69	0.12	20	1.31	0.07	0.51
Mixed – animal	64	40	1.68	0.12	29	1.43	0.09	0.22
Mixed – wind	9	74	2.18	0.19	54	1.99	0.20	0.10
Outcross – animal	164	50	1.99	0.17	36	1.54	0.12	0.20
Outcross – wind	102	66	2.40	0.16	50	1.80	0.15	0.10
Geographic range								
Endemic	81	40	1.80	0.10	26	1.39	0.06	0.25
Narrow	101	45	1.83	0.14	31	1.45	0.11	0.24
Regional	180	52	1.94	0.15	36	1.55	0.12	0.22
Widespread	85	59	2.29	0.20	43	1.72	0.16	0.21
Life form								
Annual	140	51	2.07	0.16	30	1.48	0.11	0.36
Short-lived	152	41	1.70	0.12	28	1.40	0.10	0.23
Long-lived, woody	110	65	2.19	0.18	50	1.79	0.15	0.08
Seed dispersal								
Gravity	198	46	1.81	0.14	30	1.45	0.10	0.28
Attached	55	69	2.96	0.20	42	1.68	0.14	0.26
Explosive	27	30	1.48	0.09	21	1.25	0.06	0.24
Ingested	67	46	1.69	0.18	32	1.48	0.13	0.22
Wind	105	55	2.10	0.14	43	1.70	0.12	0.14

Note:
N, minimum number of species in a category; *P*, proportion of loci polymorphic; *A*, average allelic richness, or number of alleles at a locus; *h*, the average gene diversity.
Source: Derived from Hamrick & Godt (1989).

Pinus resinosa is atypically low in variation compared with most other pines. Another problem is the correlation among species of several attributes (for example, between uniparental mating systems and colonizing habit and hence successional stage). Such correlations make ambiguous any claim of a direct causative link between attribute and variation pattern based on survey data alone.

A further question arises as to how representative a picture of genetic variation isozymes afford. Their major advantage is the comparability among species and their relative efficiency in handling large samples. Yet isozymes are a limited and biased sample of the plant genome. Isozyme surveys largely detect differences between amino acid sequences that alter net charge on the protein molecule. The surveys measure variation only in the structural gene of a restricted set of enzyme loci. In view of these shortcomings, evidence is needed of population genetic differences directly at the DNA level sampled from the complete range of DNA sequences.

2.2.3 Molecular polymorphism

Investigation of the kinds and levels of molecular diversity in plant populations is a burgeoning field (Clegg 1989; Schaal *et al.* 1991; Bachmann 1994). Because of fundamental differences in their function and organization, several major classes of DNA sequence need to be recognized. These are (i) single-copy genes, (ii) multigene families, (iii) hypervariable minisatellite and microsatellite sequences and (iv) the organellar genomes of the chloroplast and mitochondrion. Surveys for variation within any of these major classes can be made at the level of restriction site or fragment length differences, or comparative physical maps of several restriction sites, or ultimately complete DNA sequences. Despite the few comparative data in plants at the population level, it is already clear that each class of sequence has its own pattern of variation and rate of evolutionary change. This variety reflects variation in the evolutionary forces operating and the different functional constraints imposed upon these sequences.

Nuclear sequences

For *nuclear sequences of low copy number*, evidence of diversity levels comes from surveys of restriction fragment length polymorphisms (RFLPs) using specific probes (such as that by Gepts and Clegg (1989) in *Pennisetum glaucum* using cloned maize alcohol dehydrogenase (*Adh*) genes), or with anonymous random genomic or cDNA clones. The restriction analysis of DNA variation within and among several species of *Lycopersicon* by Miller and Tanksley (1990) is an example of the latter. They used 40 single-copy, nuclear probes from three libraries (one cDNA, one genomic library made with the restriction enzyme *Eco*RI, and another with *Pst*I which yields libraries richer in single-copy sequences). Genetic distances were computed from the proportion of fragments shared between two individuals, either from the same accession (population), or from different accessions of the species. Table 2.2, an extract of their data, shows that populations of the self-

Table 2.2. *Average genetic distance between individuals sampled from the same ('within') or from different ('between') accessions of species of* Lycopersicon

	Average genetic distance (%)	
Taxon	Within accessions	Between accessions
Self-compatible species		
L. esculentum		
cultivars	0.0	0.3
landraces	0.1	0.9
var. *cerasiforme*	0.1	0.7
L. pimpinellifolium	0.3	1.0
L. cheesmanii	0.1	0.3
L. chmielewskii	0.1	0.4
Self-incompatible species		
L. penellii	1.5	2.7
L. hirsutum	0.5	2.7
L. peruvianum	1.7	4.4

Source: From Miller & Tanksley (1990).

compatible species have an order of magnitude less variation than those of the self-incompatible species. This ratio is improved only slightly for DNA variation between populations – the self-compatible species are still markedly less variable than the self-incompatible.

For sequences with low copy number, finding polymorphism in outbreeding species (e.g. maize) is much easier than in inbreeding species (e.g. lettuce, wheat, tomato and soybean). In tomato (Paterson *et al.* 1988) and soybean (Keim *et al.* 1990), interspecific crosses with wild related species were used to detect polymorphism and develop high-density RFLP linkage maps. This constitutes an important use of the close wild relatives of inbred crops. As to the kinds of sequence changes that lead to RFLPs, it appears that small rearrangements (either insertion/deletions or inversions) are a common cause. Such rearrangements are probably more frequent than polymorphic variation arising from the gain or loss of restriction sites.

The recently developed RAPD (random amplified polymorphic DNA) technique permits surveys of variation for a fraction of the cost of Southern hybridization procedures (Clegg & Durbin 1990). The technique employs the polymerase chain reaction (PCR) to amplify a sample of sequences of arbitrary length bounded by a random primer sequence. It is usually not

Table 2.3. *The percentage polymorphism* (P), *allelic diversity* (A) *and gene diversity* (h) *of allozyme, RFLP and RAPD diversity in two species of* Populus

Species	Marker	Trees	Loci	*P*	*A*	*h*
P. tremuloides	Allozyme	118	13	77	3.3	0.25
	RFLP	91	41	71	3.3	0.25
	RAPD	102	61	100	2.0[a]	0.30
P. grandidentata	Allozyme	96	14	29	1.4	0.08
	RFLP	75	37	65	1.8	0.13
	RAPD	95	56	87	1.9	0.30

Note:
[a] The maximum possible value for *A* with presence/absence data.
Source: From Liu & Furnier (1993).

known whether the amplified sequences belong to the single-copy, multigene family, repetitive or organellar DNA fractions of the genome. A further limitation of most RAPD markers is the unknown homology between markers and their two-state phenotype with presence being dominant to absence. Despite these problems, a quickly growing number of RAPD studies of DNA variation are being made in wild and cultivated plant species.

Examples of these reports on wild species follow. Chalmers *et al.* (1992) measured the *partition* of variation among and within populations of *Gliricidia sepium* and *G. maculata*, two useful tropical leguminous outbreeding tree species. About 60% of the variation for presence of RAPD fragments occurred between populations. For populations of dioecious (outbreeding) buffalo grass (*Buchloe dactyloides*), Huff *et al.* (1993) found a similar proportion of variation in such markers between geographic regions, a further 10% of variation among populations within a region and only 30% between individuals within a single population. These values for population divergence are much higher than those detected for isozymes. This conclusion parallels that for RFLPs; for example, Zhang *et al.* (1993) contrasted the divergence (G_{ST}) values of 48% for RFLPs with that of 30% for allozymes among populations of wild barley (*Hordeum vulgare* ssp. *spontaneum*).

Liu and Furnier (1993) compared the levels of polymorphism in two species of aspen revealed by allozyme, RFLP (for random genomic probes) and RAPD markers (Table 2.3). The RAPD markers were more polymorphic than were the RFLPs or isozymes. However, RAPDs failed to detect the difference between the species in level of polymorphism (*h* and *A*) in single-copy DNA found by the other methods. On the other hand, Mosseler *et al.* (1992) found strong concordance between the patterns of isozyme and RAPD diversity in *Pinus resinosa*. Red pine had very low levels of genetic diversity for both kinds of markers, in contrast to the spruces *Picea glauca* and *P. mariana*. The authors conclude that recovery from red pine's low level of genetic diversity must be a slow process, if the bottleneck that produced it took place during Holocene glacial episodes.

At the level of complete DNA sequences, little is yet known about the partition of variation among and within natural plant populations. So far only specific alleles known to differ on other grounds (e.g. function or electrophoretic mobility of the encoded protein) have been sequenced comparatively, rather than a random sample of genes. The sequencing of such selected alleles at a locus (for example, the *Adh1* locus in maize: Osterman & Dennis 1989) has shown a variety of sequence changes, including rearrangements in the non-coding regions associated with transposable element activity. Thus in addition to the partition of variation between and within populations, the relative roles of tandem duplications, insertion/deletion events, silent base substitutions, and amino acid replacements in the various regions of the gene (introns, transcribed, and translated regions) are yet to be resolved.

Multigene families

Considerably more evidence is available on variation in the nuclear multigene family coding for (18S) ribosomal RNA (rDNA) (Dvorak 1989; Appels & Baum 1991; Bachmann 1994). Typically, the basic unit includes the 18S, 5.8S and 26S RNA sequences, with the two internal and one external transcribed spacers and an intergenic (non-transcribed) spacer. Part of the intergenic spacer comprises a number of subrepeats. Variation in the number of subrepeats gives rise to length variants in the basic unit. The unit is tandemly repeated in large blocks on one or more chromosomes. Thus an individual plant could be heterogeneous for its rDNA complement at up to four levels. Two or more types of units could occur within the one block, or could comprise two different blocks on non-homologous chromosomes, on related or homoeologous chromosomes from the different genomes in polyploids, or in segregating alternative 'alleles' in heterozygotes.

Two special modes of evolutionary change for spacer length variation are first, the generation of new kinds of units through gene conversion or unequal crossing over, and second, the rapid homogenization of the gene family within and between blocks of repeats. This latter process of homogenizing the members of a multigene family in the presence of considerable sequence divergence between species and populations is termed 'concerted evolution' and is also attributed to gene conversion or unequal crossing-over. In addition, Dvorak (1989) has suggested that changes in the prevalence of the various blocks of the 18S–26S family in polyploids evolve jointly with changes in blocks in the 5S family. This would be an example of coadaptation in the Darwinian sense. Further, the number of repeats in a block can respond to environmental stress making it one of the more labile elements in the plant genome (Cullis 1985). Somatic variation of rDNA has also been found among the asexual progeny of *Taraxacum officinale*, suggesting that somatic processes can be important sources of variation in apomictic plants (King & Schaal 1990) (see also section 3.1.3).

Studies by Sytsma and Schaal (1990) of the extent of population divergence for rDNA in the *Lisianthius skinneri* complex serve to emphasize another trend. Their data indicate a pattern of rDNA variation with almost total homogeneity within populations, but fixed for alternative repeats among populations. In contrast, population divergence for rDNA loci in *Pinus sylvestris* in Finland is only 0.14; however, this value of G_{ST} is still much higher than 0.02 for allozymes (Karvonen & Savolainen 1993). Thus it appears that rDNA loci may diverge between populations to a greater extent than do allozymes. As these genes are involved in the fundamental cellular process of protein synthesis, their variation patterns could be particularly important to differential adaptedness and thus to conservation strategies.

Hypervariable sequences

A hypervariable class of dispersed repeat sequences or minisatellites occurs in plant genomes (Dallas 1988). They provide powerful genetic markers for studying individual genotypic identity and aspects of clonal biology. The tandem arrangement of the set of repeats renders the set liable to frequent mispairing and unequal crossing-over, resulting in changes in size of the set. Because of their excess variability, they have rarely been used to estimate variation *per se*. Their potential lies more in tracing current or recent identity by descent among individuals or subpopulations, particularly in testing the isolation of fragmented populations. They also provide extensive markers for linkage mapping in crosses between two parents that are closely related.

Wolff *et al.* (1994) have compared the population genetic structure of allozyme variation with that of minisatellite variation, detected by the human M13 probe, in three species of *Plantago*. The three species differ in breeding system. The relative levels of microsatellite variation between

Table 2.4. *Gene diversity* (h) *and population divergence* (F_{ST}) *for allozymes and minisatellite dissimilarity and divergence in a selfing, a mixed mating and a self-incompatible species of* Plantago

Species	Outcrossing rate	Populations	Allozymes (h_P)	Allozymes F_{ST}	Minisatellites dis	Minisatellites F_{ST}
P. major	<0.05	5	0.031	0.22	0.19	0.65
P. coronopus	>0.34	4	0.041	0.07	0.50	0.22
P. lanceolata	1.0	4	0.115	0.04	0.40	0.14

Source: From Wolff *et al.* (1994).

populations (F_{ST}) and between species (dissimilarity) corresponded with the levels of allozyme variation, reflecting the major influence of breeding system on genetic diversity (Table 2.4).

Simple sequence repeats or microsatellites, analysed by PCR amplification of sequences containing tandem arrays of short repeats (e.g. a dinucleotide unit $(AT)_n$), offer an additional array of markers. Loci with more than ten repeats ($n > 10$) are likely to have many alleles. For example, Wu and Tanksley (1993) detected multiple alleles (between 5 and 11) at all eight microsatellite loci in 20 accessions of rice (*Oryza sativa* and four related species).

Organellar sequences

Reports of intraspecific variation in organellar DNA sequences in plants have recently appeared but information on population structure of this variation is still limited. Yet levels and patterns are likely to differ from those of variation in nuclear sequences. The reasons for such contrasts are the predominantly uniparental inheritance of organelles, and the lack or extreme rarity of recombination between variants. Presumably, other features, like the high number of copies per cell and specialized genome organization and function of organelles, also regulate the nature and extent of changes accepted.

Despite the slower rate of evolution of the chloroplast genes compared with nuclear genes, intraspecific variation in *chloroplast DNA* (cpDNA) has been found in over 50 species of plants (e.g. *Hordeum vulgare, Lupinus texensis, Zea mays, Heuchera micrantha, Pinus banksiana* and *Glycine* species) (Soltis *et al.* 1992). Both changes to restriction sites and altered fragment lengths due to insertions or inversions have been detected. The magnitude of intraspecific variation, expressed as an average percentage sequence divergence per nucleotide between two individuals, differs widely among species, ranging up to 0.07% (in *Heuchera* and *Salix* species), and approaching the level of cpDNA divergence between congeneric species.

With uniparental inheritance and the lack of recombination, the population structure of cpDNA variation resembles that for nuclear markers in selfing species. For example, *Tiarella trifoliata* is one of many species to show strongly developed geographic patterns of cpDNA variation (Soltis *et al.* 1992). Byrne and Moran (1994) found an average sequence divergence of 0.016% within ten populations of *Eucalyptus nitens* and divergence between populations of 0.056%. On both a haplotype (plastome) and a nucleotide basis, the divergence (G_{ST}) was remarkably high (0.78). In such species with maternally inherited cpDNA, the restriction of migration of cpDNA variants to the seed dispersal phase of the life cycle encourages population divergence.

An intensive survey of restriction site characters in the cpDNAs of four related diploid species of *Glycine* found considerable polymorphism (Doyle *et al.* 1990) and divergence among some of the plastomes within the species. In part this divergence was geographic. A cladistic analysis of site changes showed a limited congruence between plastome type and taxonomy based on morphology. Similar conclusions have emerged from several other analogous studies. In part, the limited concordance is probably due to hybridization and introgression among the taxa, and to the maintenance of ancestral cpDNA polymorphism. It appears that in these cases, morphological divergence (largely coded by nuclear DNA) is only partially correlated with organellar DNA divergence. Such lack of correlation constitutes additional genetic discontinuities among populations.

Variation in *mitochondrial DNA* (mtDNA) is of major interest as a source of cytoplasmic male sterility for hybrid plant breeding. In maize, the male sterile Texas T cytoplasm is completely associated with susceptibility to the southern corn leaf blight disease, caused by *Bipolaris maydis* (Levings 1990). The gene responsible (*T-urf13*) apparently arose by complex rearrangement. In contrast to animals, most intraspecific variation of mtDNA in plants is structural rather than in primary sequence (Palmer 1988).

Breiman *et al.* (1991) assayed mtDNA variability in four wild diploid relatives of wheat from populations spanning their range in the Middle East, using eight well-defined mtDNA primers. They estimated the nucleotide divergence among samples of the outbreeding species *Aegilops speltoides* to be 0.027, whereas it was 0.001 in *Ae. longissima* and zero in *Triticum*

Table 2.5. *Mitochondrial DNA haplotype diversity in five species of* Pinus *within and among north American populations, based on RFLPs for cytochrome C oxidase I or II probes*

Species	Populations	Probe	Diversity (h_P)	Divergence (F_{ST})	Source
P. banksiana	8	*CoxII*	0.09	0.50	1
P. contorta	8	*CoxII*	0.21	0.70	1
P. attenuata	4	*CoxI*	0.08	0.86	2
P. muricata	10	*CoxI*	0.03	0.88	2
P. radiata	5	*CoxI*	0.11	0.83	2

Sources: (1) Dong & Wagner (1993); (2) Strauss *et al.* (1993).

monococcum and *T. tauschii*, three autogamous species. These levels of variability correlate directly with nuclear variability, indicating that mating system differences affect both nuclear and organellar variation patterns.

Like cpDNA variation, mtDNA variation can be strongly structured geographically. The mtDNA variation in five North American species of *Pinus* serves as a good example (Table 2.5). In all cases, the values of F_{ST} exceeded those typical for allozymes in conifers. Strauss *et al.* (1993) suggest that because of this differentiation between populations, organellar DNA variation may be very useful in detecting distinctive populations and setting conservation priorities.

Overall, the population structure of molecular variation attests to greater proportionate divergence between populations (*h*, section 2.2.1) than does allozyme variation. In the first instance, this divergence is because DNA variation is less prone to cryptic variation beyond the limits of electrophoretic resolution. But of more consequence, these sequences reflect the action of additional modes of evolutionary change, such as unequal crossing-over, concerted evolution, uniparental inheritance and migration limited to that in seeds. Such modes encourage genetic divergence between lineages while fostering homogeneity within lineages.

2.3 MAJOR DETERMINANTS OF VARIATION

Genetic diversity within any species of land plants is the sum of variation interacting with three kinds of factors. First are the abiotic ecological forces of climate, location and soil. Biotic interactions including competition, symbiosis, parasitism and predation form the second group of determinants.

The last group includes species characteristics such as population size, mating system, mutation, migration and dispersal. All three kinds of factors influence the genetic composition of populations in both a directed and a stochastic fashion, through natural selection or random events. For example, *Pinus sylvestris* in Finland ranges over many degrees of latitude with populations that differ genetically so that each is more adapted to the day length and seasonal length of its original habitat than at other latitudes (Mikola 1982). Yet annual variation in climate at any one site could elicit shifts in allele frequencies that from the perspective of next year's climate would be essentially random.

Similarly, the response of a plant population to a pathogen combines both systematic and stochastic elements. Race-specific and non-specific resistance fluctuate in response to, but one step behind, the prevalence and racial composition of the pathogen. Likewise, species-specific attributes, such as the ways in which new variants arise (mutation) or are assorted into genotypes (recombination), the ways in which the gametes are brought together (mating system) or new propagules move to new sites (migration), all have both deterministic and stochastic aspects.

Thus the debate that was an early focus for studies of isozyme variation as to whether the variation is neutral or selected (Nevo 1983) is largely moot. Furthermore, no specific fraction of the standing variation within a species or a population should be described as 'selected', with the remainder thought of as 'neutral'. This proportion varies over generations, as will the particular loci strongly selected, and the periods during which such selection acts.

2.3.1 The physical environment

Genetic differentiation in response to environmental variation within the spread of a species is a classic finding in plant ecology (Turesson 1922; Clausen *et al.* 1948). The environmental variables eliciting population divergence include temperature, light, day length and moisture regimes, parental rock, soil physical and nutritional status, slope, aspect and drainage. Ecological races are extensively discussed in texts on plant evolution, and the ecotype concept was very influential in early writings on plant genetic resources (Bennett 1965).

The divergence between ecotypes is clearly adaptive. Transplant experiments, particularly between populations within a geographic region, have usually found a strong relative advantage for the resident population compared with the introduced one. Bradshaw (1984) has noted that this relative fitness of the alien population is commonly half that of the native one. Local transplantation within a site has been less frequently studied, although spatial patterns of genetic diversity are found in natural plant populations (Epperson 1989).

The genetic basis of ecotypic divergence is likely to be very complex. In a cross between two ecotypes of *Plantago lanceolata*, Wolff (1988) found that F_2 individuals carrying a majority of allozyme alleles from a hayfield population performed better when transplanted to a hayfield than did their sibs with a majority of alternative alleles from a pasture population. At the pasture site, plants with a majority of pasture-derived alleles performed better. Adaptive differentiation between the two ecotypes apparently segregated at many loci scattered throughout the genome.

Yet while such adaptation is multigenic, any major contribution from a single locus or gene complex is of great significance in the use of wild genetic resources. This is more likely to be the case when ecotypes have developed in response to localized ecological stresses. For example, Macnair (1983) found that copper tolerance in *Mimulus guttatus* is determined primarily by a single major gene. Such major genes are not dissipated by recombination following gene flow, as would be tolerance based solely on multigenes. Thus multigenic adaptation permits limited influx of genes and the potential benefits of migration within the framework of a metapopulation.

In considering how plant species might respond to climate change, Bradshaw and McNeilly (1991) argue that species will behave in a conservative manner and migrate to a more favourable habitat rather than evolve. This implies that the conservation of species declining due to such change may require assisted migration, and that contiguous reserves should lie along ecological gradients to allow natural migration during long-term changes. They argue that most species have a limited range and are in a state of 'genostasis' with further evolution at the margins of their present niche limited by lack of adaptive variation.

Does this mean that the genetic variation within the species has only a minor role to play in the survival of such temporal trends? Clearly for those species with pronounced ecotypes, it has been important in adapting to a range of environments in space, and presumably will be important for adjustments in the future. However, equally clearly it is no panacea for the problem of long-term environmental deterioration.

An example of ecotypic variation on a *geographic* scale is that of variation in sensitivity to UV-B radiation. Krupa and Kickert (1989) have collated the published data from either glasshouse or field testing of a wide range of crops (77 kinds) and wild species (34 herbaceous plants and 20 forest trees) for response to increased UV-B radiation. Based on biomass accumulation, they rated about 40% of species of all three groups as 'sensitive' to UV-B. In several cases, reports differed for ratings, and one factor that accounts for disparities is intraspecific genetic variation. Evidence for such variation has been reported in soybeans and rice (Teramura *et al.* 1991). In wild species, Barnes *et al.* (1987) compared UV-B induced photosynthetic damage in plants from high latitudes with conspecific (for *Plantago lanceolata* and *Rumex acetosella*) or closely related plants from low latitude alpine sites. In general, plants collected from equatorial alpine sites, where solar UV-B irradiance is high, showed no UV-B induced damage. Significant accumulation of UV-absorbing leaf pigments occurred only in populations from higher latitudes, but this differential response was insufficient to protect them as well as their equatorial conspecifics. Recent measurements indicate that the actual UV flux ratio (UV-B to total irradiance) has increased at a rate of 1.2% per year in alpine areas because of stratospheric ozone depletion (Teramura *et al.* 1991). Such steady temporal changes might be expected to select for tolerant species and tolerant ecotypes within species, in conjunction with the migration that Bradshaw and McNeilly envisage. (See section 10.2 for a community perspective on climate change.)

In a review of microhabitat or *microgeographic* differentiation, Huenneke (1991) stressed the increasing evidence 'that in a natural population much genetic variation is related to natural selection arising from small-scale environmental heterogeneity'. The clearest evidence is at the level of the whole genotype in the growth patterns of ramets of a single clone. However, such a fine scale of subtle differentiation is beyond the focus of deliberate conservation. It is possible, although unproven, that a repertoire of such variation in a population is important for its future evolution. Ensuring that local population sizes are appreciable (but with little deliberate attention to where each individual is growing) will perforce provide a sample of such variation.

2.3.2 Biotic interactions and host–pathogen systems

Interacting with the physical environment are diverse biotic forces that both shape and are shaped by the genetic variation present within and between populations. The local population is host to a whole range of organisms – microorganisms, pathogens, pollinators, dispersal agents, herbivores. Its members are neighbours to a variety of coexisting plant species making up the community (see Chapter 8). These species exert a dynamic influence on the size and structure of the population with potential to exert strong selective pressure. As well, the local physical environment affects these cohabiting members of the community, thus indirectly and additionally affecting the species in question.

As Haldane (1949) recognized, pathogens can exert a selective pressure on host species. Such pressure will lead to the increase in frequency of a new gene or gene complex that conditions resistance to the current biotypes of the pathogen. In turn this places selective pressure on the pathogen for new sources of virulence. In addition, genes for host resistance and genes for pathogen virulence may have additional (pleiotropic) effects on their carriers – the resistance gene is said to have a fitness 'cost'. Person (1966), Jayakar (1970) and Leonard (1977) have all shown that given such costs, the frequency dependence of the interaction sets in train a complex dynamic coevolutionary system that promotes the retention of variation. However, high fitness costs associated with resistance and virulence are not the only means to maintain variation. The long-term demographic and genetic effects of local extinction and migration on host–pathogen associations, dispersed in a metapopulation framework, are an alternative mechanism (Thompson & Burdon 1992).

A rapidly growing body of empirical evidence confirms that natural plant populations are often polymorphic for response to pathogens. Burdon (1987) tabulated evidence from eight species ranging from annual herbs to forest trees, and reviewed detailed evidence in species of *Avena*, *Glycine* and *Trifolium*. Genes that mediate interaction with pests, pollinators, seed dispersers and symbionts, and that entail costs as well as benefits to the plant population, are similarly likely to be polymorphic.

Interest surrounds the shape of the frequency distribution of resistant responses. Is the frequency distribution of infection type responses normally distributed, with the most frequent class in the population being the one with an intermediate reaction, or is it skewed to higher levels of resistance, or of susceptibility? Can inference be drawn from this distribution as to the recency and severity of the pathogen as a selective force, and its biotype diversity (Burdon 1980, 1987; Dinoor & Eshed 1984)? This question is important in sampling for sources of disease resistance and in deciding whether diseased sites should be preferred. Clearly the origins of the test isolates of the pathogen, whether local or exotic, is one factor involved. Others are the genetic basis of resistance and virulence, the level of pleiotropic effects, and the breeding and recombination systems in both partners (Burdon 1987). As a consequence, inferences as to the significance of interaction based on the pattern of variation can only be tentative.

2.3.3 Mating system and historic effects

As determinants of genetic variation, the physical and biotic environments act upon variants that are directly involved in meeting environmental challenges. Such genes are clearly of major conservation interest. The third set of evolutionary forces are the biological properties of the species that affect the genome as a whole, and hence the level and patterns of genetic variation generally. They are the pollination and mating systems, population size and structure, dispersal patterns within populations and migration rates between them, mutation rates and mutator systems. Of these, the mating system is one that varies greatly among plant species. Hamrick and Godt's analysis of allozyme data (Table 2.1) highlights its prominence in determining the allelic richness and gene diversity of populations.

With uniparental inheritance for organelles, however, one might expect that plant mating system *per se* would have little effect on the levels and patterns of cpDNA and mtDNA variation. However, preliminary evidence indicates that population diversity for organellar DNA is less, and divergence is greater, in autogamous than in outbreeding species. Fenster and Ritland (1992) found that the level of cpDNA diversity in the selfing species *Mimulus micranthus* was half that in its more outbreeding relative, *M. guttatus*. In parallel, the selfer had one-quarter the level of allozyme diversity.

Such a trend arises from the *interaction* between mating system and historic bottlenecks in population size. Autogamous species are more able to recover from severe bottlenecks, or colonize new sites with few founders (see Box 5.2). Schoen and Brown (1991) among many others have stressed that the patterns of diversity in inbreeding species are highly non-uniform. Large portions of a species' geographic distribution may have little or no diversity in contrast to the remainder. Examples include the valley ecotype of *Avena barbata* (Clegg & Allard 1972) and the northwestern European populations of *Hordeum murinum* (Giles 1984). In these cases, the values of diversity are unrelated to current local population sizes.

The effect of population size, separate from mating system, on the level of genetic diversity has proved harder to discern. Ellstrand and Elam (1993) summarized ten studies in outbreeding species and noted that positive associations between size and diversity were present in seven of them. Figure 2.1 illustrates two of these examples, for allelic richness in *Scabiosa columbaria* and *Salvia pratensis* (Van Treuren *et al.* 1991) and data from *Eucalyptus albens* (Prober & Brown 1994). Detecting a relationship between population size and genetic diversity can be difficult if (i) experimental sampling strategies vary, (ii) present sizes poorly reflect historic effective population sizes (section 5.4.1), or (iii) electrophoretic recognition of distinct alleles falters as allelic richness (section 2.2.1) increases.

The effects of mating system and of fluctuations in population size are felt over the whole genome. They affect population biology and population survival, as is discussed further in Chapter 5. Here we point out that populations are variable for genes that affect the population dynamics and reproductive behaviour. Such genes are termed 'modifying genes' in

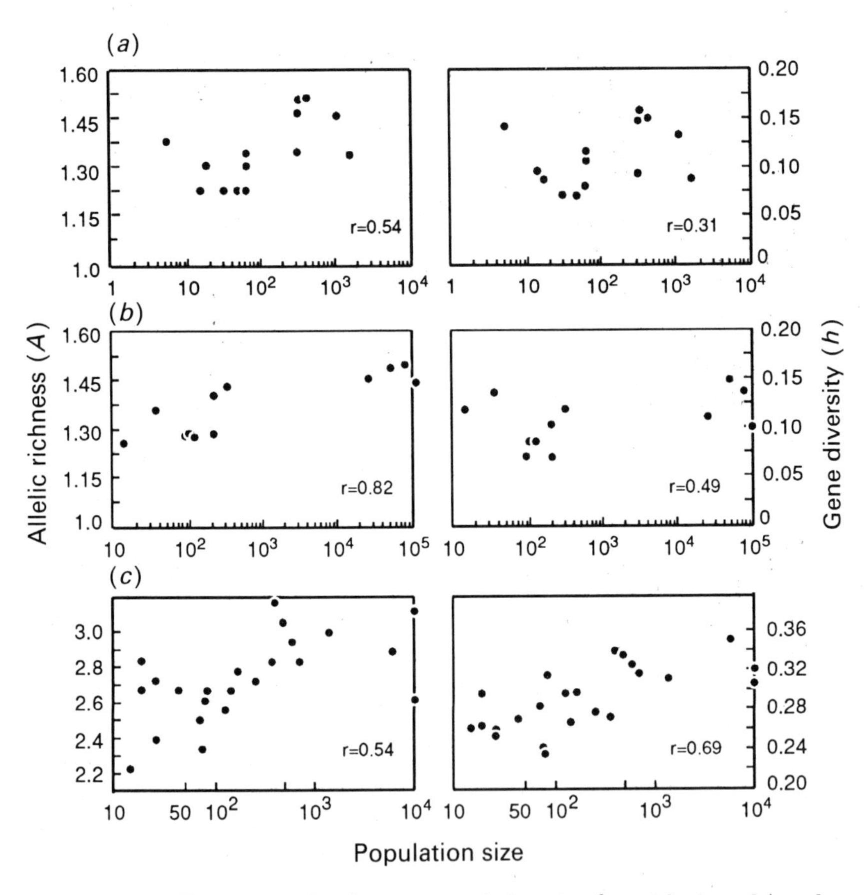

Figure 2.1. The relationship between population size (logarithmic scale) and allelic richness (A) or gene diversity (h) in populations of (a) Salvia pratensis, *(b)* Scabiosa columbaria *and (c)* Eucalyptus albens. *Redrawn from (a,b) van Treuren* et al. *(1991) and (c) Prober & Brown (1994).*

population genetics. A clear example of this class is the genes that affect mating systems such as self-incompatibility loci, the gene complexes controlling heterostyly, nuclear and cytoplasmic male sterility. The determination of such genes in natural populations is of interest not only for the conservation biology of species, but also for those wishing to manipulate mating systems in cultivated plants.

2.4 USE OF WILD GENETIC RESOURCES IN PLANT IMPROVEMENT

Wild species related to cultivated plants have contributed to the breeding of most crop plants (Harlan 1976; Stalker 1980; Prescott-Allen & Prescott-Allen 1988). Wild relatives are commonly used as sources of genes, but they can also provide disease-resistant rootstocks in horticulture, experimental resources, character models, pathogen differentials, etc. One example among many of a crop richly benefiting from its wild relatives is sugarcane. Modern sugarcane (*Saccharum officinarum*) cultivars are complex derivatives of artificial species hybrids, to which the wild species, *S. spontaneum* in their pedigree contributes yield, vigour and disease resistance (Simmonds 1976).

2.4.1 Resistance genes from wild relatives

The principal interest in wild relatives is the wealth of genes for resistance to parasites they have acquired over long periods of coevolution (Lenné & Wood 1991). In many cases, including the cereals and their rusts, the centres of diversity of the ancestral species and of their important parasites coincide. Other cases with a lack of coincidence arise in many ways, including (i) absence of the alternate host that many fungal pathogens require for the sexual phase of their life cycle, and (ii) domestication of the host in a region lacking the pest or pathogen.

A classic example of the latter is the tobacco–blue mould interaction. Of the 55 species of *Nicotiana* tested for resistance to the blue mould fungus, *Peronospora tabacina*, all but one of the 19 Australian endemic species possessed resistance. In contrast, all but one of the 35 American species, including the putative parental species of *N. tabacum*, were susceptible (Wark 1963). When the fungus spread throughout the tobacco growing areas of Australia, United States, Europe and the Near East, resistance was obtained from the Australian species *N. debneyi* and *N. goodspeedii*, and transferred to *N. tabacum* despite strong sterility barriers (Wark 1970).

Most of the resistances obtained from wild species are under the control of *major genes*, especially in the cereal rusts. However, Watson (1970a) noted that even in wild species, where major genes conferring specific resistance – i.e. resistance to specific biotypes of a parasite – have been found and extensively used, a genetic system of minor genes for non-specific resistance may be present. As evidence he quoted *Solanum demissum* as a source of both specific and non-specific resistance to *Phytophthora infestans*. Indeed, Turkensteen (1993) controversially argues that breeding for 'durable' resistance should be done in the absence of major R-genes so that polygenic resistance from wild species can be selected. Another example is the wild oat (*Avena sterilis*) which has non-specific resistance to crown rust (*Puccinia coronata*) effective against all races in Israel in addition to the specific resistance genes that have been extensively studied (Murphy *et al.* 1967; Dinoor 1975).

Are specific resistance genes derived from wild species more effective than those from domesticates? Watson (1970a) made two points. First, genes from wild species 'are often effective against all strains of the organism in the area where the work has been done'. Second, resistance genes transferred from wild species are not necessarily more durable than those from cultivated sources. For example, the specific resistance to *Puccinia recondita* in *Aegilops umbellulata* transferred to wheat by Sears (1956), and the resistance to *P. striiformis* derived from *Ae. comosa* (Riley *et al.* 1968) can be overcome in cultivars of hexaploid wheat. Since wild genes do not necessarily possess greater durability (Lenné & Wood 1991), their most effective and economical use is in combination with each other, i.e. in the form of multiple resistance.

Resistance gene deployment

Whatever the potential pool of resistance sources, it is finite. Therefore the rate at which resistances are overcome through pathogen evolution is of concern for the future of agricultural productivity. In past experience, each system of resistance has provided instances of both long-term stability and fairly rapid breakdown. However, there are substantial differences in the relative frequencies. Specific single-gene resistance in homogeneous cultivars constitutes the most vulnerable of the resistance systems, although there are examples of stability (see Watson 1970b). At the other end of the scale, non-specific, polygenic resistance (Van der Plank 1975) may be less vulnerable, but there are examples of fairly rapid, as well as of slow and gradual, decline in efficacy. It is likely, though an over-simplification, that race non-specific resistance, being potentially more stable, is the more desirable system. But it must be recognized that single-gene resistances, singly or in intra- or intergenotypic combinations, are much easier to manipulate in breeding programmes, and remain the principal defences against many pathogens that attack most of the crops that feed the world.

There are a number of approaches that aim to increase the effectiveness and the stability of resistance systems based on specific resistance. Multilines are composite varieties consisting of lines carrying different resistance genes, each of which is either resistant to all biotypes ('clean approach') or to single, but different biotypes ('dirty approach') (see Marshall 1977). Both kinds of multilines make heavy demands on resistance sources and breeding effort. However, *varietal mixtures* are an important alternative that has been implemented, for example, in barley to control mildew (Wolfe 1985; see section 3.2.2). These mixtures have the major advantage of utilizing existing cultivars and hence show little yield loss relative to the best pure-line cultivars. Disease levels in varietal mixtures are frequently less than half those occurring in their component pure lines alone (Burdon & Jarosz 1989).

Such approaches enhance rather than diminish wild relatives as sources of resistance. In a growing number of crops the resistance genes stored in related wild species are the main, in some instances the only resources available when the known domesticated sources are fully used. This was the case in resistance to some pathotypes of crown rust until resistance was obtained from *Avena sterilis* (Murphy *et al.* 1967). In other instances, resistance has so far been found only in wild species. Two diseases of the sunflower (*Helianthus annuus*), sunflower rust and downy mildew, are a case in point. Wild sunflower species are native to North America and were brought to Europe by early explorers. However, the improvement in oil content that made possible sunflower exploitation as a major crop is recent and took place in Russia. The sunflower rust (*Puccinia helianthi*) is prevalent wherever sunflowers are grown, but resistance is lacking in the domesticate. Downy mildew (*Plasmopara halstedii*) is a seed- and soil-borne disease of sunflower that causes heavy crop losses in the USA and elsewhere. A source of resistance to the rust, derived from a natural cross of a cultivar with wild *H. annuus*, also confers resistance to downy mildew through separate genes. Resistance to rust occurs in at least five species of the genus and that to downy mildew in eight (Seiler 1992).

More extensively than in any other economic plant, wild relatives have been used in breeding tomatoes resistant to the main pathogens – *Cladosporium*, *Fusarium*, nematodes, *Septoria*, *Stemphyllium*, tobacco mosaic virus and *Verticillium*. Cultivars with multiple disease resistance have been produced (Rick 1982).

2.4.2 Sources of new adaptations

In addition to pathogen resistance, wild relatives of crop species have donated a diversity of economic traits for crop improvement. Specific examples are cytoplasmic male sterility from *Helianthus petiolaris*, and fertility restoration from wild *H. annuus* into sunflower; and several components of fruit quality from *Lycopersicon chmielewskii* and *L. cheesmanii* into tomato (Rick 1982).

The use of closely related wild and weed species to extend the adaptive range of crops is increasingly under trial, given the need to adapt crops to more marginal environments. Zohary *et al.* (1969) drew attention to 'the wild diploid progenitors of wheat [which] constitute large gene pools largely unexplored and untapped by plant breeders'. The three diploids differ in ecological adaptation and in many other respects. Hybridization of *Triticum tauschii* (the donor of the D genome) with tetraploid wheat has tapped into novel wild diversity (Lagudah *et al.* 1993). Autecological studies of wild tomato species have revealed a diversity of adaptations that are of considerable interest to tomato breeders (Rick 1973). They include tolerance of an extreme range of conditions – from waterlogging to drought – in *Lycopersicon esculentum* var. *cerasiforme*; extreme resistance to desiccation in *Solanum pennellii* and *L. chilense*; and near-complete freedom from insect predation in *L. hirsutum*. In an attempt to widen the genetic base of the potato (*Solanum tuberosum*), a 'Neo-Tuberosum' strain of *S. tuberosum* ssp. *andigena*, the putative parent species of *S. tuberosum*, was developed by repeated mass selection for yield under Northern European conditions and controlled pollination (Simmonds 1966). The programme has ultimately resulted in enhanced germplasm for breeders and at least one hybrid (Tuberosum-× Neo-Tuberosum) cultivar (Hermsen 1989).

Weed relatives of crop species may have a special place as genetic resources for ecological adaptation to less favourable environments. So far they have been little considered in such a role. One exception is a project at ICARDA (the International Center for Agricultural Research in Dry Areas) to investigate *Hordeum vulgare* ssp. *spontaneum* as a source of tolerance to climatic stress (Ceccarelli *et al.* 1992).

Finally, wild relatives furnish model systems for phenotypic characters which, even if they are excluded from genetic transfer, might still be copied. For example, the wild relatives of linseed have seed oil with a range of fatty acid profiles (Green 1984). In particular, species of the section Linastrum have high linoleic and low linolenic acid; *Linum tenuifolium* averaged 81% and 4%, respectively. Although this showed that edible oil quality could be attained in the genus, the species cannot be used directly as genetic resources because of strong reproductive barriers. However, the finding provided a model for mutation breeding and linseed mutants that mimic the character have been obtained (Green 1986; see also section 3.2.7).

Cotton (*Gossypium hirsutum*), in common with all species of the genus, has glands in growing aerial tissues that contain gossypol, a natural pesticide. In cotton and its near relatives, these glands are already present in the seed, and the gossypol must be removed in processing the seed for stock feed. Seed of the Australian wild relatives (e.g. *G. sturtianum*), in contrast to the crop, lack mature glands and have low gossypol levels. Glandless-seed with glanded-plant would be a valuable combination to transfer to, or mimic in cotton (Altman *et al.* 1987).

2.4.3 The need for collecting wild relatives

Most of the wild and weed relatives of domesticates have not been collected, studied and conserved to an extent commensurate with their promise. Even for many of the principal crops, the number of wild accessions in major germplasm collections is still small, although the last decade has seen increased collecting. As well as provision for conservation *in situ* (section 6.2.4), representative collections are needed *ex situ* for ready observation and use by biologists and plant breeders.

Most of the species in question are not exposed to rapid extinction. However, there is concern for the continued existence of representative gene pools, covering the full range of each species (Marshall 1989). Prescott Allen and Prescott Allen (1988) reviewed the conservation status of many wild relatives and stressed threats of habitat loss, over-exploitation from harvesting, predation and competition from introduced species. Sections 4.4 and 6.2.4 address conservation methods for wild relatives in more detail.

The number of wild species that are either closely or distantly related to domesticated or exploited species is very large, certainly in the thousands. Further, some species are widely distributed and show marked geographic differentiation (e.g. *Lycopersicon pimpinellifolium*: Rick *et al.* 1977). How are priorities to be assessed among this taxonomic and geographic diversity? Several criteria are available to take into account.

1 The degree of relationship of the species to the cultigen, based on the Harlan and de Wet's (1971) concept of crop gene pools (Box 2.1). Chapman (1989) suggested that closer relatives should have higher priority.

2 The degree of endangerment of the species, or portion of the gene pool. Marshall (1989) suggested that threatened taxa merited special effort.
3 The extent to which previous collections represent the phylogenetic diversity of the genus and the ecological and genetic diversity of the species.
4 The value of and likelihood of finding significant genetic resources for improvement of the cultivated species.

BOX 2.1. CROP GENE POOLS AS DEFINED BY HARLAN AND DE WET (1971)

Primary gene pool (GP1): the true biological species including all cultivated, wild and weedy forms of a crop species. Hybrids among these taxa are fertile and gene transfer to the crop is simple and direct.

Secondary gene pool (GP2): the group of species that can be artificially hybridized with the crop, but where gene transfer is difficult. Hybrids may be weak or partially sterile, or chromosomes pair poorly.

Tertiary gene pool (GP3): including all species that can be crossed with difficulty (e.g. requiring *in vitro* hybrid embryo culture), and where gene transfer is impossible or requires radical techniques (e.g. radiation-induced chromosome breakage).

In the case of barley, GP1 comprises the crop species *Hordeum vulgare* ssp. *vulgare* and its progenitor, *H. vulgare* ssp. *spontaneum*; GP2 includes *H. bulbosum*; GP3 includes all other species of the genus (Bothmer *et al.* 1991).

Developments in genetic engineering may have reduced the importance of the first criterion because gene transfer is now possible without requiring hybridization. However, closer relatives still merit priority as they are more likely to be alternative hosts of crop pathogens and hence to develop resistance.

2.5 SAMPLING STRATEGIES

Programmes for the conservation of biodiversity at all levels necessarily entail sampling (Brown 1992). This is especially true for the conservation of allelic diversity of wild species because it is neither possible nor justifiable to conserve all allelic variants of all wild species. Conserving wild genes and, indeed, collecting and using them, are sampling processes.

Genetic diversity may be conserved *in situ*, i.e. in populations in their natural habitat within the communities of which they form part. It may also be held in *ex situ* germplasm collections. In either case, the key parameter for gene conservation is the allelic richness of the population or sample. For simplicity in what follows, we will refer to samples, with *ex situ* procedures in mind. The same principles also apply to *in situ* conservation of populations, as will be seen in Chapter 6.

2.5.1 Sampling theory of the infinite neutral alleles model

If the patterns of genetic diversity of the populations to be sampled were known in detail, an unambiguous optimal strategy would follow immediately. However, such information is limited or unavailable ahead of sampling. Indeed, the actual number of alleles in a population (the allelic richness) is a difficult parameter to estimate because its value in a sample steadily increases with increasing size of that sample. Instead, it is possible to calculate the theoretical effect of sampling on allelic retention, based on hypothetical distributions of allele frequencies. For this purpose the neutral allele model of Kimura and Crow (1964) is used as its sampling theory has been well developed (Ewens 1972). Thus, for example, the number of selectively neutral alleles (k) in a sample of size S random gametes from an equilibrium population of size N at a locus with mutation rate u is approximately

$$k \approx \theta \log_e[(S+\theta)/\theta] + 0.6$$

where $\theta = 4Nu > 0.1$ and $S > 10$. This formula shows that the expected number of alleles in a sample increases in proportion to the *logarithm* of the sample size. In contrast, the effort required to sample the individuals at a site is in direct proportion to the sample size. Thus there is a diminishing return (in terms of collecting new alleles) per unit of cost for increasing sample size which becomes progressively more wasteful of resources.

This conclusion derives from using the neutral infinite allele model under equilibrium conditions to specify genetic variation. However, the conclusions broadly apply in other cases (heterotic models, non-equilibrium situations, empirical distributions) where they have been checked (Brown 1989b; Brown & Briggs 1991). The neutral allele model is a useful benchmark because of its relative simplicity, but more testing of the generality of its sampling behaviour is needed.

With the above as theoretical background, there are several basic sampling questions to address. How many individuals should make up the original sample to form an accession in a collection? How many accessions are needed for each region of a species distribution, and for the whole species? When should samples be random, or when should they be systematic, stratified, or orientated to particular phenotypes, or when should repeated sampling procedures be adopted? Such questions are of two general kinds – those related either to the sufficiency of effort for some purpose, or to the optimal allocation of limited resources.

2.5.2 Sufficiency of sampling procedures

Under *sufficiency*, the concern is with the total amount or number of items required to attain a given objective or achieve a specific purpose to a reasonable level of certainty. It is necessary to clarify the biological basis of an objective in order to specify and defend it. For example, in determining the number of individuals for each sample from the field, Marshall and Brown (1975) argued that the objective should be to collect at least one copy of each allele that occurred in the population with frequency >0.05 with 0.95 level of probability. The sample size required for this is 59 random unrelated gametes (assured by 50 individuals). A general formula for the sample size (S diploid individuals with an inbreeding coefficient F) to be 95% certain of obtaining one copy of an allele with population frequency (p) is

$$S \approx 3/\{(F-2)\log_e(1-p)\}$$

While this size of sample is adequate for including a single copy of all the common alleles in the population, it may be inadequate for accurate studies of their population frequency (Brown & Moran 1981). For example, a sample of about 400 gametes is required to estimate allele frequency within a 95% confidence interval of 0.1 (assuming the worst case of an allele with population frequency of 0.5). In the case of genetic resources for crop improvement, however, the accessions are intended primarily for study and use in the future, at which time their increase may be justified. Hence a single copy of an allele is notionally all that is needed now.

With respect to geographic distribution, alleles fall into four classes. The '*locally common*' alleles occur at an appreciable frequency – greater than 5 or 10% – in only very few localized populations. They may be rare or absent elsewhere. Second are alleles that are *common* (in population frequency) and *widespread* (geographically) with several common occurrences. The third class are only ever *rare* in frequency ($<10\%$) but are *widespread* in the species. The final class consists of alleles that are both *rare* and *localized*. Locally common alleles merit a higher priority than the two classes of rare alleles because they presumably contribute substantially to the current adaptation of populations (Marshall & Brown 1975).

Deciding upon the ideal number of populations to sample is more difficult than the individual sample size. This number is often set by the available resources, the extent of material in the field, or the need to sample other regions or species. A cardinal principle is that the target region be divided into distinctly different environments (using climatic, edaphic and vegetational variation) and sufficient sites chosen to cover the ecological range. (See section 6.2.1, pp. 153 to 157 for further discussion of stratification.)

2.5.3 Optimal allocation of sampling effort

Allocation questions deal with the optimal allocation of a fixed level of limited 'resources' in order to maximize a desirable product. For example, Marshall and Brown (1975) considered the question of the number of sites to sample in any single collecting mission as a trade-off with the number of individuals per site. The desirable product or basic parameter to maximize is the total number of locally common alleles in the collection. Hence the optimal strategy that maximizes this number is to collect samples of sufficient (minimum) size from as many sites as possible.

Sampling procedure (whether random, orientated to particular phenotypes, structured, stratified or repeated) needs to be considered. On the one hand, random sampling is generally a robust and statistically desirable procedure. On the other, the biologist usually has some information about the species to structure the sample and increase its diversity. Stratified random sampling is the logical compromise when the stratification depends upon major biological features known ahead of sampling. Latitudinal range is one such feature, as Mikola's (1982) study of geographic patterns of morphological variation in *Pinus sylvestris* in Finland shows.

Two components form the procedure, the stratified or selected component, and the random or stochastic component. They provide a conceptual link with the two kinds (systematic and stochastic) of forces acting on plant gene pools (see section 2.3; Brown 1992). The two components of sampling match two kinds of purposes: a random component of unspecified variation for indefinite future needs, and a stratified, selected component, for defined ecological conditions.

One problem with much of the sampling theory developed for *ex situ* conservation is that it assumes an even spread of diversity among populations. Since sampling frequently has to take place in the absence of information of diversity, this assumption is justified. However, as more data emerge about the geographic patterns of diversity, more efficient strategies will be ones that use such information. Indeed the assumed evenness of variation is generally violated by inbreeding and asexual plant species. Ultimately the best way around this problem is two-stage sampling, where diversity studies of the first round of samples are used to improve the second round (Jain 1975).

2.6 CONCLUSION – THE GENES OF CONCERN

A vast array of diversity at the gene level confronts the conservationist. This array includes the genes that differ among plant species and that account for their differential adaptation. To this is added intraspecific variation, on which the evolution and future survival of today's species and their descendants depend. Claims are commonly made that not only individual species, but their genetic resources must be conserved. In response to such claims, the question is, just how much of this diversity can be coped with in conservation programmes?

Modern techniques for surveying diversity at the DNA level have revealed samples of this diversity and the patterns they display. These data have pointed to considerable divergence between plant populations – a theme that unites this evidence with that of the classical work on ecotypes. The patterns result from a combination of ancestry and ecology. Although it is impossible to specify unequivocally where gene conservation priorities lie, clearly the divergence between populations stresses the need for ecological diversity in sampling. In this way a richer heritage, and one more likely to contain the resources for the future, will be conserved.

3

The genetic diversity of cultivated plants

3.1 EVOLUTION OF GENETIC DIVERSITY

3.1.1 Introduction: The evolutionary continuum

All domesticated species derive, directly or indirectly, from wild species, some of which were among the much larger numbers of species used by hunter-gatherers for unknown millennia. The ancestors of many domesticates have been identified, although those of some important crops, among them maize and common wheat, are still subject to research and argument. Wild species, including most forest and pasture species and some tropical tree fruits, are used and extensively planted without having undergone the more or less drastic morphological and physiological changes that are associated with domestication of many crops. Thus there is an evolutionary continuum linking the prehistoric pre-domesticates with the present-day cultivars; and an ecological continuum linking wild and semi-domesticated species with those which have been modified beyond the point of no return to natural conditions.

These relationships are relevant for two reasons. First, wild species continue to be taken into some form of domestication. With the increasing pressures for intensification of land use and for new industrial raw materials, species exploited in their natural state and others as yet not used at all are taken into cultivation and, ultimately, into full domestication. Second, wild progenitors and other close relatives of crop species – the secondary gene pools (Harlan & deWet 1971) – have assumed an important role as genetic resources used by plant breeders. Gene transfer by genetic engineering is opening up tertiary (distantly related) gene pools, a process whose only constraint seems to be the rapidly expanding techniques of gene transformation and the regeneration of transformed cells into whole plants.

Cultivation grew out of food gathering which imperceptibly led to elements of domestication. Yen (1985) views this transition as a form of 'intensification of plant gathering'. This may involve a gradual 'domestication of the environment' by measures such as water control in New Guinea. In the lives of gatherers 'the advantages of growing plants on purpose are not conspicuous at the beginning and the differences between gathering and cultivation are minimal' (Harlan 1975a). White (1989) remarks on the narrow dividing line between 'wild' and 'domestic' strains of rice and yam, of which prehistoric domesticators in southern Asia may initially have been unaware. But once cultivation became a part of the food gathering process, it set up new evolutionary dynamics which inevitably led to a progressive transformation of the plant.

In biological terms, the dividing line may be similarly diffuse. The oil palm (*Elaeis guineensis*) grows on the edges of West African forests. It was utilized extensively as a source of oil and palm wine, and varieties were selected that favoured one or other product (Harlan 1975a). It is from such materials that high-yielding cultivars and hybrids have been produced, in Malaysia and elsewhere. Where is there a cut-off point between wild and domesticated? Indeed, in the light of the extensive use of wild species in current plant breeding, the question loses some of its relevance.

As is evident from its title, this chapter is largely focused on cultivated plants. *In situ* conservation of wild plant species used by humans, either directly, or as genetic resources in crop breeding, is the subject of Chapter 6. However, aspects of conservation that are common to all plants, such as the preservation of seeds or of tissue culture, and aspects of organization and management applying to all plants used by humans, are included in Chapter 4.

3.1.2 Domestication

Domestication is the foundation stone of the biodiversity of crop species. It is from this foundation that landraces[1] emerged and diversified, and they in turn were the foundation stock for the modern cultivars developed in the last hundred years. Much is still obscure in what has been a process which in some instances extended over centuries. But for an increasing number of species some of the elements are being unravelled gradually: the wild progenitor, the site or region, the timing, the singleness or repetitiveness of the event. For others, like the progenitors of many minor yet important

[1] Landraces are the crop varieties of peasant farming; see section 3.2.1

species which help to sustain farming communities in the tropics, their origin will forever remain obscure.

In a definition that accommodates a wide range of genetic and/or cultural adaptations, Harris and Hillman (1989) delineate domestication as 'human intervention in the reproductive system of the plant, resulting in genetic and/or phenotypic modification'. They suggest three distinct, though not mutually exclusive, pathways of domestication.

1. Very rapid genotypic change, resulting in inability of the plant to survive in the wild. Examples are the domestication of cereals and grain legumes in South-West Asia.
2. Gradual genotypic change, eventually resulting in inability to survive in the wild, through a process of 'incremental ennoblement' (Wilkes 1989). An example is the evolution of maize.
3. 'Plastic' phenotypic change without (immediate) genotypic change. An example is the experimental modification of wild yam (*Dioscorea rotundata*) by Chikwendu and Okezie (1989). This pathway may apply to many root and tuber crops, and to some fruit species whose 'instant domestication' was achieved through the adoption of vegetative propagation.

The three pathways are distinguished by their prevailing breeding system. Species in the first group are inbreeders, those in the second are outbreeders, and those in the third are vegetatively propagated.

We add a fourth, poorly defined group of species which do not properly fit into any of the three categories, possibly because of lack of information. They include vegetable and fruit species developed and maintained in peasant cultivation in the tropics, many of which are derived from wild species that are unknown or have vanished.

1. *Rapid domestication by adaptation syndromes*

Prototypes are the 'founder crops' of Near Eastern agriculture. They include the cereals einkorn and emmer wheats and barley, and the grain legumes peas, lentil, chickpea and bitter vetch (Zohary 1989). Domestication of wheat and barley appear to have followed, not preceded, cultivation. According to archaeological evidence the first cultivated crops had a brittle rachis, natural selection favouring the dispersal facility it provided. Selection of the non-brittle rachis depended on husbandry methods, including time of harvest relative to maturity, and harvest tools. In an experimental model of domestication of einkorn, computer simulation indicated that domestication could theoretically have been achieved in 20–30 years (Hillman & Davies 1990).

Genetically, the acquisition of a non-brittle rachis was a more or less simple process in the cereals, in most instances conditioned by one or two genes. It has been obtained in experiments in every grass species in which selection has been attempted, usually with great ease, e.g. in *Andropogon hallii* by harvesting seed a month later than the usual time (Harlan *et al.* 1973). This favours genotypes with a non-brittle rachis. While in most cereals selection for tough rachis was involuntary, in sorghum, according to Harlan (1989) it must have been deliberate since at harvest plants are handled plant by plant.

Once cultivation was established, other adaptations evolved in response to selection pressures associated with harvesting and with seedling competition. These resulted in larger inflorescences, uniform ripening, increase in the proportion of seeds set, and an increase in seedling vigour resulting from the larger size of grain and from more rapid and synchronous germination through loss or reduction of germination inhibitors (Harlan *et al.* 1973).

In the grain legumes it was the loss of seed dormancy which was the critical change for domestication to eventuate. Dormancy is strongly expressed in the wild lentil (*Lens orientalis*), and presents a selective advantage through retaining a reservoir of seed in the soil. In contrast with the cereals, the development of non-shedding was not a crucial element of domestication. If threshed on the spot, losses through pod dehiscence can be reduced through early harvesting; but with transport to the homestead the importance of non-dehiscence was enhanced. Selection for non-dehiscence could have been a rapid process. With indehiscence based on one gene, a mutation rate of 10^{-4} and a rate of seed loss of 50%, indehiscence could have dominated the gene pool in under 20 years (Ladizinsky 1989).

2. *Gradual genotypic change*

According to a recent review on the domestication and evolution of maize by Wilkes (1989), the earliest cultivated maize (3500–2300 BC) was virtually identical with the earliest material found (5000 BC), which was arguably a wild maize. The earliest introgression from teosinte – a close relative of maize – came about 1000 years later, and continued ever since.

Molecular evidence has confirmed the now widely held view on the interrelations between maize, teosinte and the related genus *Tripsacum*. Dennis and Peacock (1984) found that the knobs located on the chromosomes of these species had as a main component of their knob heterochromatin, a 180 bp repeat sequence. *In situ* hybridization experiments showed that this

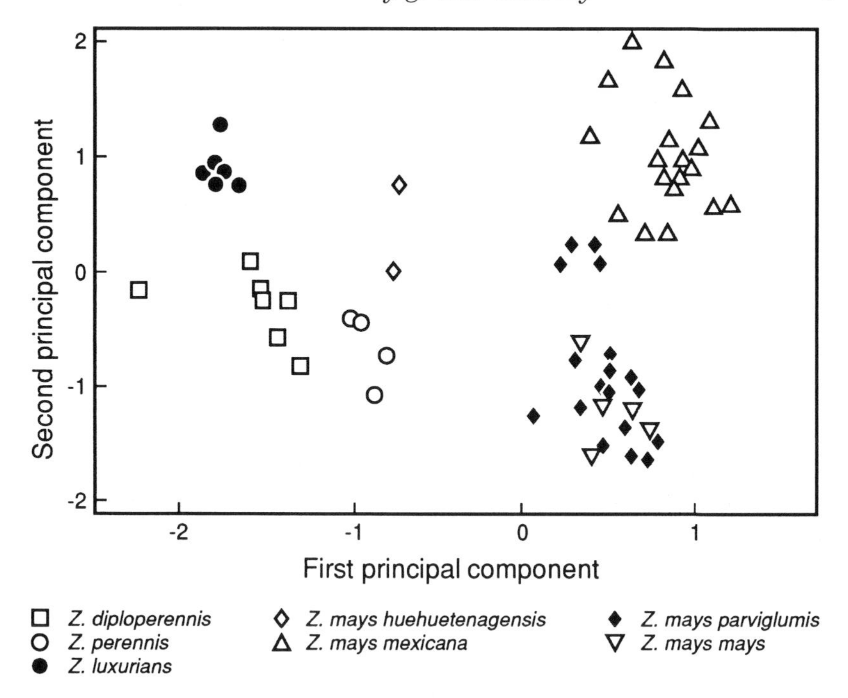

*Figure 3.1. The close relationship between maize (*Zea mays*) and teosinte (*Zea mays parviglumis*) shown by principal component analysis of isozymic variation in* Zea *(from Doebley* et al. *1984).*

sequence was also present in terminal knoblike heterochromatin of diploid *Zea diploperennis* and tetraploid perennial teosinte, *Z. perennis*. However, it was absent in three species of the related genera *Coix* and *Sorghum*. Maize and teosinte share additional 202 bp and 235 bp segments which are absent in *Tripsacum* and the perennial teosinte. While the evidence demonstrates the close relationship of maize and teosinte, it fails to establish the derivation of maize from teosinte, as against an hypothetical wild maize.

Studies of isozymic variation and chloroplast (cp) DNA of maize and its wild relatives, reviewed by Doebley (1990), provide significant information on their relationships and on the origin of maize. They clearly demonstrate the close relationship between maize and teosinte, *Z. mays* ssp. *parviglumis* (Figures 3.1 and 3.2), suggesting the latter to be the ancestor of maize. Molecular evidence also suggests the Balsas River valley in southern Mexico as the area of domestication. There is, as Doebley (1990) emphasizes, no convincing evidence for multiple origins of maize.

3. *Domestication without drastic genetic change*

In this third group, Harris and Hillman (1989) placed 'plastic' phenotypic change, without genetic modification, as a prototype. In such cases domestication amounts to the manipulation of pre-existing responses of the plant to specific environmental conditions. We include in this category other instances of domestication of wild species taken into cultivation with little if any immediate genetic change. There are two types.

(i) 'Plastic' phenotypic change not involving directional genotypic change, domestication amounting to the manipulation of pre-existing responses of the plant to specific environmental conditions. Wild species of

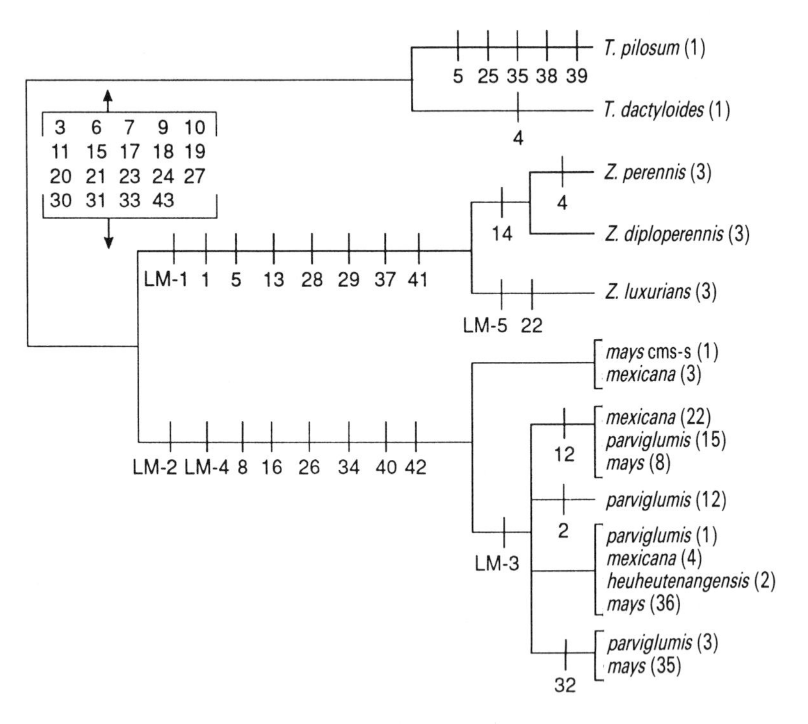

*Figure 3.2. A Wagner parsimony phylogenetic tree showing the close relationship between maize (*Zea mays*) and teosinte (*Zea mays parviglumis*) based on cpDNA variation. Two species of* Tripsacum *were used as outgroups to root the tree; identification numbers for specific mutations appear along the branches; numbers in parentheses are the number of populations in which that cpDNA type was found. [Reprinted by permission from* Economic Botany **44** *(3, suppl.): 6–27, J. Doebley, Copyright 1990, The New York Botanical Garden.]*

yam (*Dioscorea* spp.) have been used in many parts of the tropical world, for food or for poison. They are readily reproduced. For example, the Australian aborigines, who had no horticulture, planted the heads back at harvest time (Harlan 1975a).

Recently, Chickwendu and Okezie (1989) tested experimentally the attaining of a domesticated phenotype through transformation of the environment. Tubers of wild African yam (*Dioscorea rotundata*) were collected in their forest environment, and setts taken from them planted in a range of cultural regimes. While the wild tubers had a dense network of coarse roots and strong shoots, both with strong thorns along their length, after five seasons of vegetative propagation under cultivated conditions, the shape of the tubers began to approximate that of domesticated *D. rotundata*. The roots were greatly reduced as were the thorns, the fibre content of roots was much reduced and the starch content increased. The authors argue that this experiment may contribute to an understanding of the domestication of yam and possibly of other tropical root and tuber crops.

The question arises whether some form of genotypic variation generates opportunities for adaptive evolution in asexually reproducing plants. In a recent study of genetic variation in asexual lineages of dandelion (*Taraxacum officinale*) King and Schaal (1990) discovered significant restriction site variation in rDNA among sibling clonal lines derived from sibling plants. They suggest that clonal lineages accumulate genotypic variation which may become the basis of adaptive changes. Cultivation in specific environmental conditions induces heritable morphological changes in flax, including weight and height of plants. The changes are associated with molecular alterations, e.g. in the copy number of genes coding for the 18S and 25S ribosomal RNAs and for the 5S RNA (Cullis 1985). Specific and repeatable changes in restriction fragment patterns, arising by DNA rearrangement, were also detected with a 5S RNA probe (Schneeberger & Cullis 1992). Therefore we cannot assume that the mode of domestication postulated for yam is absolutely without any genetic change.

(ii) Selective clonal domestication. A shift from sexual to vegetative reproduction is a rapid and effective way of establishing a domesticate. In this way important characteristics, such as a distinctive size or shape of the fruit, can become fixed and adopted as the standard and hallmark of domesticated races. This, according to Zohary and Spiegel-Roy (1975), was the case in the domestication of the ancient fruit species of Near Eastern origin – olive, grape, date, pomegranate and fig. All of these species are cross-pollinated and extremely heterozygous, hence sexual reproduction yields a multitude of mostly undesirable types. Vegetative reproduction is simple in all of them, and outstanding individuals were probably selected to initiate clones which may have been maintained for centuries. The authors suggest that in five or six millennia under cultivation these species may have undergone very few sexual cycles. Such instant domestication may have occurred in many temperate as well as tropical fruits. Parthenocarpic, seedless forms of the breadfruit (*Artocarpus altilis*) are common in eastern Polynesia, with a fertility cline from the western Pacific. A similar transformation may have occurred in the banana (*Musa*) and in the New Guinea banana, *Australimusa* (Yen 1985).

However, sexual reproduction remains as a means for the release of allelic diversity and for recombination. This is well established in the sweet potato (*Ipomoea batatas*). It produces seed which may germinate in unexpected places. Diversity among seedlings has been demonstrated, and a seedling origin for some of the numerous cultivars is highly probable (Yen 1974).

4. *Vegetable and fruit crops of tropical peasant farming*

As stated previously, this poorly defined group includes a large number of vegetable and fruit species used in peasant agriculture in the tropics. Their wild ancestors are largely unknown or have vanished. These species are of importance to rural households and communities but do not reach world markets. They originated, and are cultivated, in tropical Asia, America or Africa where they have not been subject to modern plant breeding. For the time being their genetic resources are relatively safe. The vegetables and fruits are sources of vitamins and minerals that are of special importance in diets based on rice, potato, cassava, taro or sweet potato. Some, like the bean crops, are valuable sources of protein.

Many of these plants are derived from wild plants which no longer exist. The Andean crops are old domesticates, with archaeological records of up to 5000 years. In South-East Asia, the archaeological record is scant. There can be little doubt that these ancient crops, many of which are harvested individually and are consumed in the cultivator's household, would have been subject to a good deal of selection for size, shape, flavour, colour, etc.

Three small books on vegetables, root and tuber crops, and fruits grown in South-East Asia provide information on some of the plants grown by peasant cultivators for home consumption and local markets (IBPGR 1980, 1981a,b). Vegetables like the bitter gourd (*Momordica charantia*), the yard long bean (*Vigna unguiculata*) or the winged bean (*Psophocarpus tetragonolobus*) are of more than local importance, as are many fruits like mangosteen (*Garcinia mangostana*) or rambutan (*Nephelium lappaceum*). The 'potato yam' (*Dioscorea esculenta*), a native of Indonesia, is cultivated in countries as far apart as New Guinea and West Africa. It has never been found in the wild state (D. E. Yen, personal communication).

A group of Andean crops, the 'lost crops of the Incas', has recently been highlighted under this heading by the US National Research Council (1989). All of them were domesticated in the Andean region, many of them thousands of years ago, from wild ancestors that are no longer identifiable. Some of these crops, important staples in Inca farming, were suppressed by the conquerors but kept in being by peasant farmers. Some, including potatoes, peppers, tomatoes, squashes and lima beans, have spread worldwide. Others, like oca, tamarillo and passionfruit, have been adopted in particular countries – oca in Mexico and New Zealand, tamarillo and passionfruit in New Zealand and elsewhere. Others again have remained confined to their Andean home. Yet some of these have promise for wider distribution, as emphasized by the National Research Council (1989).

Why are the wild ancestors of so many of these crops apparently missing? Extinction through habitat deprivation, gene flow from cultivated forms into the original wild populations, or over-exploitation of the wild populations are possible causes. Assuming close similarity in morphological and physiological characteristics, in ecological requirement and in retention of mating compatibility, mutual integration would be the most likely outcome. However, the possibility of a polyploid origin, with or without hybridity, needs to be examined where appropriate.

3.1.3 Single versus multiple domestication

Whether the domestication of a species was a single event or whether it occurred repeatedly, in time and/or in space, is not only of historical, but of evolutionary significance, limiting or expanding the genetic base. Ladizinsky (1985) contends that many seed-propagated crop plants exhibit a founder effect due to their derivation from small groups of plants that became isolated from their wild progenitor by a series of genetic changes, including major mutations or polyploidy. Similarly, Pickersgill (1989) suggests that founding populations of domesticates may represent only a small proportion of a progenitor's genetic diversity (founder effect), or variation may be randomly reduced by genetic drift.

The occurrence of multiple versus single domestications is associated with the breeding system and hence with the pathway of domestication (cf. the preceding section). There is widespread agreement that domestication of seed-propagated species, involving major genotypic and morphological changes (pathways 1 and 2) is a rare event. Evidence comes from genetic and cytological observations on domesticates and their progenitors. The Near Eastern founder crops (see page 41 above) have as their major distinction from their wild progenitors the loss of seed dispersal mechanisms (brittle rachis or dehiscent pods, respectively). This involves a single major gene change, with the exception of barley which has two genes for brittleness. Had domestication been a multiple event, a diversity of mutant genes might be expected. Further, in peas, lentils and tetraploid wheat only one of the chromosomal variants of the ancestral species is represented in the domesticate (Zohary 1989).

Domestication of different, though related, ancestral species can result in multiple, though distinctive, domestications. This has, for example, occurred in chile pepper (*Capsicum*) with at least three domestications derived from distinct progenitors. They resulted in clearly distinguished species – (*C. pubescens*, *C. baccatum* and the *C. annuum–chinense–frutescens* complex (Pickersgill 1989)). This is confirmed by isozyme analysis (Doebley 1989). Other examples are independently domesticated species of *Gossypium* and *Chenopodium* (Pickersgill 1989).

In contrast with the seed-propagated crop plants, multiple domestication must have been a common event in vegetatively propagated species (pathway 3). Vegetative reproduction in root and tuber crops and in many fruit species is a simple operation, and so is what Zohary and Spiegel-Roy (1975) have called 'instant domestication'. Hence in such species there is the potential for great diversity. However, in the long-lived fruit species this is scarcely realized, for the simple reason that once a good strain is recognized, its long-term retention is likely. One may presume that many wild species were used for long periods prior to being taken into gardens or fields, and that cultivation, and hence domestication, was initiated by many intelligent cultivators; multiple domestication would have been inevitable. In many instances the wild progenitors have disappeared, possibly because they merged with their domestic descendants. It is therefore not possible to ascertain the changes brought about by domestication, and hence whether strain differences arose before or after domestication.

3.1.4 Genetics of domestication

Two genetic events are of particular significance in the domestication and subsequent evolution of crop plants. These are major gene mutations (MGM) and polyploidy. The crucial role of MGMs in the domestication of many crops – in particular of the cereals – has emerged in the preceding sections. Some MGMs provided the umbrella for the unconscious and/or deliberate selection of genes fitting a plant to the principal elements of cultivation – sowing and harvesting (Ashri 1989). As Jain (1989) points out, many adaptations, for example to temperature or photoperiod, are controlled by major genes. However, many other ecological adaptations are polygenic.

Lester (1989) notes that under domestication many morphogenetic characters have evolved that are based on mutations to recessive alleles. Characters of survival value based on dominant alleles, such as seed dispersal, dormancy, disease resistance and alkaloids that are distasteful to herbivores, have been lost. Evolution under natural conditions results in the slow accumulation of adaptations favouring survival, whereas domestication may lead to rapid changes which are due to the loss of effective products of particular genes.

Polyploidy, by combining the genes of several species, was a radical evolutionary event that greatly expanded the diversity of many crops. In wheat, the first polyploidization, resulting in the tetraploid *Triticum dicoccoides*, occurred prior to domestication. Of great importance for the establishment and expansion of the polyploid wheats was the evolution of the genetic system that restricts pairing in meiosis to chromosomes of the same ancestral diploid species. This is due to the action of the *Ph* gene, situated on chromosome 5B. Reference has already been made to the second step in polyploidization, the origin of the hexaploid wheats (see Box 3.1).

3.1.5 Geography of domestication

Vavilov (1926, 1949–50, 1951), a geneticist and plant geographer, explored the agricultural flora in many of the less developed and largely mountainous parts of the world, where the indigenous crop varieties had not yet given way to cultivars selected by plant breeders. On the basis of the geographical distribution of genetic diversity in different crop species he identified areas as centres of genetic diversity, which he believed were centres of origin. All of them were located in mountainous regions with an ancient agricultural civilization (Figure 3.5). It was soon recognized that the 'Vavilov centres' were indeed centres of diversity, but that sites of domestication and diversity did not necessarily coincide. However, the concept of centres, or regions of genetic diversity proved of great significance in the study of the evolution of crop species and in the assembly and utilization of genetic resources. Figure 3.5 presents the Vavilovian centres[2].

Subsequently several authors proposed modifications, largely on the

[2] For an account of the origin of the centres concept, see Harris (1990).

BOX 3.1. DOMESTICATION OF THE PRINCIPAL CEREALS

The possibility cannot be excluded that the cytogenetic events which led to the establishment of polyploid domesticates are not uncommon, though domestication itself may have been a rare or unique event. *Hexaploid wheats*, including common bread wheat (*Triticum aestivum*), are descended from tetraploid emmer (*T. turgidum* ssp. *dicoccum*) crossed with the wild diploid *T. tauschii*. This could have occurred only after domesticated emmer arrived in the *T. tauschii* habitat (extending eastward from the southern end of the Caspian Sea), since the current distributions of *T. tauschii* and wild emmer do not overlap. Zohary and Hopf (1988) suggest that hexaploid hybrids very probably were produced several times, involving different varieties of the two parent species, and state that such hybridization still occurs in wheat fields in Iran. Recent evidence, however, indicates a narrow base for the *T. tauschii* component, the D genome of the hexaploid. Lagudah and Halloran (1989), using seed esterase and protein markers, examined a large number of accessions of *T. tauschii* from its main area of distribution. They found that three accessions belonging to the variety *strangulata*, derived from an area south east of the Caspian, corresponded closely to the D genome constitution of selected hexaploid subspecies, and to representative hexaploid wheats in a germplasm collection. This evidence tends to support a monophyletic origin.

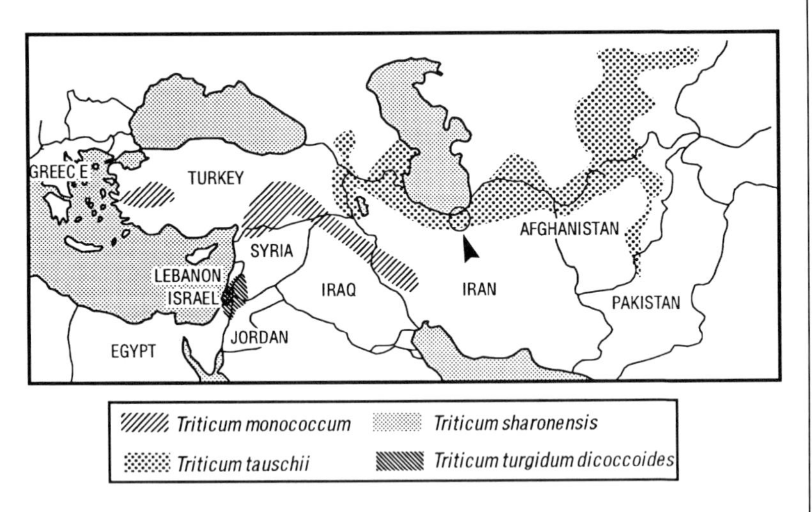

Figure 3.3. Present-day distribution of the ancestral species of wheat (redrawn from Lagudah 1986). The area in which the variety strangulata *is found is circled.*

BOX 3.1. (*cont.*)

Most authors conclude that maize originated in Mexico (e.g. Wilkes 1989; Pickersgill 1989), with an early established secondary centre in the Andes; morphological characteristics, chromosome knobs and mitochondrial DNA exhibit similarities as well as substantial differences between these centres. Kato-Y (1984) concluded that maize had been domesticated at several independent centres in Mesoamerica. However, molecular evidence clearly indicates a single area of origin in southern Mexico (see section 3.1.3).

Asian (*Oryza sativa*) and African rice (*O. glaberrima*) have a common origin, though geographically separated since the break-up of Gondwana; they share the A genome (Chang 1976b). African rice has its primary centre in the upper Niger River area. Asian rice diversified over a vast area in South, South-East and East Asia.

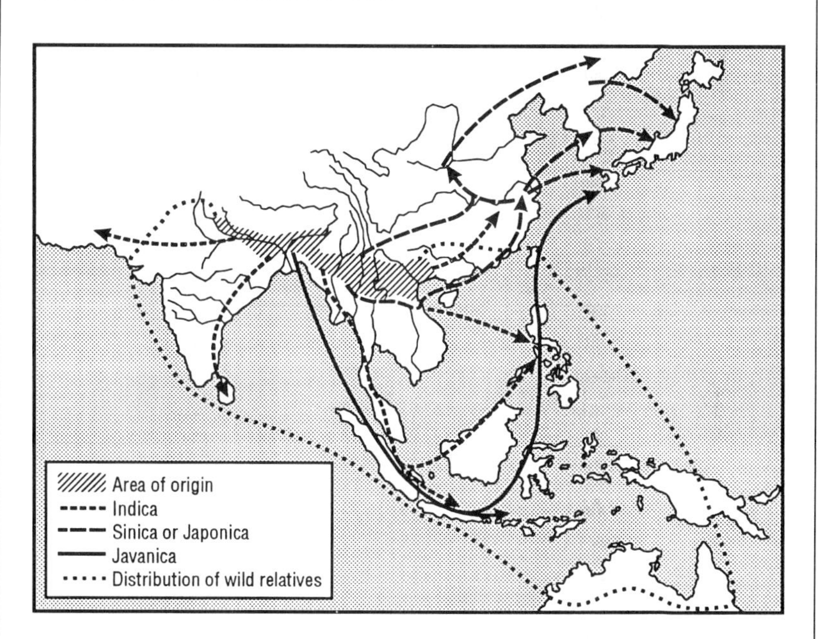

Figure 3.4. Distribution of wild relatives and spread of ecogeographic races of Oryza sativa *in Asia and Oceania (from Chang 1976a; reprinted by permission of Kluwer Academic Publishers).*

There are two levels of uncertainty in the domestication of Asian rice (*Oryza sativa*), the first relating to the origins of the major constituent branches, *indica* and *japonica*, the second to the domestication process

BOX 3.1. (*cont.*)

itself. Regarding the first, the majority of authors, including Oka (1974) and Chang (1976a) assume a monophyletic origin of *O. sativa* from *O. rufipogon,* with subsequent differentiation of northern (*japonica*) and southern (*indica*) subspecies and *javanica* as an offshoot from *indica*. On the other hand, Second (1982), on the basis of extensive isozyme surveys, proposed two distinct pathways, one from *O. rufipogon* in China, resulting in *O. sativa* ssp. *japonica,* the other from *O. rufipogon* in South or South-East Asia, resulting in the *indica* subspecies. The second uncertainty relates to the site or sites of domestication. Much of the evidence pertains to sites and timing of early cultivation, but not to domestication, due to the scarcity of archaeological findings. Chang (1976b,c) suggests that rice was domesticated independently at many sites in a broad belt extending from the Ganges plains, across Burma, northern Thailand and Vietnam to South China. Its dispersal routes are indicated in Figure 3.4. Oka (1988) also assumes multiple origins in a vast region stretching from India to China, and points to the need for solid evidence. This could come from more extensive archaeological investigations and from studies of characteristics like seed retention and uniform germination in wild rice populations which could have served as founding stocks for domestication.

basis of their own experience. Recognizing that some centres of genetic diversity did, but others did not, fit into the geographical concept of a 'centre', Harlan (1971) suggested that there were 'centres' and 'non-centres', the latter extending over vast areas of great physical and cultural diversity. Thus he recognized a Near East Centre, and an African non-centre across Africa between the Sahara and the Equator; a North China centre and a non-centre including South-East Asia and the South Pacific; and a Mesoamerican centre and a South American non-centre covering a vast area including the northeast and west of the continent.

3.1.6 Dispersal and diversification, gene pools and cultivars

Dispersal by humans played a major role in diversifying the gene pools of crops. Archaeological evidence shows that many domesticated species spread fairly rapidly throughout the known world. Wherever they went they were modified by the environment and by the cultural methods adopted by different civilizations. Secondary centres of diversity evolved in areas with environments favouring genetic diversification. For example, Ethiopia became a secondary centre for the Middle Eastern founder crops including barley, emmer, peas, lentils and several others.

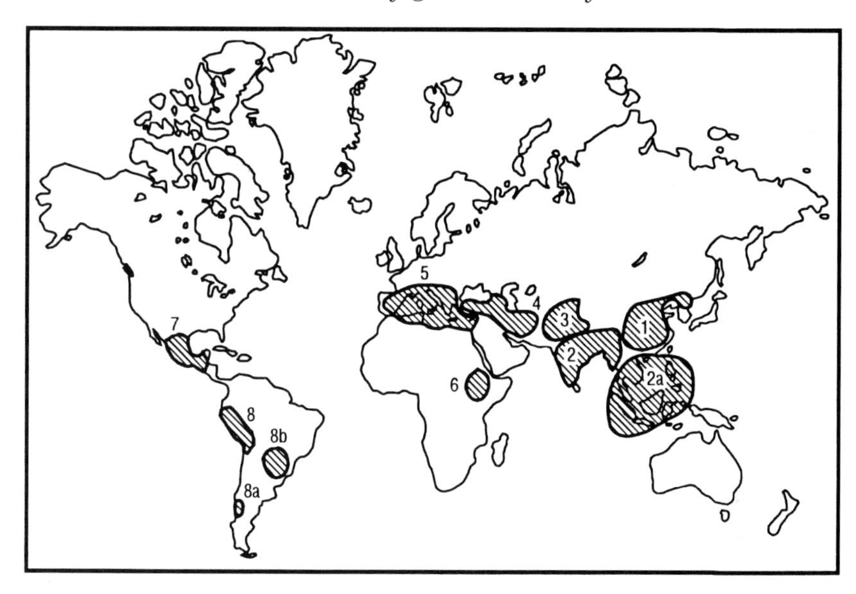

Figure 3.5. The eight centres of origin of crop plants as proposed by N.I. Vavilov in 1926. (Reprinted with permission from J.R. Harlan, Agricultural origins: centers and noncenters Science **174**: *468–74. Copyright 1971, American Association for the Advancement of Science.)*

Along with dispersal, frequent or sporadic recombination within primary gene pools (Harlan & de Wet 1971) occurred, as did mutations conferring resistance to parasites encountered and/or evolving in various environments. Introgression from wild and weedy relatives – the secondary gene pool – enriched the gene pools of crops and broadened the scope for the selection of adaptations. Such introgression could still be observed in recent times (Harlan 1951; Zohary & Feldman 1962). Thus there emerged distinctive local races, adapted to the many variants and interactions of natural and cultural environments to which crop species were gradually exposed. We have come to call them 'landraces' or 'primitive cultivars', in distinction from the 'advanced cultivars', the products of scientific plant breeding in the last 100 years, initiated by selection from landraces.

There is a marked difference in the nature of the adaptation of landraces and of the advanced cultivars of the last 100 years. Landraces evolved, presumably by a combination of natural and deliberate selection, under conditions of traditional, low-input cultivation, and adapted to the particular environment in which they were grown. Such localized adaptation resulted in diversification between landraces which is still recognizable in local names for ancient cultivars of crops or fruits (e.g. grape vines). In contrast, advanced cultivars are bred for high performance usually under intensive, high-input cultivation which tends to level out environmental differences of site and season. Environmental levelling is paralleled by genetic levelling: modern varieties are deliberately selected for performance over a range of environments, thus reducing the need for specific local adaptation. Genetic levelling within cultivars is even more drastic: the heterogeneity of landrace populations is replaced by the relative homogeneity of near-pure lines, hybrids, vegetative clones or blends. Clearly, advanced cultivars, like landraces, are ecospecifically adapted, though on a different scale. Over vast areas a modern cultivar can take the place of the multitude of landraces which had been in use until early this century. In some species of livestock, for example in poultry, this process has gone even further. In consequence genetic resources have been placed at serious risk.

3.1.7 The dynamics of genetic diversity

From the plants gathered by pre-agricultural hunter-gatherers to those by which we live now, levels of genetic diversity have undergone dynamic changes at specific and intraspecific levels. The trend has been for a decrease in the number of species used, accompanied by an increase in intraspecific diversity. However, in the last century, as we saw in the preceding section, the impact of intensive agriculture and scientific breeding also resulted in drastically restricting the intraspecific diversity of crop species. The emerging pattern derives from a progressive transition from a generalized natural ecosystem to a specialized artificial ecosystem, involving a drastic reduction in environmental diversity (Harris 1969).

During the transition period from hunting–fishing–food gathering to domestication of plants and animals, some 10,000–8000 years ago, a great variety of plant and animal species were used, with the distinction between ‘gathering’ and ‘cultivation’ gradually emerging. At Tepe Ali Kosh, in southwest Iran, Helbaek found in excavations dating from the early beginnings of domestication (7500–5600 BC) seeds of 40-odd plant species and bones of about 35 animal species; two of the former and two of the latter belonged to domesticated species. The total number of species greatly exceeded that used today in the same region. During the following millennium the inhabitants of Tepe Sabz, also in southwest Iran, used at least seven domesticated plant species (including nine distinct types) and three domesticated animal species. As the number of domesticated species increased, the number of wild species used for food decreased (Flannery 1969).

With the spread of agriculture in Afro-Asia and in Mesoamerica, plants, animals and humans, as Darlington (1969) observed, evolved in close interaction. Selection produced crops from wild plants or from weeds. The latter, as colonizers of disturbed environments, were already pre-adapted for cultivation (Hawkes 1969). Others, like the hexaploid wheats, originated from the combination of wild and domesticated species. The number of cultivated species increased with the rise of civilization and the differentiation of human needs, and with the spread of settled agriculture into regions which made their own contributions, such as the numerous tropical fruit and vegetable species originating in South-East Asia and Meso- and South America. Quoting Mangelsdorf (1966), 'during his history, man has used at least three thousand species of plants for food and has cultivated at least one hundred and fifty of these to the extent that they have entered into the world's commerce'. But, Mangelsdorf continues, there has been a tendency to reduce the number of crops to those which are the most efficient; and 'today the world's people are actually fed by about fifteen species of plants'. It should be noted, however, that this process applies in the main to staple food crops. A number of wild species have been taken into cultivation even in the last hundred years – as pasture plants or for soil conservation, as industrial or medicinal plants, or as ornamentals; and, as Brücher (1968a) pointed out, there is as yet a reservoir of potentially useful indigenous plants in South America and this is also the case in tropical regions of Asia.

As the number of major food plants declined, variation within species increased. This has been the case especially with the ancient staple food crops, which spread early in their history from the region in which they originated into other regions where they acquired further diversity. In this way the gene pools of our major crop plants were greatly enriched. When wheat reached China 3000 years after its domestication in the Near East (Darlington 1969), many types evolved which are peculiar to that region. Some, very likely evolved under irrigation, have ears with unusually large numbers of spikelets, and spikelets with unusually large numbers of grains. Wheat was carried across the Sea of Japan; and one may guess that it was this same character combination which, possibly centuries later, reappeared in a modern Japanese wheat, Norin 10, to find its way into Vogel's record-yielding hybrids (Vogel *et al.* 1963) and ultimately into the high-yielding Mexican varieties which helped to introduce the 'green revolution' in countries of Asia and Latin America. The process of intraspecific diversification continued throughout recorded history. The momentum was maintained by social and technological changes and by migrations, the most momentous being the post-Columbian exchanges across the Pacific and

Atlantic oceans. With the advent of scientific agriculture, plant introduction greatly extended the transfer of genetic materials on a world-wide scale. The extension of Old World gene pools had marked effects in the 'new' continents, North America and Australia, as had the New World potatoes, tomatoes, maize, and many more, on Old World agriculture.

In recent times scientific plant breeding has had the effect of restricting genetic diversity everywhere. In Western Europe, pure-bred cultivars began to replace landrace populations from the middle of the last century, and some fifty years later few if any of the landraces of major crops remained in what we now call the developed countries. Indeed, this progression has been a major part of 'development', and is so now in the developing countries. Introductions, selections from indigenous varieties, hybrids between them, and recently varietal composites are rapidly replacing the landrace populations which remain. The new cultivars have one feature in common – a high degree of genetic homogeneity. The drastic increase in productivity which resulted from the replacement of traditional forms of agriculture was as essential as it was inevitable. However, it placed in jeopardy the reservoirs of genetic diversity on which the continuing evolution of cultivated plants to a large degree depends. This dilemma will be discussed in later sections.

3.2 GENETIC RESOURCES TODAY

The evolutionary processes outlined in the preceding section have resulted in the great diversity of crop genetic resources which are now recognized. This diversity ranges at the evolutionary level from wild ancestors to advanced cultivars, at the ecological level from components of primeval ecosystems to those of high-input agriculture and horticulture, and at the genetic level from populations to genes. A functional classification was introduced by the International Biological Programme (IBP 1966; Frankel & Bennett 1970b).

- Landraces
- Advanced cultivars
- Wild relatives of domesticated plants
- Wild (i.e. non-domesticated) species used by humans

To these we now add

- Genetic stocks
- Cloned genes

Wild species are considered in Chapter 6, genetic stocks and cloned genes in Chapter 4.

3.2.1 Landraces

'Landraces', in the words of J. R. Harlan (1975b), 'have a certain genetic integrity. They are recognizable morphologically; farmers have names for them and different landraces are understood to differ in adaptation to soil type, time of seeding, date of maturity, height, nutritive value, use and other properties. Most important, they are genetically diverse. They are balanced populations – variable, in equilibrium with both environment and pathogens, and genetically dynamic.' The genetic diversity of landraces thus has two dimensions: among sites and populations, and within sites and populations. In the main, the former is generated by heterogeneity in space, the latter in addition by heterogeneity in time, i.e. by short-term variation between seasons, and by longer-term climatic, biological and socio-economic changes.

Generally, landraces evolved at low levels of cultivation, fertilization and plant protection, subject to selection pressures for hardiness and dependability rather than for productivity. That genetic diversity within landraces provided some protection against climatic extremes and epidemics is plausible, though we lack experimental or observational evidence; and disastrous epidemics have occurred throughout recorded history. Yet in less adverse conditions, subsistence agriculture was capable of producing at least enough to keep the cultivator alive until the next crop, and, as D. E. Yen (personal communication) points out, of providing a 'social surplus' for ritual needs, exchanges, etc.

The components of landraces often are deliberately manipulated by farmers. This is evident in maize crops in Mexico which are composites of deliberately selected types of great diversity; sometimes including the 'weed' teosinte, thus continuing the process of introgression which started with the domestication of maize (Wilkes 1989). Harlan (1975a) describes a similar procedure in the selection of seed stocks during the sorghum harvest in Africa.

Under adverse conditions landraces can still outyield selected cultivars. This was demonstrated in recent experiments conducted in Syria in which populations derived from eight Syrian landraces of barley and four control modern cultivars were tested in three environments, including highly productive, drought, and drought–saline conditions. In the two latter environments, the majority of the landraces outyielded the control cultivars by a considerable margin (Weltzien & Fischbeck 1990). It should, however, be noted that none of the cultivars used as controls had been selected in the adverse environments in which the landraces succeeded.

It does not follow that landraces were selected solely for survival of crop and man. In fact it is clear that selection for a multitude of characters took place, in the field and in the cultivator's home. This is obvious in vegetables and fruits where size, health and flavour are readily assessed. But it is also true of field crops – with selection of barley for beer, for food and for feed, of maize for flour, for popping and for religious rites, of glutinous and non-glutinous rice, of Pacific root crops for storage and processing qualities. And, as Jain (1989) points out, it seems very likely that throughout the history of agriculture, humans observed and selected novelties for colour, plant form, sex form, seed hairs (for fibre), etc., capitalizing on gene mutations. Indeed, species which were prone to produce such mutants, e.g. through the presence of active controlling elements (*Zea mays*, *Antirrhinum*) may well have enjoyed preference. More importantly, it appears that some selection for productivity also did occur, even in the small-grain cereals. The fact that the first advanced cultivars, whether selected from European landraces in Europe or in North America, or from indigenous landraces in Asia, produced greatly improved yields, goes to show that landraces contained high-producing components.

In favourable environments, landraces of the major crops are now mainly used as donors of genetic components which in some way enhance the biological or economic adaptedness of a crop plant, or the quantity or quality of a plant product. Such components may be well defined, simply inherited and easily transferred, such as many resistances to diseases and pests. They can be adaptations to specific environments; or they may be coadaptations inherited as supergenes, or held together by linkage, by the breeding system, or by selection pressures.

In less favoured and indeed in *marginal environments*, landraces of major as well as minor crops have remained the predominant form of cultivation (Table 3.1). Under such conditions very little breeding work has been done. Indeed, many national and international plant breeding establishments question the need for specific adaptations to marginal environments and place their emphasis on adaptability. Improvements over a range of environments are expected to extend to variable and marginal ones. However, this has not been confirmed by increased production by farmers in disadvantaged environments. Nor is it supported by experimental evidence: crossovers of regression lines indicate specific adaptations to both

Table 3.1. *Percentage distribution of durum wheat varieties in the countries of the ICARDA region*

Country	Indigenous varieties or landraces (%)	Improved cultivars (%)	New high yielding varieties (%)
Afghanistan	90	0	10
Algeria	85	0	15
Cyprus	0	90	10
Ethiopia	100	0	0
Iran	100	0	0
Iraq	25	65	20
Jordan	20	80	0
Lebanon	30	65	5
Libya	0	55	45
Morocco	50	50	0
Saudi Arabia	100	0	0
Syria	64	29	7
Tunisia	0	65	35
Turkey	10	40	50
Pakistan	88	0	12

Source: Reprinted with permission from Srivastava & Damania (1989), Cambridge University Press.

favourable and marginal conditions (Figure 3.6). For a discussion of the extensive literature on 'crossovers' see Evans (1993).

Work at ICARDA (the International Center for Agricultural Research in the Dry Areas, located in Syria) has demonstrated the value of local landraces both on farmers' fields and in plant improvement. At ICARDA the best pure lines are used in crosses among themselves or with non-landrace material, or as mixtures of pure lines 'which are a sort of improved landrace'. Emphasis is placed on 'selection of parents and segregating populations being located in environments climatically and agronomically similar to farmers' conditions' (Ceccarelli 1994).

It should be recognized that ever since Vavilov, landraces have been regarded not only as genetic resources, but also as sources of information on the evolution and genetic differentiation of crop species. Much less is known about the population structure of landraces, as it has evolved in different species and in different environments. In the section which follows we examine the population structure of landraces at two levels: within landrace populations, and among ecologically, genetically or agronomically related

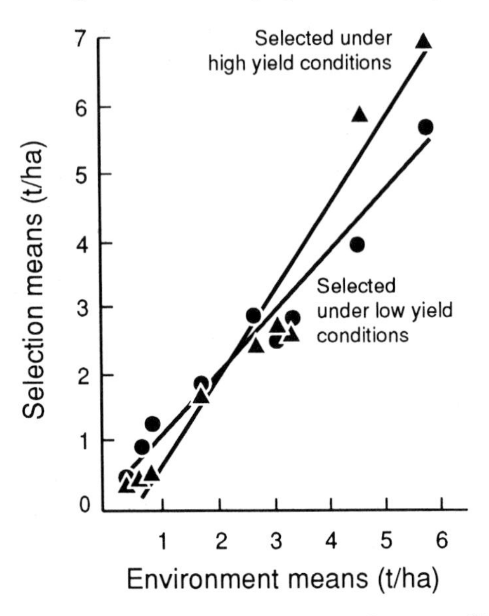

Figure 3.6. Linear regression over environmental means of barley genotypes selected for high grain yield in high yielding conditions (▲) or for high grain yield under low yielding conditions (●) (from Ceccarelli & Grando 1991; reprinted by permission of Kluwer Academic Publishers).

groups of landraces. The representation of this diversity in germplasm collections is discussed in section 4.5.1.

3.2.2 Landraces – population structure

Within populations

Heterogeneity and interactions: We have already seen that landraces are recognized as heterogeneous populations reflecting the natural and social environments in which they have evolved and diversified, subject to natural selection and, in some crops more than in others, to a measure of mass selection[3] by cultivators. An understanding of the population structure of these plant associations, reaching back to the earliest phase of settled agriculture, would be of considerable interest. Here we address two questions: (i) does heterogeneity itself contribute to the success of a landrace; and (ii) do the components of a landrace population interact in a

[3] Mass selection is the selection of a phenotypically superior group.

mutually beneficial manner? Crop evolutionists tend to answer both questions in the affirmative. Harlan (1975b) regards a landrace as 'an integrated unit [whose] components have adjusted to one another over the generations'. This interpretation is widely accepted in spite of the fact that evidence to sustain it is partial or indirect (see below).

An answer to the first question, whether heterogeneity *per se* is adaptive, comes from studies of the epidemiology of pathogens in varietal mixtures and multilines[4]. The question is whether components of genetically diverse populations have a disease buffering effect in reducing the spread of pathogens. A review of the experimental evidence by Burdon and Jarosz (1989) showed that in the majority of observations this had been the case. In over 100 observations, infection in two-component mixtures averaged 25% below that of the more susceptible component in monoculture. Wolfe (1985) observed that the overall infection in varietal mixtures approximates that of the resistant component grown alone. He also found that varietal mixtures are more generally effective than multilines (which are near-isogenic except for the resistance gene) because of the higher level of genetic heterogeneity of the former.

While the second question, whether components of landraces are interactive, has not been explicitly examined, we can derive indirect evidence from experiments on the yields of varietal mixtures and their components. Marshall (1977) reviewed the effects on yield in a large number of experiments with mixtures. In 328 varietal mixtures representing eight crop species, 289 resulted in yields corresponding to the mean yield of the components, four were lower and 37 higher than the mid-component. Clearly, there is no convincing evidence of positive interactions being widespread. Nor is there evidence of an increased stability of mixtures over that of the best performing components. However, very different conclusions resulted from a study by Allard and Adams (1969) of interactions in F_{18} of a composite cross derived from intercrosses of 31 barley varieties of highly diverse origins. Eight randomly selected lines were tested for their effects on each others' yields. In contrast with varietal mixtures included in the experiment, the composite cross population had been exposed to mutual selection for a long period of years; as the authors' results indicate, with marked effect. Of the small number of lines in the test, two yielded substantially more in competition with sister lines than in pure culture, and none substantially less. Selection for 'ecological combining ability' resulted in lines that are good competitors and at the same time good neighbours. It is

[4] Varietal mixtures are synthesized from individual lines selected for performance and interactions. Multilines are composed of lines that are near-isogenic except for a resistance gene.

worth noting that the best two lines yielded almost as well as the best commercial cultivar.

From the experiments with varietal mixtures we conclude that: (i) there is a positive effect of heterogeneity in buffering the spread and increase of pathogen infections; and (ii) interactions among components resulting in increased productivity could not generally be established. It must, however, be recognized that varietal mixtures do not – like landraces – result from selection over many generations; nor are they likely to generate and retain heterozygosity. But the fact remains that it has generally been possible – probably in all crops – to select, within landraces, individuals with an all-round performance surpassing that of the parent landrace. This suggests that poorer components limit the productivity of the population as a whole. This may even apply in adverse conditions where landraces do best. Reference may be made to a previously reported experiment on barley under drought–saline conditions in which local landraces outyielded improved cultivars (Weltzien & Fischbeck 1990; see section 3.2.1). One may, however, hazard the guess that cultivars bred from the successful local landraces might have outperformed their parental populations.

These observations cast some doubt on the dogma of the interactive structure of landraces and its *short-term* benefits. However, they do not touch upon the long-term benefits derived from landraces in having a population genetic structure which accommodates an indefinite range of genetic diversity; perhaps much of it of little immediate value, but a store for adversity and opportunity alike. It may well be that it is the *weakness* of selection to which landraces have been exposed, rather than its intensity and direction, which maintained and preserved their essential *long-term* quality. Had individual selection and progeny testing been invented prior to the nineteenth century, it might not have been possible to rescue and preserve the genetic diversity of crops in the twentieth.

Landraces of vegetatively reproduced species represent a special case. While seed-reproduced plants – and especially the inbreeders – store genetic diversity within and among populations, vegetatively reproduced plants as a rule are highly heterozygous: hence, in addition, preserve allelic diversity at the level of the individual. Since the majority of such species have the capacity for sexual reproduction, this stored diversity is released on selfing or crossing.

Among populations

Variation in the population structure of landrace populations has been studied in a large number of species, using isozyme techniques (for a list of

BOX 3.2. ISOZYME ANALYSIS OF LANDRACES

The use of isozyme markers makes possible an analysis of multigene diversity at the individual plant level. An example is a study by Geric *et al.* (1989) of representative accessions from the maize germplasm collection which had been assembled in Yugoslavia from old landrace populations established in the last two or three centuries. The 276 landrace populations studied were formed into 18 groups, mainly on morphological characteristics. The average number assayed per accession was 20. The analysis involved 12 enzyme systems controlled by 21 loci.

All loci were found to be polymorphic, though some only rarely, with two to eight alleles per locus. All 18 groups were genetically heterogeneous. There were no common features in the allozyme patterns of populations within groups – scarcely surprising in view of the composition of the groups. It was, however, possible to identify populations with a common ancestry, and this was confirmed by cluster analysis which was carried out within groups, but not over the material as a whole. Overall, variation within populations exceeded that among populations. The results confirmed that the commonly used germplasm of corn breeders is narrowly based, whereas exotic races, old varieties, isolated populations and wild relatives are rich sources of genetic variation.

This report contains the largest assembly of allozyme data on record. It calls for a more extensive analysis. For example, insight might be gained from following up ecogeographical and historical affinities in the analysis of isozyme patterns.

Isozyme analysis has been extensively applied in the study of wild relatives of crop species which is referred to in Chapter 2.

species, see Simpson & Withers 1986). Here we consider isozyme variation among populations with comparable ecogeographic background.

Landraces of barley collected in Iran had the authenticity of population structure which maintenance in germplasm collections almost inevitably tends to infringe (Brown & Munday 1982). The authors sampled 25 spikes each in 12 barley fields situated in four distinct districts of Iran. The outstanding result was the high level of variation at the 25 loci surveyed. In the 12 landraces the mean number of alleles per locus ranged from 1.16 to 1.64, and the proportion of polymorphic loci from 0.16 to 0.44. No heterozygous plants were identified. Thus, in spite of the absence (or low frequency) of outcrossing, a high degree of diversity is maintained. One may ask what is the significance of this diversity: is it adaptive and maintained by natural selection, or is it there because of the weakness of selection? The observations made in this survey are relevant to the comments made in the preceding section on the short and long-term significance of the genetic diversity within landrace populations.

Samples of two composite crosses of barley (CC21 and CC34) were included in this survey, and results of a previous study of wild barley (*Hordeum spontaneum*) in Israel, were also available for comparison. The landraces were intermediate for genetic diversity, with the wild barleys the highest and the composite crosses the lowest. However, such comparisons have their limitations. In a study of wild and cultivated barley collected in neighbouring locations in eastern Mediterranean countries, isozymic variation of cultivated barley equalled that of wild collections (Jana & Pietrzak 1988).

A second example involves a cross-fertilized species, with an isozyme analysis of 34 races representative of Mexican maize (Doebley *et al.* 1985). Ninety-four collections were analysed for 13 enzymes coded by 23 loci. Variation was extremely high, with 72% of the total variation occurring within collections and 28% among collections. The average number of alleles per locus was 7.09. This exceeds the allozymic diversity of most species so far examined. Some loci were near monomorphic, while *Glu1*, with 18 alleles, was the most polymorphic. Diversity among races was not marked, but overall, three geographically (and ecologically) defined groups could be discerned.

Conclusions

So far we have considered landraces as entities in their own right: products of domestication, exposed to a diversity of climates, to attack by pests and diseases, to introgression from more or less related genotypes, to the will of perceptive or fumbling cultivators. With a genetic structure formed by all of these, they retained in spite of, and perhaps because of this multiplicity of impacts, the high level of population diversity that has been testified in all extensive examinations. For thousands of years landraces sustained humanity's slowly rising numbers and, as we have seen, still do so in less favoured physical or social environments. Starting from the middle of the nineteenth century, landraces became the foundation materials of modern cultivars and, until recently, the principal gene resources of modern plant breeders. They have retained a prominent place in many breeding programmes, even of major crops. An example is the number of landraces or parents of unknown origin which entered CIMMYT popular wheat lines as progenitors: as Figure 3.7 shows, over the last 30 years there has been a steady rate of increase (Byerlee & Moya 1993).

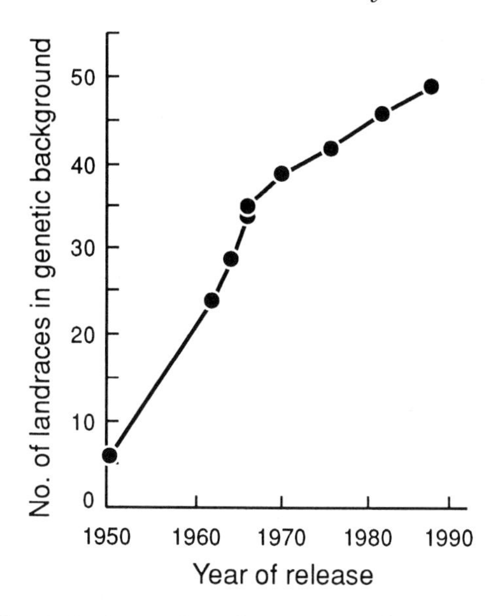

Figure 3.7. Trends in the number of progenitors of popular CIMMYT lines, calculated by tracing all ancestors to the original landraces or to parents of unknown origin (from Byerlee & Moya 1993).

3.2.3 Landraces – specific contributions

We shall now review a few examples of the contributions that landraces have made to modern plant breeding as donors of specific genetic components. However, we must recognize that today, landraces are no longer the only breeding resources. In recent years they have been giving way to breeders' populations and to crop-related wild species, especially in breeding for resistance to diseases and pests (cf. Chapter 6).

Germplasm collections[5] have been screened for a wide range of characteristics, some of which were discovered at very low frequencies. The world collection of landraces of safflower (*Carthamus tinctorius*) established by Professor P. F. Knowles, was screened in Israel for resistance to a range of diseases, including safflower rust (*Puccinia carthami*), the leaf spot diseases *Ramularia carthami* and *Cercospora carthami*, and phyllody, which is a

[5] Germplasm collections are assemblies of genotypes or populations representative of cultivars, genetic stocks, wild species, etc., which are maintained in the form of plants, seeds, tissue cultures, etc. Individual samples are called 'entries' or 'accessions'.

gradual reversion of florets to vegetative organs, induced by a mycoplasma disease transmitted by a leaf hopper. Resistant lines were identified for all these diseases. Resistance to rust was prevalent among lines from the Middle East, while lines from India–Pakistan and East Africa were more susceptible, which may be due to racial diversity in the pathogen. In *Ramularia* the situation was reversed, resistance being centred in parts of India where the disease is common, as against the Middle East where it is absent. A number of lines had resistance to two or more of the diseases. There was good correspondence in the geographical distribution of some disease resistances and of some morphological characters with which they are associated (Ashri 1971).

A survey of a large section of the barley world collection led to the discovery of two sister lines, collected in Ethiopia in 1923, both with a high protein content, and one of them also with a high lysine content. The latter, named Hiproly, had shrivelled grain and low yield, but its nutritional quality was transferred to high-yielding varieties (Munck 1972). A corresponding search for high lysine in the world sorghum collection yielded two high-lysine lines (Singh & Axtell 1973). Stolen (1965) tested a collection of 630 varieties of barley for tolerance of low pH in the soil. A small proportion were moderately tolerant, and one did significantly better at low than at high pH.

There is no country with a greater range of environments than India, nor a greater diversity of farming systems and crops. Even prior to the advent of high-yielding varieties of wheat and rice, cultivars bred in India reduced the genetic diversity of the principal crops. But to this day India retains extensive reservoirs of ancient diversity in farmers' fields in many parts of the subcontinent, but especially in mountainous and tribal areas where physical, ecological and social barriers have slowed the advances of modern technology. Exploration in the Himalayan foothills of northeast India yielded large numbers of primitive rice cultivars with resistance to diseases and pests, including blast, bacterial blight, tungro virus, gall midge and stem borer (IARI 1972; Seetharaman *et al.* 1972). Similarly, exploration in maize growing areas in the northeastern Himalayan region yielded cultivars resistant to leaf blight, downy mildew and corn borer (Singh 1974). Singh stresses the immense variability in the maize populations found in this region. As many as 15 individual races of maize have been identified and fitted into the race system developed by Mangelsdorf. They included highly promising material: some races equalled or even exceeded high-yielding hybrids or composites used as controls (Singh 1977).

In recent years, landraces have been complemented by wild relatives as sources of resistance. But the literature on genetics and breeding for resistance to diseases and insect pests – see, for example, a comprehensive review of the latter by Khush and Brar (1991) – is full of examples of landraces as donors of resistance genes. In many instances they may not be identified as such by the breeder for whom the status of a donor has little relevance. Yet there can be little doubt that the majority of resistance sources used by breeders in the past were landraces or their derivatives, obtained from germplasm collections which consisted largely of such materials.

3.2.4 Landraces – adaptations and coadaptations

From such well-defined genetic elements which affect a particular structure or function we turn to genetic contributions which have more generalized effects on the adaptedness and efficiency of genotypes. Most prominent among these is tolerance of climatic stresses such as drought, heat or cold, or edaphic stresses such as salinity, alkalinity or mineral toxicity.

Reactions to climatic stresses are highly complex phenomena involving various organs and processes, with effects differing with the time of incidence and the intensity of the stress. Screening for tolerance may therefore necessitate a number of complementary approaches, including field tests in stress environments, laboratory or glasshouse tests designed to reproduce the effects of natural stresses under controlled conditions, or laboratory tests developed from physiological or histological research into the effects of the stress. All these were applied by Russian workers in screening many of the collections in the Vavilov Institute of Plant Industry for frost and drought resistance. Among 19,000 accessions of winter and spring wheat, some 200 were highly resistant to frost and 150 to drought. As might have been expected, highest frost resistance is not found in the centres of genetic diversity for this crop, but in wheats from countries of the former USSR, Canada and USA. The drought-resistant types came from Central Asia, India, Syria, Chile, Mexico and, somewhat surprisingly, from Canada (Dorofeev 1975).

The germplasm collection of the International Rice Research Institute (IRRI) was extensively screened for drought resistance in mass field screenings. The largest proportion of resistant types was found in upland rices from Africa and South America, in hill races from Laos, and among early ripening varieties from Bangladesh. Field tests were supported by greenhouse techniques which separated the effects of early (vegetative) and late (reproductive) stresses. They also made it possible to discover associations between plant characteristics and drought resistance. Plants of drought-resistant varieties have more and thicker deep roots than susceptible ones, and this is now being used in screening techniques (IRRI 1975).

Combined with drought conditions, salinity is a major factor restricting yields in Middle Eastern countries. An extensive search for salt tolerance in wheat was carried out by Sayed (1985) in a collection of over 5000 accessions, mostly from semi-arid and arid regions. All lines were tested at the seedling stage. The 442 salt-tolerant ones were tested during a full life cycle. Seedling tolerance was found to be indicative of life-cycle tolerance. There was considerable diversity between regions, less between countries, but a good deal within countries – highest in Egypt and USA, where also the highest tolerance levels were found. Tetraploid wheats were more tolerant than hexaploid and diploid ones.

3.2.5 Landraces today and tomorrow

Many factors are relevant to the role of landraces in agriculture and biology: the physical environment (latitude, altitude, climate); the social environment (cultural, political, educational, technological); the agro-ecosystem (whether monoculture, multicrop, complex rotation); the species, and its utilization (diversity of plant parts used). It is therefore scarcely possible, or useful, to attempt generalizations on such issues as the retention or replacement of landraces, or *in situ* versus *ex situ* conservation, without consideration of the ecological system to which the generalizations apply. In some agricultural ecosystems landraces have virtually disappeared, but are actively exploited by plant breeders. In contrast, some rural communities largely depend on landraces of traditional crops that have not been subjected to plant breeding. There are, of course, many intergrades and combinations. Here we examine the place of landraces in the two extremes outlined above, which for simplicity, we call 'modern' and 'subsistence' systems. We also examine the role of landraces in production systems which are designed as 'sustainable'.

Landraces in 'modern' agro-ecosystems

Though still prevalent in some less favoured environments, the remaining landraces of crops produced for the market should be replaced by plant breeders' cultivars. These may be 'pure line' cultivars, hybrids, multilines, varietal mixtures, etc. *appropriate for the social, economic, agronomic and crop-specific conditions and opportunities.* The superiority and dependability of plant breeders' cultivars over local landraces has been proved in numerous experiments and by growers' experience over long periods. The

BOX 3.3. COADAPTATIONS AND GENE COMPLEXES

Of particular interest for an understanding of the evolution of landraces and of their contribution to future crop breeding, are their *coadaptations*, i.e. their complementary adaptations. Darwin, in *The Origin of Species*, emphasized the common occurrence of coadaptation in domesticated animals and cultivated plants as strong evidence of selection. He had in mind the coordination of many traits for some single adaptive purpose discernible at the level of the individual organism. Frankel and Bennett (1970a) suggested that 'the enormous diversity of gene complexes determining adaptation and productivity, assembled and incorporated over centuries of cultivation under different environments, must be recognized as the outstanding and unrivalled characteristic of primitive varietal populations'. Other authors have suggested that coadaptation occurs at several other levels; ranging from the level of tightly linked loci forming a switching supergene (Ford 1971); or of inversion heterozygosity in *Drosophila pseudoobscura* with mutually selected gene sequences (Dobzhansky 1970); to that of harmoniously interacting genes in the local gene pools of a population (Mayr 1963).

In barley landraces, Qualset (1975) examined the frequencies of combinations between eight characters in the Ethiopian collections noted previously. The characters were heading time, seed and ear traits, and barley yellow dwarf virus (BYDV) resistance. Over three-quarters of the combinations showed association, significant at or below probability of 0.05. Such associations are on a broad geographic scale. For example, populations from higher elevations were late flowering and had higher virus resistance, suggesting that both characters are needed jointly to achieve adaptation to higher elevations.

Often the genetics and selective basis of such complexes of characters are poorly understood. Genes can be held together in a population by linkage, by uniparental breeding systems (apomixis or self-fertilization), by isolation from other populations, by selection or by combination of any of these. The relative roles played by these forces, and particularly the strength of epistatic selection, need much more investigation. Allard (1988) has recently reviewed the extensive evolutionary changes seen over the generations of certain barley composite cross populations. The changes were in allelic frequencies for isozyme marker loci and disease resistance genes and morphological traits. Analyses of joint frequencies of alleles at several loci, that is of *multilocus structure*, showed that the dynamics of each marker locus reflects not only its own individual effect on adaptedness, but also an 'all-genome' effect. This effect is due to the reduced recombination between genes that results from the close inbreeding in the population. The relationships between marker multilocus genotypes and morphological traits were complex with substantial

BOX 3.3. (*cont.*)

higher-order interactions. From these studies, one might infer that a polymorphic landrace population also contains genotypes in which alleles at different loci have positive synergistic effects on fitness. Yet this proposal remains contentious and hard to prove (Hedrick 1983). This very difficulty of proof suggests to us that complexes which are polymorphic within populations will in general be difficult to sample, to conserve and to use in cross-breeding.

In contrast, the complexes which most if not all members of a particular population share are of major importance. A good example is the syndrome of characters that contribute to the productivity of the 'high yielding varieties' of wheat with Norin 10 ancestry. In addition to short height, there are increases in grain weight and in the number of grains per ear. All of these can be seen as adaptations to conditions of ample water supply and high fertility. The syndrome was derived from a Japanese variety which, in spite of its somewhat mixed ancestry, seems to point to China as a likely source of the syndrome. Wheat had been grown in China under irrigation for thousands of years, and Chinese landraces embodying features of the syndrome were common some 50 years ago, as was evident from a collection of Chinese wheats assembled in New Zealand.

present and future role of landraces is to serve as sources of genetic materials for plant improvement. Such materials can be either direct selections of individuals for the establishment of a cultivar; or they can be specific genes (or characteristics) to be incorporated in existing genotypes or cultivars. The former has been and is the first step in the genetic improvement of any crop. Today, however, wherever plant improvement has taken root the major agricultural crops are past this stage. Where this has not been the case, as in some parts of Asia or Africa, there is now a tendency to replace indigenous landraces with introduced 'high-yielding varieties' rather than with locally selected ones. This may result in the loss of valuable indigenous germplasm if it is not incorporated in germplasm collections.

Landraces in 'subsistence agriculture'

Reference has already been made to landraces of vegetable and fruit crops in traditional peasant farming (section 3.1.2). Typically, such landraces are maintained as populations, although group or individual selection may occasionally take place. As a rule production is for family use and for local markets. Some of these crops, as previously pointed out, have reached wider markets, sometimes through introduction to other, more production-geared socio-economic environments.

There is no question that the horticultural crops of small-scale tropical agriculture are important in the economy, nutrition and social well-being of large rural populations. This is likely to continue, even if the production of staple crops like rice, cassava or beans, were proportionately to increase. Nor are landraces likely to be displaced by locally adapted advanced cultivars, due largely to the scarcity of professional plant breeders. However, competition is likely to come from commercial imports of vegetable seeds – with potentially disastrous consequences for indigenous species and their genetic diversity.

The retention of landraces of major crops, especially in marginal environments, is discussed in section 4.2.

Landraces in 'sustainable agriculture'

Landraces play a part in planning for 'sustainable agriculture', based on traditional systems of agriculture. The plans emphasize stability of production and preservation of genetic diversity, rather than production levels. Sustainability is to be achieved by the interactions of a diversity of species (multi-cropping), by nitrogen-fixing leguminous crops, and by partial inclusion rather than elimination of weed species in the ecosystem. The aim is to generate equilibria between pests and their predators and to control undesirable weeds (Gliessman *et al.* 1981).

Discussions on sustainable agriculture tend to focus on subsistence farming in tropical coastal lowlands (Gliessman *et al.* 1981) or highlands (Altieri & Merrick 1987). These traditional systems seem in tune with economic and cultural community standards, and with limitations imposed by farm size and climate. Traditional varieties or landraces often are the only planting material, because of the species diversity, the ecological diversity, and the small scale of production. Landraces evolved in such systems are more dependable for growing in traditional ways than would be exotic modern cultivars.

Systems of sustainable farming have been suggested for broader scale cash-cropping, with the idea of using landraces because of their yield stability and built-in population-based defence against parasites (Altieri & Merrick 1987). However, 'with respect to disease and insect resistance, genetic diversity alone is an inadequate defense unless that diversity includes genetic resistance to the organisms in question' (Brown 1983). Further, yield stability (lower variance in yield) of heterogeneous populations is often purchased at the price of lower mean yield. Indeed, the least productive genotypes tend to have the lowest variance in yield.

Modern cultivars have a role to play in developed, sustainable systems. Breeding for resistances reduces the need for pesticides; their productive capacity has been greatly raised and continues to rise (see, for example, Austin *et al.* 1980). Clearly the use of landraces for broad-acre planting would entail substantial yield losses relative to cultivars selected for productivity in specific agro-ecosystems. A return to the famines, such as those in India prior to the green revolution, is unthinkable.

3.2.6 Advanced cultivars

Modern, or advanced cultivars are (more or less) homogeneous populations derived from selection by plant breeders. They emerged in the middle of the nineteenth century when plant breeders in France and England discovered the advantage of individual selection and progeny testing, by comparison with mass selection which had been practised previously. Johannsen (1909) provided the theoretical background, landraces the raw material. Early crossbreeding anticipated the rediscovery in 1900 of Mendel's laws of inheritance. Selection from landraces continued well into the twentieth century, to be joined and replaced by crossbreeding in all major crops.

Successively, the earlier products of rapidly developing plant breeding activities became obsolete. Many remained in existence in germplasm collections. This is how some 'obsolete cultivars' survived, although most of the earliest ones have vanished through lack of interest or attrition of repeated reproduction. Yet these obsolete cultivars, though lacking the population diversity of landraces and their multitude of adaptations to diverse environments, are not without interest. Exposed to more rigorous selection for performance than landraces had ever been, they had been bred prior to the widespread use of fertilizers, intensive cultivation, or irrigation. They may possess adaptive characteristics which later became redundant, yet may be needed again for less favoured environments or low-input agricultural systems. Further, resistance to currently rare biotypes of pathogens may again be required should these biotypes build up in the future. Another reason for retaining representative collections of obsolete cultivars is that many of them have been evaluated and documented for many characteristics associated with performance. It is, however, relevant to note that in yield tests of old and contemporary English wheat cultivars under high and low fertility conditions, the recent cultivars outyielded the obsolete ones in both situations (Austin *et al.* 1980). In view of their large number, the question arises, which to preserve? Judicious culling is inevitable, though along what principles?

The most popular genetic resources of plant breeders today are current or recently released cultivars. They are the biological benchmark and the

Table 3.2. *Major sources of broadening germplasm; percentage of respondents using each source*

Crop	Elite unadapted %	Landraces %	Related species %
Cotton	59	76	24
Soybean	9	85	0
Wheat	100	25	15
Sorghum	13	88	0
Maize	48	70	4
All crops	34	71	8

Source: Reprinted by permission from *Economic Botany* **38**, 161–78. D. N. Duvick, copyright 1984, The New York Botanical Garden.

Table 3.3. *Major sources of needed pest resistance; percentage of respondents using each source*

Crop	Elite adapted %	Elite unadapted %	Landraces %	Related species %
Cotton	41	65	35	35
Soybean	79	70	33	24
Wheat	95	95	50	50
Sorghum	63	75	38	38
Maize	83	61	39	17
All crops	76	72	39	17

Source: Reprinted by permission from *Economic Botany* **38**, 161–78. D. N. Duvick, copyright 1984, The New York Botanical Garden.

genetic base for further improvement, not only for the breeder who produced them, but for others, in the same country or elsewhere. In addition, as a survey of 100 American plant breeders in 1981 showed, a surprising number used elite (i.e. advanced) cultivars for 'broadening the germplasm' they are using; but of course not exclusively (Duvick 1984) (Table 3.2). In four of the five crops, more than two-thirds of the participants use landraces, but only cotton breeders include related species. In breeding for pest resistance, again a large proportion of breeders find resistance sources in advanced cultivars (Table 3.3); but up to 50% resort to landraces and wild relatives. However, breeders appear to agree that elite varieties rather than landraces contribute the best sources of tolerance of climatic stress.

Genetic vulnerability became of widespread concern as a result of the 1970 epidemic of southern corn leaf blight (*Helminthosporium maydis*) which destroyed some 15% of the corn crop in the United States. Two major factors were responsible: uniformity for major genes controlling susceptibility, and the extent of the area planted to a susceptible variety. Accordingly, the greatest risk occurs when widely used varieties are exposed to a build-up of a pathogen to which they lack resistance (National Research Council 1972). Such vulnerability is becoming a widespread phenomenon in many crops due to the recent narrowing of the genetic base. In rice it has two independent sources. The first is the use of the same gene, *sd1*, for short plant stature by breeders in several countries, and its subsequent spread in the cultivars derived from the early hybrids such as IR8. The second source of vulnerability is the extensive use of cytoplasm derived from China (Tjina) in the breeding of semidwarf cultivars; and the identity of the cytoplasm in the hybrid rices widely grown in China (Chang 1984).

There is a widespread realization that broadening of the genetic base is a real need if genetic vulnerability is to be reduced and further progress to be achieved. For maize this was recently expressed by Goodman (1988), who noted that extensive isozyme surveys had shown the close relationship, amounting to pedigree duplication, in US hybrid maize. The genetic base of this crop is even narrower than previously assumed, with little evidence of the entry of exotic germplasm into the gene pool.

3.2.7 Life-support and under-utilized and new crops

The distinction between life-support and under-utilized crops is by no means strict. The former are plants which help to sustain human life in 'stress-prone habitats under temperature, soil and water stress and in emergency situations' (Paroda *et al.* 1988). Under-utilized crops, like the former group mostly tropical and subtropical, are not limited to specific environments. Their distinguishing feature is a potential for expansion of what is currently a limited area of cultivation. Some had been more widely used in the past but have given way to more remunerative alternatives. Both groups have attracted scientific and popular attention and have become targets in the search for new crops or new products from old ones.

Life-support species include a wide array of wild and domesticated species with actual or potential uses stretching from vegetables and fruits to medicinal plants. An international workshop on life-support species (Paroda *et al.* 1988) dealt mainly with stress conditions and with life-support species in India, but the observations and conclusions apply to comparable conditions in Asia, Africa and Latin America. The discussion which follows refers to contributions to this workshop.

Life-support species, in the wider sense, are any species grown under stress conditions, including even cold-resistant rice (Vergara 1988). In the narrower and more commonly used sense, they are species that are restricted to, and mostly endemic in, stress conditions of various kinds – drought, waterlogging, heat, cold – or on poor or problem soils such as saline, acid or alkaline soils. Shankar (1988) points out that a distinction needs to be made between a regularly occurring stress and an occasional incidence, such as years of drought when, thanks to improved communications, drought-relieving supplies can be brought in. Zakri and Ghani (1988) suggest that the economic significance of life-support species is of a lesser order than their social significance. Yet their cultivation might extend into other – mostly unfavourable – environments in consequence of population increase and soil degradation (Randhawa 1988).

Domestication or improvement of some species may open up or enhance their cultivation. However, as Smith (1988) points out, this is an expensive process, hence the cogent need for selection of the most promising target species. This should take into consideration the social and economic relevance of a species, as well as the prospects of its expansion into new, including more productive agro-ecosystems. If, as is proposed by several contributions to the workshop referred to above (Paroda *et al.* 1988), this involves the collecting and preservation of the genetic diversity of the full array of such species, their large number makes discrimination inescapable. Yet the small number of priority species nominated indicates the difficulty of choice.

'Under-utilized' species (National Academy of Sciences 1975) are species which are believed to have a potential for expansion beyond their current area of cultivation and/or form of utilization. Some, like a number of the 'lost crops of the Incas' (National Research Council 1989) have been widely cultivated in the past and have given way to agronomically or commercially more attractive alternatives. The main impetus for revivals or promotions of under-utilized crops is to diversify and stabilize agricultural production. For this there are agronomic and economic reasons. Alternative or additional crops should make a system more sustainable through modifying and diversifying crop rotations, improving the soil structure and the nutrient status (e.g. by the inclusion of nitrogen assimilating crops), and through reducing the build-up of pests and parasites which occurs in monocrop or

short-rotation agricultural systems. The economic advantage of diversification is the stabilization of farm incomes that depend on major commodities subject to price fluctuations.

The promotion of under-utilized crops is the objective of a recently established autonomous, independent centre, the International Center for Underutilized Crops (ICUC). It has established working groups to identify priority species on a regional or commodity basis. The centre is to foster the improvement (genetic and agronomic) of target species, establish a database and organize training activities. Once a target species is identified, major genetic changes are likely to be needed to make it agronomically and economically attractive to growers and to consumers of the product. This need is recognized by ICUC. Even for an established crop a seemingly small genetic change resulting in a new product is likely to require a major effort.

An example of such a procedure is the recently performed transformation of linseed oil, the oil contained in the seed of flax (*Linum usitatissimum*), into an edible oil. Used in the production of paints, linoleum, etc., linseed oil has been progressively displaced by synthetic materials, whereas the demand for edible oil is increasing. Linseed oil is not suitable for use as an edible oil because linolenic acid – its principal oil component – is subject to autoxidation which is the cause of 'off'-flavours. Replacement of linolenic acid by linoleic acid would solve the problem; linoleic acid is one step ahead on the metabolic pathway to linolenic acid. Green (1986a, b) succeeded in obtaining two induced mutations, located on different chromosomes, which when combined resulted in lowering linolenic acid to 2% and increasing linoleic acid to 48%. The new type of flax, named Linola, will be a valuable component in a rotation based on wheat.

An under-utilized crop, considered a prospect for large-scale adoption, is kenaf (*Hibiscus canabinus*). After many years of costly development, including the design of technologies and machines, there is a prospect of success, at present realized on a minor scale in Texas. The aim is to use this fibre crop in paper production, as a replacement for shrinking timber resources.

A cooperative project to 'Improve Conservation and Use of Under-utilized Crops of Regional and Local Importance' has been initiated by workers in South, South-East and East Asia, with coordination from the Asia–Pacific group of the International Board for Plant Genetic Resources (IBPGR 1993). The target species are buckwheat, sesame, safflower and minor millets (*Panicum, Setaria, Eleusine,* etc.). The species have been grown in substantial areas for long periods, mainly as subsistence crops, without attempts at developing their production potential. This programme is designed not only to preserve the genetic resources but to make them available for use and plant improvement. *In situ* conservation of landraces is being considered as part of this project (see section 4.2).

In recent years, emphasis has been given to 'new crops' (Janick & Simon 1990). For the most part these are under-exploited ones being given new emphasis. The likely result is an increasing diversification of the food supplies – mainly vegetables and fruits – in developed countries, with some economic benefits to developing countries which produce many of them. But these are limited prospects. As Janick (1990) points out in the 'Afterword' to the symposium *Advances in New Crops* quoted above, 'fueled by prices of $1 to $3 a pound, we can point to such successes as kiwifruit and broccoli'. But crops with a value of 5–10 cents per pound have a great difficulty in getting established; hence the prospect of displacing established crops, with their infrastructure of growers, processors and marketing agencies, quite apart from the accomplishment of genetic and cultural adaptation to various environments, is slight indeed.

It may well be that in the quest to upgrade and adapt new species and under-utilized ones the main requirement is for imagination, hard work and patience of individuals, rather than the multispecies testing which has been suggested. The opposite is likely to be the case in the search for new resources for *industrial use*. Here extensive knowledge of plants and their components, and extensive, systematic searching and testing, seem essential conditions for success; and one may ask whether opportunities are likely to be greater among under-utilized species than in the vast reservoir of unutilized ones. The US Department of Agriculture conducted an extensive screening programme for components of interest to chemical industries (White *et al.* 1971). Several species emerged with sufficient promise to encourage further examination. Four potential oil seed crops may serve as examples of promising species tested and developed in subsequent years: cuphea (*Cuphea* spp.), vernonia (*Vernonia galamensis*), jojoba (*Simmondsia chinensis*) and meadowfoam (*Limnanthes* sp.). Oils derived from these species differ widely in chemical composition and industrial application (Carlson *et al.* 1992). The authors believe that these species may be among the new crops for the 21st century. What they, and most of the potential 'new crops' now being tested and developed have in common, are their botanical diversity and wide range of products, far exceeding the circle of staples on which past agricultural production has been based.

Although, for reasons given earlier in this section, the emphasis appears to be on agricultural crops, in the short term success is more likely in the various *horticultural industries*. A recent example is the Chinese gooseberry or kiwifruit (*Actinidia chinensis*). Developed in New Zealand this century, it has spread to many countries world-wide, a delightful addition to the diet of the well-fed rather than new sustenance for the hungry. On a smaller scale, some of the tropical 'life-sustaining' vegetables and fruits might be developed for export, as 'specialty' or 'boutique' crops, opening limited opportunities for small-scale rural producers, and for the diversification of food resources of the well-to-do. On the positive side, opportunities for new species for the floricultural trade are virtually without limit.

In the area of major food crops, the chance of finding new species that would help to sustain the population pressure of the 21st century, is scant. It is not easy to go against the historical and evolutionary trend that has narrowed the great diversity of food crops to the few that can be produced to the best effect, and which have been transformed to sustain our ever growing numbers. One can scarcely help feeling that if there are to be new and better species we may have to construct them ourselves.

4

The conservation of cultivated plants

4.1 CONSERVATION IN PERSPECTIVE

In this chapter we are concerned with the conservation of plant species, populations and genotypes which are actually or potentially useful to humans. To this qualification we must add two others which derive from the first. Since we cannot determine what characteristics will be required in the future, or even which of the current economic species will continue to be used, a reasonable strategy of conservation would include (i) types likely to be useful in the near future, (ii) representative samples of the genetic resources for use and study by future generations. Further, it follows from the qualification of usefulness that a time scale of concern (see Chapter 1) could be formulated in accordance with reasonable expectations of future uses. This has some tactical implications. For example, in view of the uncertainties of future requirements on one hand, the possibility of new methods for generating and using genetic variation on the other, a period of 50–100 years seems an adequate time scale of concern for the preservation of crop genetic resources. Such a commitment on behalf of the future is on a normal human scale, in contrast to long-term nature conservation which inevitably carries the burden of the awareness that commitments by any generation are subject to the vagaries of decision making by each subsequent one. Another implication of a relatively short time scale of concern is the realization that the regeneration of stored seeds which are losing vitality – perhaps the most burdensome operation in seed preservation – may possibly be avoided altogether (see section 4.4.6).

In an earlier chapter we recognized four classes of genetic resources with distinctive genetic and ecological characteristics (section 3.2). In germplasm conservation concern is focused on two categories with distinctive requirements for conservation – wild species and domesticates. Wild species are best conserved in their natural habitats within the communities of which they form a part. This applies to those wild species which are regarded as actual or potential genetic resources. In general the principles of nature conservation apply, with some qualifications discussed in Chapter 6. Only when such communities or some individual species within them are threatened may some form of protection be necessary, either under natural conditions as in forestry reserves (see section 6.2.1), or in specially designed 'genetic reserves' (see section 4.2); or *ex situ* (section 4.3). However, the genetic resources of the majority of wild species used by humans can be regarded as reasonably safe in at least a proportion of their natural habitats, although in some instances there is a need for protection, in others for continuing watchfulness.

By way of contrast, all domesticates require positive measures for their maintenance and preservation. Few if any of the most vulnerable genetic resources – the landraces of major crops remaining in cultivation – can be regarded as safe unless they are adequately represented in a well-conducted germplasm collection, i.e. *ex situ*. All genetic resources maintained in whatever form of *ex situ* preservation require continuing supervision, maintenance and protection, in some instances involving considerable expenditure. Hence efforts to limit the burden, by cooperation on one hand and rationalization on the other (section 4.5.2), are to be supported. However, landraces of some minor crops and of many horticultural and forestry species, and of some crops grown in marginal environments, are preserved in cultivation, i.e. *in situ*, by peasant farmers (section 4.2).

The life cycle, the size and cost of the unit of preservation, and the method of preservation, largely determine the scale and effectiveness of operations. For example, fruit species, preserved as trees, are expensive to maintain and are subject to virus and insect attack, restricting both scale and effectiveness. Preservation as seeds, pollen or tissue culture allows an increase in scale, reduces costs and expedites distribution.

Here we make a special plea for the preservation of accessions or stocks which have been extensively used or evaluated in investigations of any kind, overriding differences of use, purpose or biological status. They include materials used in cytogenetic, evolutionary, physiological, biochemical, pathological or ecological research on one hand, accessions evaluated for their agronomic or breeding propensities on the other. An example of the latter group is given by the wild species collected for their potential as forage species in other parts of the world. It is not unusual for such accessions to be examined and evaluated for a number of years, only to be rejected and lost if not found entirely suitable. Yet such material might be of use in other environments. While it is true that if sites are well documented the material could be collected again, the expense of collecting is considerable, to which must be added the value of previous evaluation which strictly applies only to the material on which it was carried out.

4.2 CONSERVATION *IN SITU*

Both domesticates and their wild relatives can be preserved *in situ*, i.e. in an environment in which they are subject to continuing selection pressures. *In situ* conservation is therefore described as dynamic (Figure 4.1). In domesticates, landraces cultivated in their home environment are an example of dynamic conservation. Once landraces are displaced by improved cultivars as part of a change in the agronomic and social system, *in situ* conservation of landraces, as a deliberate measure, would meet with difficulties. Attempting to retain 'primitive farms' in a radically changed social and technological climate would, except in the short term, be self-defeating. However, based on long experience in Iran and Turkey, Kuckuck (in Bennett 1968, pp. 32,

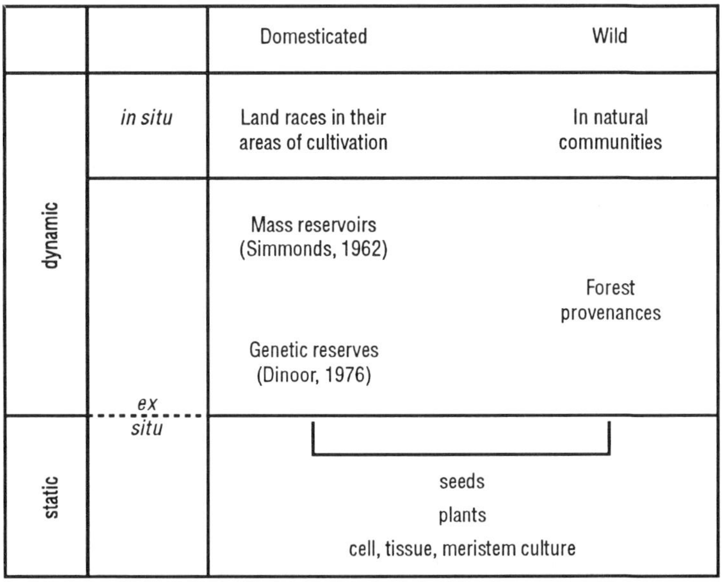

		Domesticated	Wild
dynamic	*in situ*	Land races in their areas of cultivation	In natural communities
dynamic	*ex situ*	Mass reservoirs (Simmonds, 1962) Genetic reserves (Dinoor, 1976)	Forest provenances
static	*ex situ*	seeds plants cell, tissue, meristem culture	

Figure 4.1. Conservation of domesticated and wild species of economic significance (reprinted with permission from O.H. Frankel and M.E. Soulé, Conservation and evolution, *1981, Cambridge University Press).*

61) proposed the establishment of what might be called 'crop reservations' – areas of 0.5 to 1 ha in size where a local crop variety would be maintained under the supervision of a local agricultural officer. The areas would be subject to changes in the environment brought about by agricultural development – fertilizers, cultivation, etc. – and to genetic change mediated by natural hybridization, mutation and natural selection. They would thus form 'mass reservoirs' (see next section), with opportunities for gradual adaptation to changing environments and with genetic self-renewal through mutation and introgression. Frankel (1970a) questioned the practicability of maintaining a number of small isolates, with all the inherent tactical difficulties, let alone the inevitable loss of identity. Proper maintenance might be available in experiment stations, which, however, would be technically *ex situ*.

A new approach to *in situ* preservation of landraces, spearheaded by ethnobotanists, has been fostered in recent years (e.g. Oldfield & Alcorn 1987). It involves preservation in peasant cultivation as a dynamic, and, as the authors claim, safe and inexpensive form of long-term conservation. Cultivated within the system in which they have evolved or become adapted, landraces remain exposed, and may adapt, to impacts from their environment – new species or races of parasites, introgression from wild or weed relatives, changes in the climatic, edaphic or socio-economic environment and, to varying degrees, to forms of selection by cultivators.

It is, however, a fact that in all the major, and in many minor crops, landraces have been replaced in the 'broad acre' farming of many developing, as well as developed countries, by plant breeders' cultivars, with landraces persisting in arid, mountainous and other stress environments. Altieri and Merrick (1987) restrict their examples of persistent landraces to subsistence farming in isolated mountain or forest environments, and to crops peculiar to subsistence farming. As Brush (1989) points out, improved varieties have been readily adopted in areas of relative homogeneity; whereas in marginal farming areas with environmental, economic and cultural heterogeneity, landraces have remained.

Other authors maintain that in some more favourable environments landraces have been retained as part of a deliberate management strategy. This is based on what could be called the greater 'resilience' of landraces to biological or climatic impacts. The population structure of landraces is presumed to include components such as pest resistance genes, which enable the population to survive and to produce a modest return, as against the pure-bred cultivars which may be wiped out by an epidemic.

In a wide-ranging review of past, present and future conservation strategies, Brush (1989) contrasts the emphasis on *ex situ* conservation of landraces with the ideas expressed by ethnobotanists and others (cf. the three preceding paragraphs) which emphasize conservation through continuing cultivation of landraces within their agro-ecosystems. The adoption of *ex situ* conservation was caused by the rapid spread of high-yielding varieties (HYVs) of wheat and rice, with the result that in these, and in due course in many other crops, landraces would be totally displaced by improved cultivars. *Ex situ* conservation was therefore advocated as a safeguard against what was called 'genetic erosion', the loss of genetic diversity represented by landrace populations. Brush (1989) points out, however, that the idea of retaining landraces in cultivation gains strength from the realization that in many marginal agro-ecosystems landraces have not been extensively replaced, and that, as already mentioned, some landraces have been retained elsewhere for a variety of agronomic and socio-economic reasons, such as food flavours, or the need for long straw for roofing or for bedding of stock. He concludes that the role of landraces in existing ecosystems is not adequately understood, and that quantitative assessments by ecologists, ethnobotanists and social scientists should replace the anecdotal evidence elicited by an earlier generation.

In distinction from the ethnobotanists whose main concern is with 'under-utilized' and 'subsistence' crops (see sections 3.2.5 and 3.2.6), Brush (1991) discusses the continuing cultivation of landraces of major crops. Examples include potatoes in Peru, maize in Mexico, and rice in Thailand. Brush contends that the replacement of traditional varieties is not a simple linear process resulting from economic, social and agricultural changes. Over substantial areas in developing countries landraces are holding their own alongside 'modern' varieties. In the case of potatoes, the reasons are culinary appeal and higher prices on the market, against modern varieties with higher yields and, rather surprisingly, resistance to disease and stress. In the maize and rice examples the advantages are less marked, and the reasons are likely to be ecological and cultural. Traditional maize varieties are retained in a transitional area where both traditional and modern varieties are grown, but, one may guess, the available modern varieties are not adapted to the marginal subsistence farming.

Brush (1991) suggests that the (dynamic) *in situ* cultivation could be of continuing value to the (static) *ex situ* collections, which in itself would justify international support. He is opposed to direct subsidies to farmers; yet in the long term an advantage to the farmer is hard to see. It is likely that farmers would be better served if in place of 'modern' varieties, bred for favourable agricultural environments, farmers – in what the author states often are marginal environments – had available varieties *bred for their conditions*.

Brush is undoubtedly right that germplasm collections and their users stand much to gain from landraces preserved *in situ*, especially since this occurs prevailingly in regions of great environmental diversity. Yet it is difficult to cast aside the feeling that it is the farmer who foots the bill, whether he knows it or not. One can perhaps take comfort from the thought that not only the developed countries, but also a rapidly growing number of developing ones, served by modern plant breeders, take advantage of germplasm collections. It should be noted that the need for breeding varieties for marginal environments is increasingly recognized: see, for example, Ceccarelli (1994) and CIMMYT (1991; section 3.2.1).

Two final comments are appropriate. First, it should be recalled that when 'improved varieties' first appeared in Europe a hundred years ago, they swept away the landraces, which, in all the major and many minor crops, rapidly disappeared. Second, these improved varieties were not the 'high yielding varieties' of our day, but individual selections from landraces based on progeny testing and, later, hybrids among them. Such an approach, applied today, could greatly improve productivity and resistance to parasites in areas where landraces still prevail, combining ecological adaptedness with genetic improvement. Nor need such cultivars impose the environmental burden of intensive applications of fertilizers or pesticides which could endanger fish or wildlife supplies. It may well be the case that in such conditions *it is not the superiority of landraces* which is the cause of their retention, *but the lack of local plant breeders* who could produce *'improved varieties' in tune with the agro-ecosystem*.

4.3 POPULATIONS *EX SITU*; MASS RESERVOIRS

When populations of domesticated or wild species are maintained in habitats other than those in which they evolved or became adapted, they are exposed to selection pressures prevailing in their adopted site. Such exposure may be incidental, as in many introductions of crops, forages, or forestry species which are regularly reintroduced from the country of origin (e.g. eucalypt species from Australia). On the other hand exposure to, and selection in new environments may be the main purpose of the operation, resulting in populations adapted to different habitats.

A specific proposal relates to the use of mass reservoirs as a form of dynamic genetic conservation. Mass reservoirs are populations derived from crosses between large numbers of diverse parental types, maintained for a number of generations (see, for example, Harlan 1956). Simmonds (1962) proposed the use of mass reservoirs as more appropriate for long-term conservation than what he called 'museum collections' of individual accessions. His reasons were the high rate of attrition and the absence of continuing natural selection in long-term storage. However, attrition has been largely overcome by current technology (see section 4.4). That mass reservoirs have a place as 'an adjunct to plant breeding' (Simmonds 1962) has been asserted by some, questioned by others. Here we are concerned with their role in conservation. The question is whether genetic variation is maintained more effectively than in collections consisting of individual accessions. Marshall and Brown (1975), using as evidence the loss of from 50 to 70% of the original genetic variation for height and heading date in barley Composite Cross V, argued that mass reservoirs 'are of little value in preserving variation, potential or expressed'. Nor would growing mass reservoirs in many different places – to counteract genetic depletion – reduce the effect by comparison with maintaining individual lines in storage. Further, they quoted evidence to show that different composite cross populations, grown at the same site, far from maintaining their differences in allelic frequencies, were approaching the same end point after a number of years, thus tending to retain the same fraction of variation. Frankel (1970b) questioned whether exposure to current selection pressures may adversely affect the frequency of specific characteristics and even more so of coadapted complexes (see section 3.2.4). Further, collections maintained in the form of individual entries make it possible to study separate gene pools and to relate their characteristics to their original environments. The pragmatic anonymity of bulk methods is in itself an argument against their exclusive application.

Recently, Marshall (1990) resumed the discussion of mass reservoirs as a form of dynamic conservation, in the light of the current argument on the need for providing self-renewing sources of host resistance to balance genetic changes in parasite virulence. Marshall argues that the large germplasm collections now available for most crops provide a substantial safety factor for the future. Further, many crops – including oats, barley and rice – have associated and cross-compatible weed species which, as Burdon (1987) pointed out, are likely to respond to challenges from new parasite races. Genetically heterogeneous species possess reservoirs of genetic diversity enabling them to interact with their parasites. A group of species that have neither a reservoir of heterogeneity nor associated weed relatives, includes 'pure line species, hybrids and vegetatively propagated species'. Even here Marshall is doubtful of a likely success for mass reservoirs, as against selection in landraces, mutagenesis, or genetic engineering.

Another form of dynamic conservation was proposed by Dinoor (1976).

He suggested that landraces (or possibly composites of landraces?) be grown in what he termed 'nature reserves' (a rather misleading term since they would require cultivation) or 'genetic reserves', under conditions of continuing evolution and, more significantly, of coevolution with pathogens. By comparison with the *in situ* preservation of landraces which have been, or are about to be, displaced, the populations proposed by Dinoor could be raised wherever there are expectations of host–pathogen evolutionary changes, presumably in the centres of diversity of the pathogen (see Leppik 1970), or wherever a concentration of pathogen diversity has been ascertained. Such an experiment could be an aid in plant breeding or in phytopathology, but could not be regarded, as Dinoor suggested, as a method of genetic conservation, if for no other reason than that the number of strains that could be included would inevitably be small (Frankel 1978).

4.4 STATIC CONSERVATION

In contrast to the dynamic conservation of populations discussed in the two preceding sections, the conservation of seeds or of individual plants aims to preserve as far as possible the genetic integrity of individuals or populations. This means that while frequencies of genotypes or alleles may undergo inevitable changes, gene and allele erosion are minimized and recombination with alien material is avoided. The purpose is to retain genes and gene assemblies which have evolved in landraces, in wild populations and in modern varieties, as well as in various genetic stocks which are or may be required for research or breeding.

It should be emphasized here that static and dynamic forms of conservation are in no way mutually exclusive; rather they are complementary. Endangered species may have to be moved to the safety of *ex situ* conservation, and, generally, *in situ* conservation may be complemented by *ex situ* opportunities, for convenience as much as for safety. Clearly, *ex situ* facilities provide ease of access for study and use, and the convenience of a stable reference source under direct control.

4.4.1 Conservation of seeds

Conservation is both safest and cheapest if life processes are reduced to a low level. This is the case in species which can be preserved in the form of seeds. Some species have seeds which, by currently used procedures, can only be kept alive for short periods, and some plants produce no seeds. However, the majority of seed-reproduced crop species can be stored over long periods of years without substantial loss of vitality and, as far as existing evidence goes, without substantial genetic damage.

The advantage of long-term storage over the alternative procedure of regular regeneration at intervals of three to five years, is the avoidance or reduction of natural selection, genetic drift in small populations, natural hybridization, destruction by parasites or climatic rigour, or loss through human error; it is also less expensive.

Roberts (1975, 1989) reviewed the problems of long-term storage of seeds which can be maintained under conditions of low moisture content and low temperature. Such seeds have been called 'conventional', 'orthodox' or desiccation tolerant, in contrast to 'recalcitrant' seeds in which a decrease in moisture content below relatively high levels (between 12 and 31%, depending on the species) infringes viability. The majority of species have orthodox seeds, but many forestry and fruit species and tropical crops, e.g. oak, chestnut, mango, cocoa, coconut and rubber, have recalcitrant seeds.

4.4.2 Long-term storage of orthodox seeds

The long-term storage of orthodox seeds is now well worked out and widely applied (Box 4.1). The longevity of seeds is subject to water content and temperature being maintained at levels which have been established for a large number of species. Roberts (1975) developed what he termed the three basic viability equations which describe the relationships between temperature, moisture content and seed viability. Oxygen pressure is ignored since in open storage it is that of the atmosphere; in sealed conditions it soon drops to zero due to respiration. The equations have been used to develop viability nomographs for a number of species for which the required experimental evidence is available. They facilitate predictions on the storage life of species, subject to seed viability, seed moisture and temperature.

4.4.3 Cryopreservation of seeds

Storage at the temperature of liquid nitrogen has yielded encouraging results with orthodox seeds, and difficulties, especially in cooling and thawing, have proved less than anticipated (Sakai & Noshiro 1975). Experiments at the National Seed Storage Laboratory indicate that, for orthodox seeds, storage in liquid nitrogen (LN) is distinctly promising and problem free. Moisture contents need to be strictly controlled; and physical damage to seeds of sensitive species may be avoidable by modifications of moisture content or by re-warming procedures. No genetic effects have been noted,

BOX 4.1. SEED STORAGE CONDITIONS

The FAO Panel of Experts on Plant Exploration and Introduction formulated standards and procedures for storage installations used for long-term seed conservation (FAO 1975). The 'preferred', or desirable, conditions are storage at −18 °C or less in air-tight containers at a seed moisture content of 5–2%. Sealed containers are necessary to maintain the low moisture content of seeds through periods of equipment breakdowns; also, this method is easier to operate than humidity control of the storage space. The recommended storage conditions are a workable compromise between optimal scientific standards and economic and functional practicability.

After its establishment in 1974, the International Board of Plant Genetic Resources (IBPGR) reviewed and elaborated the proposed storage procedures in the light of increasing experience. Its Committee on Seed Storage organized meetings of experts (Dickie *et al.* 1984), a handbook on seed technology (Ellis *et al.* 1985) and a handbook on seed handling (Hanson 1985).

The formulation of standards for seed storage conditions stimulated the design of appropriate *seed storage facilities*, from installations designed to accommodate collections of 100,000 accessions or more, to cabinets housing a few hundred. Two standard conditions are recommended; long-term storage, with a temperature of −18 °C or less, and medium-term storage with a recommended temperature of 3 °C, but ranging between 1 °C and 6 °C. Some of the largest long-term storages are at −10 °C. The capacity for long-term storage on a global scale facilitates the preservation of a large proportion of the world's unique and valuable germplasm. This is supplemented by a much greater volume of widely distributed medium-term storages.

In many institutions a long-term storage is associated with a medium-term storage which holds seed for evaluation and for distribution to breeders and others. Two reports on the design of storage facilities, including the cost factor, show that storage facilities are not necessarily cost prohibitive and technically elaborate, but readily manageable for institutions in most developing countries (Cromarty *et al.* 1982; IBPGR 1985).

and up to five years of LN storage had no effect on the frequency of chromosome aberrations. The estimated comparative annual cost of storage in LN versus conventional storage per accession over a 100 year period, is $0.42 for LN, against $1.65 for storage at −18°C. The saving is largely due to the avoidance of monitoring and regeneration which are unnecessary for LN-stored seed (Roos 1989).

4.4.4 Recalcitrant seeds

As already mentioned, recalcitrant seeds cannot be dried without loss of viability. In addition, many are also sensitive to low temperatures, and many are large, greatly exceeding the average weights of orthodox seeds (Roos 1989). Various approaches have been tried to discover long-term storage conditions for recalcitrant seeds, but without success. Roberts *et al.* (1984) regard storage in liquid nitrogen as the most promising prospect, but as yet it has not been fully successful. A review of the current status of work on recalcitrant seeds, accompanied by a bibliography, was produced by Chin and Pritchard (1988).

4.4.5 Loss and monitoring of viability of seeds

Loss of viability of seeds tends to be associated with chromosome breakage or aberrations resulting in bridges and fragments. This has been known for many years (Roberts *et al.* 1967; Abdalla & Roberts 1968, 1969). Clearly, the conditions for seed survival are also the conditions for chromosome stability. Gene mutation appears to be subject to similar conditions (Abdalla & Roberts 1969).

Monitoring of seed viability starts even prior to the entry of an accession into storage. Indeed, the first viability test is a major determinant of storage life, and low viability may necessitate regeneration prior to entry. The purpose of monitoring is to establish the need for regeneration of accessions in storage. The frequency of regeneration depends on (i) the initial viability, (ii) the rate of loss of viability and (iii) the regeneration standard (i.e. the viability level at which regeneration is to take place). The regeneration standard is based on a compromise between what is optimally desirable and the cost and biological risk of frequent regenerations. Lowering the standard incurs the risk of accumulating lethal and non-lethal deleterious mutations and of selection in populations (Roberts 1984). Allard (1970) observed that after a drop in germination to half the original level 'the genetic composition of the entry is drastically changed due to differential survival of genotypes'. For some species there is sufficient available information to predict the timing of monitoring with adequate reliability, while for others it is subject to judgment based on past performance. It should be noted that for each test there is a requirement of up to 400 seeds, which may constitute a severe drain on seed resources.

To reduce the burden of periodical germination tests, the Vavilov Institute of Plant Industry routinely tests 5% of accessions in any one group (e.g. by place of origin, year of collection or of last regeneration, etc.) resorting to testing further accessions in the group when test samples show a drop in vitality (Budin, personal communication) (Box 4.2).

BOX 4.2. REGENERATION STANDARDS

The determinant for the timing of regeneration is the regeneration standard. IBPGR adopted a germination percentage of 85% as the 'preferred standard', which Roberts (1984, 1989) strongly upholds. The National Seed Storage Laboratory also regards this standard as desirable, with a minimum standard for regeneration when a drop to 65% can be predicted or has actually occurred (Roos 1989).

Some curators regard the 85% standard as unnecessarily high, and a further drop of 20% as acceptable. This may be based on the consideration that collections of many of the major crops are prevailingly used as sources of individual genes of major effect rather than of gene assemblies, or entire genomes. However, it is highly desirable to maintain a high standard of integrity, for two specific reasons. First, some accessions may be used directly, without a deliberate genetic change. Such plant introduction is common practice in forage, pasture and forestry research, as well as in horticulture. Second, the end purposes and requirements of future plant breeders are not foreseeable; nor is the use of base collections confined to plant breeders but includes other plant scientists for whom genetic integrity may be an essential requirement.

4.4.6 Regeneration and multiplication

Regeneration is the reconstitution of a pure line, a clone or a population by raising progeny. It is implicit in the concept that the progeny is as true a replicate as possible of the original. The extent to which this is attainable is subject to the status of the original – whether a pure line or a population of genotypes – and to the breeding system. In a self-fertilized plant a population may consist of homozygotes which can be readily reproduced, whereas in an outcrossing plant the heterozygous components have segregating progenies. The process of regeneration is identical with the process of multiplication for the production of seed for replenishing seed supplies required for testing and evaluation, and for distribution to plant breeders and others; hence the two are considered together.

In the regeneration or multiplication of seeds two requirements need to be met. First, the breeding system must be controlled. This involves the prevention of outcrossing and, in cross-fertilized species, the control of fertilization within entries, whether by isolation or by controlled crossing (cf. Clark 1989). Second, the effects of natural selection must be minimized if regeneration takes place in an environment other than the original one. Hence it is often claimed that regeneration should take place in the environment in which the entry had evolved. This is often impracticable, always difficult and expensive, and altogether unnecessary provided that survival is maximized, i.e. *that all or most of the components of a population survive* and reproduce. This is most readily achieved in an environment in which all components of a species – or subspecies – can be successfully grown, as is the case for wheat in the South Island of New Zealand, or for *indica* rice in the Philippines. In less ideal circumstances regeneration may have to be shared between institutions in different climate zones. Many wild species require specific conditions for germination and seedling growth, necessitating some measure of environment control. Where mild winter temperatures fail to trigger flowering of winter cereals, there may be a need for vernalization, which is a pre-planting period of imbibed seeds at 3–5 °C.

Another problem is the avoidance of natural selection among components of accessions. It is minimized by the choice of an *optimal environment for survival* (see the preceding paragraph) and by control of parasites and pests. To reduce selection among sister plants in an accession, *aliquot amounts of seed should be taken from all survivors* to produce a composite sample.

Breese (1989) produced a monograph on the principles and tactics of regeneration and the hazards that need to be avoided or counteracted. Special attention is paid to the maintenance of effective population size and to the avoidance of both genetic drift and the effect of natural selection (genetic shift): see Table 4.1. Random genetic drift, causing the loss of rare, hence non-adaptive alleles, is not as harmful as genetic shift which may adversely affect the frequency of adaptive alleles. Both can be countered by an effective population size of at least 30, or preferably 50 or more.

In a large germplasm collection regeneration is a continuing task, but it need not be a heavy burden at any one time. With an expected minimum viability of between 30 and 50 years, regeneration of any one accession would be a rare event within the notional time scale of concern of one century or thereabouts. Hence the number due for regeneration, if reasonably spaced out, need never be unduly large.

4.4.7 Preservation of wild species as seeds

As Thompson (1975) pointed out, the preservation of wild species as seeds has some advantages over maintaining living specimens. It is difficult to

Table 4.1. *Regeneration of heterogeneous populations: some factors causing genetic shift*

Stage	Factors	Minimize by
Germination	Differential genotypic	
	(i) Longevities	(i) Regenerate before germination falls to <85%
	(ii) Dormancy	(ii) Artificially break dormancy
Seedling and vegetative stage	Differential genotypic survival due to:	
	(i) interaction with climatic and soil factors	(i) Regenerate in regions where a large proportion of a species can be successfully grown, or under controlled conditions
	(ii) susceptibility of diseases and pests	(ii) Protect by fungicides, pesticides, etc.
	(iii) competition	(iii) Grow at low densities
Reproductive phase	Differential production of flowers, pollen and seed	(i) Maximize production from individual genotypes
		(ii) Equalize inflorescences before pollen shed
Harvesting	Differential maturities and seed shattering	(i) Harvest heads individually at appropriate stage
		(ii) Harvest equal quantities of seed from maternal parents

Source: After Breese (1989), with permission.

preserve an adequate representation of a species as living plants, especially of large and long-lived ones, and seed preservation may be at least a useful adjunct in a conservation programme. Further, seed preservation is relatively safe from environmental hazards and from changing policies of successive administrations. Distribution also is easier in the form of seeds. Last but not least, it is cheap in terms of space and maintenance.

What information there is on storage requirements of wild plants points to a considerable diversity. Harrington (1970) listed a large number of species in which seed germinated after many years of burial in soil, and others which remained viable for more than 50 years under herbarium conditions. Many annual or short-lived species with dry seeds can be stored like seeds of orthodox crop species. Others, as we have seen, are short-lived when dehydrated. For many species, as Thompson (1975) remarked, little is known about storage conditions and germination requirements, which are complicated by dormancy, hard-seededness, or requirements for a post-ripening period. Accordingly, the seed bank of the Royal Botanic Gardens, Kew, provides a variety of seed treatments and of conditions for germination tests, including a range of temperatures, diurnally fluctuating temperatures, chilling, and treatment with gibberellic acid. Storage conditions also are varied (Thompson & Brown 1972). Thompson (1976) listed criteria which may help in evaluating the suitability of taxa for conservation in seed banks, including available information on taxonomy and reproductive biology, the accessibility in the field, storage responses, and amenability to cultivation. Applying these criteria, he found *Clarkia* highly suitable, *Gentiana* and *Ulmus* intermediate, and *Shorea* quite unsuited for seed conservation.

4.4.8 Conservation of pollen

Techniques for the long-term storage of pollen are far less developed than those for seed. Yet pollen storage has the practical advantage of making stored pollen directly available for crossing. This could be of importance in the breeding of tree species which take some years to flower when raised from seed. Wild relatives of fruit species in particular are maintained mainly for their use in crosses with established cultivars, hence long-term availability of pollen might reduce the need for costly field collections of such materials.

Cryopreservation is needed for long-term maintenance of pollen; Towill (1985) reviewed the extensive literature. With moisture contents reduced according to species requirements, pollen of a number of species has been kept at cryogenic temperatures (−180 to −196 °C) for periods ranging up to six years (Alexander & Ganeshan 1989), but mostly well below one year. Cryopreservation for long-term pollen storage is unlikely to come within reach for general application until the causes of the breakdown are understood. On the other hand, freeze-drying and vacuum drying have both been successfully applied to preserving pollen of a number of species for up to 12 years, usually at a storage temperature of +5 to −18 °C (Towill 1985).

Long-term pollen storage would be particularly useful in species with seed storage problems. Encouraging results came from work by Akihama *et al.* (1978), who kept freeze-dried pollen of peach for nine years and of peas for six, without a marked loss in germination or fruit setting by comparison with fresh pollen.

4.4.9 Vegetative propagation; field gene banks

There are many plants which either do not produce seed (clonal crops), or which are not normally reproduced from seeds so as to maintain intact a highly heterozygous genotype. To either of these belong many short-lived crops which are propagated from tubers, roots, bulbs, etc., and many long-lived shrubs and trees. For these species, as an alternative to vegetative maintenance of storage organs or whole plants, *in vitro* preservation in 'test-tube gene banks' is being explored.

Storage organs of short-lived plants have a storage life which is measured in months rather than years. Potatoes, for example, can be kept at 4 °C until the next spring, or even for a further 12 months (Hawkes 1970); but most other root and tuber crops have a storage life of less than 12 months. However, the effect of respiratory inhibitors such as maleic hydrazide, while applied to stimulate sprout production, suggests that they may have a potential role in the regulation of metabolic processes and in extending storage life (Martin 1975). Other vegetative parts of plants present similar storage problems. Unrooted or rooted woody cuttings held between −2 and +2 °C have been stored for up to five years (Howard 1975; Akihama & Nakajima 1978, pp. 80–111). Thick tea roots have been kept at 5 °C with wet sphagnum moss in vinyl bags for five years, and, judging by their carbohydrate metabolism, can be expected to survive for another ten years (Sakai *et al.* 1978).

Species which are normally propagated asexually may produce seeds which are more readily preserved than are vegetative organs, such as tubers or roots. This is the case in species related to the potato (*Solanum tuberosum*), some of which produce storable seeds. However, the majority of fruits, both tropical and temperate, have seeds with a short storage life.

Germplasm of species requiring vegetative forms of reproduction is being preserved in field collections maintained in experiment stations, research institutions, botanic gardens, and by commercial enterprises such as nurseries. The maintenance of living collections of trees or shrubs presents biological and economic difficulties. The biggest single biological difficulty comes from the likelihood of virus attack. But the organizational and economic problems of space (especially for trees and large shrubs) and maintenance are considerable, with the further consequence that numbers and sizes of samples are likely to be much reduced by comparison with what is regarded as desirable in seed-reproduced crops. Against this it must be recognized, first, that vegetatively reproduced crops have uniform vegetative progenies; that multiple grafts can reduce space requirements; and that most of them are highly heterozygous, so that seeds they produce – or can be induced to produce by various means – reproduce highly diverse progenies.

4.4.10 *In vitro* conservation

By comparison with field collections, *in vitro* preservation (see section 4.4.11 below) has distinct advantages including greater safety from virus attack, lower cost, and accommodation for larger numbers. Research on plant tissue cultures began more than 30 years ago, but only in the last decade has its potential use for long-term conservation been investigated. The International Board for Plant Genetic Resources has taken an active part in initiating and supporting research on various aspects of *in vitro* conservation. For a general account and references see Withers (1989).

The principal role of *in vitro* cultivation is in the preservation of plants with recalcitrant seeds (see section 4.4.4 above) and of clonal crops, i.e. species which are vegetatively propagated and either produce no seed or, being highly heterozygous, necessitate vegetative reproduction to retain genetic identity. They include root and tuber crops with a short storage life, like potato, sweet potato or cassava, and many temperate and tropical fruit and industrial crops, such as *Citrus* spp., banana, coconut and rubber.

Two forms of *in vitro* conservation are recognized: slow growth and cryopreservation. Existing *in vitro* gene banks, including potato, sweet potato and cassava, apply techniques safeguarding *slow growth conditions*. As a rule this includes sterile conditions, controlled – usually lowered – temperature, and the application of growth retarding substances. Genetic stability needs to be strictly monitored. Techniques have been worked out for a range of species. For a comprehensive review, see Withers (1987).

Cryopreservation – storage in liquid nitrogen – is a recent development for *in vitro* plant cultures. Its advantages are a long – perhaps indefinite – storage life, and genetic stability. Much experimentation has been done on the processes of cooling and recovery and the use of cryoprotectants: for leads to the extensive literature, see Towill (1989) and Withers (1987, 1989). As yet there are no established gene banks using cryopreservation, but methods for cassava, coconut and banana have been developed (IBPGR 1989). A pilot project for a gene bank has been set up at CIAT (Box 4.3).

4.4.11 Conservation strategies in perspective

In view of the multiplicity of conservation techniques and, in many instances, of the availability of alternatives, there is a need for an overall perspective of techniques and their application (Table 4.2). Selection of the appropriate conservation strategy is a basic task in genetic conservation programmes. This may result in more than one being applied. For a more extensive treatment, see Withers (1990).

BOX 4.3. *IN VITRO* PRESERVATION TECHNOLOGY

In the establishment and maintenance of *in vitro* cultures, three processes are involved: the selection of the plant material, disease eradication, and disease indexing (Figure 4.2, from Withers 1990).

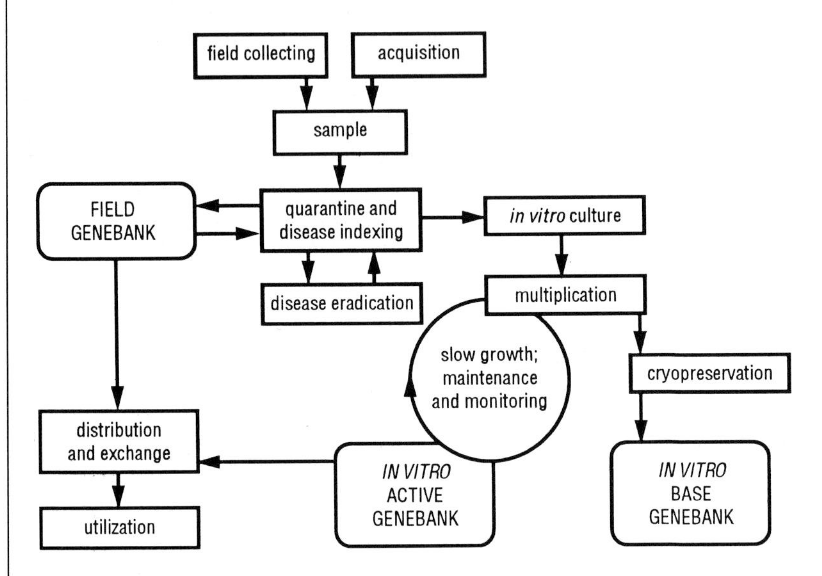

Figure 4.2. Establishment and maintenance of in vitro *cultures (from Withers 1990).*

As *plant material for establishing cultures,* organized systems, such as meristems or shoot tips, have advantages over suspension or callus cultures: greater genetic stability, possibility of producing virus-free plants, and ease of *in vitro* propagation (Stushnoff & Fear 1985). Procedures for shoot-tip culture of *Musa* (Vuylsteke 1989; Panis *et al.* 1992), for meristem-tip culture of sweet potatoes (Love *et al.* 1987) and for excised embryos of palms (Sossou *et al.* 1987) have been described in detail.

A high level of genetic stability is an obvious requirement in all forms of static germplasm preservation. On the other hand, systems giving rise to callus formation are used as sources of somatic, or 'somaclonal', variation, which has turned out to be a new source of genetic diversity, with good effect in a number of crop species. Somaclonal variation has been found to be virtually ubiquitous. For a review see Scowcroft (1984).

Cryopreservation, on the other hand, has so far been found to be highly stable, although the consequences of regeneration of cryopreserved

BOX 4.3. (*cont.*)

cultures have as yet not been widely explored. Monitoring of genetic stability is therefore a general necessity in tissue culture preservation. Morphological assessment of abnormalities is restricted by the conditions of slow growth. Chromosome abnormalities, the earliest observed indicators of disturbances in tissue culture, fail to reveal abnormalities at the DNA level. Recently, isozyme analysis, using several enzymes, has been used to identify genotypic change in tissue culture, and molecular techniques, including restriction fragment length polymorphism (RFLP), have been applied in IBPGR research projects (Withers 1989).

Virus elimination is a precondition of safe storage and distribution. Meristem culture, with or without thermotherapy, is successfully used on a wide range of species (Kartha 1986). Moreover, tissue culture is the only safe method for eliminating non-viral diseases of bananas to make them safe for transport to non-affected areas (Dale 1988).

Withers (1989) provides an overview of the present state of *disease indexing*. Recent techniques include biochemical methods such as ELISA (enzyme linked immunoabsorbent assay), and a nuclear acid hybridization technique that has gained prominence as a test for RNA viruses and viroids. A DNA copy of the viral RNA is used as a probe for the presence of the virus in the test tissue.

4.5 GERMPLASM COLLECTIONS

Reference has already been made to germplasm collections as assemblies of genotypes or populations, maintained as research materials for plant breeders, crop evolutionists, ecogeographers, taxonomists, plant pathologists, etc. Germplasm collections consist of samples of domesticated species and their wild relatives which are maintained in the form of plants, seeds, tubers, etc., or *in vitro*. Methods of preservation have been outlined in section 4.4. Whatever the method or the material – from seeds to single cells – there is a built-in capability of regenerating plants. It is for this reason that DNA is not included – at least for the time being – as a source of genotypes, rather than as a source of genes.

The earliest substantial collection of a crop plant, wheat, was assembled by Philippe de Vilmorin at Verrières, near Paris, in the mid-nineteenth century. Others followed – in England, Germany, Sweden, Australia. These collections consisted mainly of landraces obtained from plant scientists or breeders in different countries. The first systematically assembled germplasm collections resulted from the plant exploration expeditions conducted

Table 4.2. *Conservation systems for genetic resources*

Material	*in* or *ex situ*	System	Reference to section
Wild species	*in situ*[a]	Natural habitats, reserves	4.2.6
	ex situ[a]	Safeguard, availability	
Landraces	*ex situ*[b]	Genebanks	
	in situ	In cultivation	4.2
	ex situ	Low temp. storage	4.4.2
Conventional seeds	*ex situ*	Low temp. storage	4.4.2
Recalcitrant seeds	*ex situ*[b]	Field genebanks	4.4.4
	ex situ	*in vitro* genebanks	4.4.10
Clonal crops	*ex situ*[b]	Field genebanks	4.4.9
	ex situ	*in vitro* genebank	4.4.10
	ex situ	Seeds in genebank	4.4.2

Notes:

[a] *In situ* preservation is the most appropriate method for wild species and for landraces remaining in cultivation. It is more representative, involving larger population sizes than are likely *ex situ*, and it is dynamic being responsive to environmental impacts. *Ex situ* is a safeguard against *in situ* failure, and it provides an available supply for research and breeding.

[b] Field genebanks are exposed to risks from diseases (especially viruses), but are available for evaluation and for instant use in breeding. *In vitro* preservation is more secure and possibly less expensive in the long run. However, the material is not immediately available for use or research.

in the 1920s and 1930s by N. I. Vavilov and his colleagues at the USSR Institute of Plant Industry, now the Vavilov Institute of Plant Industry (Vavilov 1949–50). The aim was to obtain a world-wide representation of the genetic diversity of the crops of the USSR, and to classify the collections for morphological and physiological characteristics. This grand design has never been emulated.

Following Vavilov's lead, systematic exploration and collection of primitive and related wild plant resources intensified, especially in the areas indicated by Vavilov as centres of genetic diversity. J. G. Hawkes, himself a life-long participant, recorded 34 potato collecting expeditions in Latin American countries between 1925 and 1967 (Hawkes 1970). Collecting of other principal crops proceeded apace. More recently, an increasingly important role was played by the International Agricultural Research Centers (IARCs) which began to assume global responsibilities for germplasm of their mandated crops. This development was led by the International Rice Research Institute (IRRI), established in 1961. Systematic collecting efforts have also been undertaken by national institutions. Individual scientists were responsible for establishing comprehensive collections of tomato (National Research Council 1993), safflower (Knowles 1969), okra (Hamon & van Sloten 1989) and others.

Germplasm collections differ in origin and purpose, in size and geographical or taxonomic coverage, and in the biological status of the material. The older collections, for example some of the crop collections of the US Department of Agriculture and some European collections, contain large numbers of landraces and obsolete advanced cultivars, obtained over the years from a variety of sources, in addition to more recently acquired material derived from collecting expeditions. Recently established collections contain mainly the latter. Geographical coverage varies from global to regional to national.

It must be recognized that germplasm collections, whether large and representative, such as the rice collection of IRRI, or haphazardly acquired and maintained, contain some unique and potentially valuable genetic materials which are worthy of being identified and maintained. This is especially the case since many of the accessions may have gone out of cultivation or, like many wild crop-related accessions, are difficult and expensive to replace. It is evident that germplasm collections, individually and collectively, carry the responsibility not only for making the world's genetic resources available for all kinds of users in our time, but to preserve them for future generations.

4.5.1 Size of germplasm collections

How large need a plant germplasm collection be? Clearly no single figure can apply to all collections as they range widely in scope, function and specificity. Thus tens to hundreds may be the appropriate size for a collection of cytogenetic stocks of a species, whereas tens of thousands may be needed in a world collection of a major crop like sorghum. Collections should represent the variation that could be of use to scientists within the compass of an institution, a country, a region, or the world, depending on the function and responsibility of a particular collection. Yet it is beyond the scope of collections to include every nuance of variation.

In principle, the larger a collection, the more likely it will include the genetic information that may be useful at some future date (see Box 4.4). Since the nature of such information cannot be foreseen, it follows that representation of available genetic diversity should be as complete as possible. Yet increasing costs impose limits on the indiscriminate assembly

BOX 4.4. SIZE AND GENETIC RICHNESS OF GERMPLASM COLLECTIONS

How much does the genetic richness of a collection increase with increasing size? Much would depend on the kinds of genetic variation considered and the extent to which the origin of later samples overlapped that of earlier ones. We can turn for guidance to a simple theoretical model in which genetic variation is measured as the number of 'selectively neutral' alleles at a locus and sampling is at random from the whole species. In this case the number of alleles increases approximately in proportion to the *logarithm* of the collection (sample) size (Brown 1989b). This amounts to a relationship of sharply diminishing returns with increasing size. At the other sampling extreme, let us assume that each new sample comes from an entirely new area, completely independent of the origin of samples already in the collection. In this case the number of alleles increases in proportion more directly with the total collection size. The true situation lies somewhere between these extremes depending on the degree of overlap between succeeding groups of samples. Thus there is a marked trend towards diminishing returns with size when variation is measured on the basis of alleles per locus.

Suppose that variation were gauged not as the number of alleles per locus, but as the number of distinct multilocus genotypes. In an outbreeding species, every individual is unique, so the number of genotypes is a linear function of the total collection size. It is the power of a large collection to contain individuals with a desired *combination* of many traits, as noted by Chang (1989), that is perhaps the most telling genetic argument in its favour.

of materials. The cost of storage and maintenance of seed collections rises steadily with numbers once a basic size is exceeded. For vegetative collections this increase is particularly steep. The largest burden, in terms of effort and investment, comes from evaluation, which may involve a number of institutions (Holden 1984). 'Redundancy', i.e. near-identity or close similarity for all identifiable genes or characters, is an expensive fault of collections assembled without restraint. How is it to be avoided or reduced?

Some steps to reduce redundancy can be taken before including any new accessions in a collection. The candidate entries that are named varieties usually carry data concerning origin and parentage, phenology, resistances, performance, adaptation and various morphological characteristics. This information establishes the extent of similarity between the new entries and those already in the collection. In the case of new field collections, characterization of the target population and of its environment are valuable aids in avoiding redundant collecting, as can be the names and farmers' assessments of local landraces.

As for redundancy in a collection, the accessions can be evaluated for a range of strongly inherited, phenotypic traits ('descriptors'). The electrophoretic survey of proteins or other molecular markers can add an objective measure of genetic diversity among and within plant populations or related cultivars (Brown 1978). A multivariate analysis of all available information would then provide presumptive evidence of redundancy among accessions.

What action can be taken to reduce redundancy within a collection? Four possible approaches are as follows.

1. The entire material could still be retained, despite the evidence of redundancy. This is appropriate for accessions from regions where repeat collections are most unlikely, or for collections of great scientific or historical interest.
2. The sample size within each entry could be reduced, while keeping accessions apart. This alternative may be adopted for vegetative or tissue culture collections.
3. Redundant entries could be combined, so that the bulk sample forms a single accession with consequent saving in handling and in storage. However, screening for some characteristics may be less efficient in bulk than in separate entries, and mistakes in bulking could not be remedied.
4. A random selection from the redundant entries could be retained and the remainder discarded.

Options 2 to 4 incur a progressively increasing loss of alleles compared with option 1. Choice among them depends on the kind of collection and the reliability of the data.

4.5.2 Core collections

Reducing redundancy is one aspect of the problem of swollen collections. The more important question is, how can one organize a germplasm collection to be as representative as possible, yet not to exceed manageable limits? How can one devise better access to and use of the genetic resources in the collection? It was these questions that led to the idea of setting up a '*core collection*' or a minimal set of accessions that represents the genetic diversity present in the total collection (Frankel & Brown 1984). The accessions not chosen for the core set are retained as the *reserve collection*.

The choice of accessions for the core collection can be made after accessions have been grouped, using the available passport and characterization data (see section 4.6.2). The hierarchy of grouping begins with taxonomy (species, subspecies, cytological races). Within each taxon, the accessions are clustered according to similarity of origin (countries, states or ecological regions). Further subdivision within each geographic group will be worthwhile if based on strongly inherited characters such as genetic polymorphisms, pathogen, pest and stress resistances, habit, etc. Collections with abundant discriminating data of this type will require multivariate analysis to define the groups of similar accessions.

Having classified the collection into groups the next task is to decide the number of samples and mode of sampling within each group. The number to sample is discussed in Box 4.5. The selected accessions can be chosen at random from within the group. Account should be taken of seed availability, of data authenticity, of any previous use of accessions (such as in genetic research or as donors of resistances), and whether the combined choices include accessions with diverse expressions of any major character used in the cluster analysis. Thus, the core collection would comprise a representative selection of accessions from each of the designated groups, and hence be a structured sample of the germplasm collection. Examples of published core collections include that of the ORSTOM/IBPGR okra (*Abelmoschus esculen-*

BOX 4.5. THE NUMBER OF ACCESSIONS IN A CORE COLLECTION

How many accessions should constitute the core collection? As gauged by the sampling theory of selectively neutral alleles, a sample of 10% of the collection would retain in the core set at least 70% of the alleles of the whole collection (Brown 1989a). The proposed upper limit of 3000 individuals for any one species is sufficient to yield as many alleles as are expected to occur in the species with frequency greater than 10^{-4}.

Having fixed the size of the core collection, the next question is how to apportion this number among the various groups from the cluster analysis. These groups are likely to differ greatly in size. If the same proportion were to be taken from each group, the choice would be biased in favour of the most numerous group. Alternatively, if the same number were taken from each group the bias would operate in favour of small groups. An intermediate strategy of representing each group in proportion to the *logarithm* of its size (see Box 4.8) appears to be the best overall choice (Brown 1989b).

tus) collection (Hamon & van Sloten 1989) and of the CSIRO collection of *Glycine canescens*, the perennial relative of soybean (Brown *et al.* 1990).

In considering core sampling strategies, we have used the number of alleles per locus as a simple measure of genetic diversity. However, this measure assumes that all alleles are of equal value to the future users of the collection. Yet this is generally not the case. Other things being equal, extremely specialized alleles or genotypes highly restricted in their occurrence are less likely to be of use than are more broadly adapted ones. The latter can occur at more than one place and at appreciable frequencies. Hence the evaluation of core strategies should give weight to how well they include widely occurring alleles and locally common alleles.

The core should be subject to change, with variations in its content and size in the course of time. These changes come about from (i) the receipt of new accessions into the collection from distinctively new sources, (ii) the replacement of accessions of questionable authenticity, (iii) the revision of groupings in the light of new data and (iv) the review of users' needs, for example to ensure that strong expressions of resistance characters are included. However, one important aim of a core collection is to build up a body of knowledge on a restricted reference set of lines. Such knowledge is clearly needed to study relationships among variables. Too rapid a flux of

BOX 4.6. THE BARLEY CORE COLLECTION

A recent development of the core concept is the proposal to develop an integrative core collection for a whole crop species. The European Cooperative Programme for the Conservation and Exchange of Crop Genetic Resources (ECP/GR) plans to assemble such a collection for barley. The Barley Core Collection aims to represent optimally the genetic diversity of cultivated barley (*Hordeum vulgare*) and wild relatives and will provide well known genetic standards with a limited set of about 2000 accessions. The plan is to sample these species (or known subspecific taxa) throughout their range based on analyses of the climatic, ecological and geographical information on where the species are known to occur from observations, existing collections or herbarium specimens. The integrative core collection would be a sample of each distinct subspecific type (e.g. taxonomic – barley and wild species of *Hordeum*; physiological – winter and spring barley; use – feed, malting or human food; morphological – two-row vs six-row, etc.) from each distinctively different environment or widely separated locality. The accessions chosen by consensus among barley scientists will be assembled from numerous sources and may entail fresh field collecting.

Table 4.3. *Criteria for a good core collection*

Utility criteria
1. The core should contain no redundant entries.
2. The origins of its entries are authentic, unless no accessions of a particular taxon with such data are available.
3. It is sufficiently large to yield reliable conclusions for the collection as a whole, yet is of manageable size.
4. It is able to predict sources of useful variation.

Genetic criteria
1. The major subspecific taxa and geographic regions are represented.
2. Preference is given to including the more broadly adapted rather than ecologically specialized alleles or genotypes.
3. Within the constraints of criteria 1 and 2, genetic diversity, especially as measured by the number of alleles per locus, is maximized.

Source: From Brown (1989b), with permission.

accessions through the core would defeat this aim. Hence, altering the core should be a relatively rare process. Table 4.3 summarizes some general criteria for judging a good core collection and provides a basis for the curator for its improvement.

4.5.3 Representativeness of germplasm collections as research material

Primarily intended as resources for plant breeding and associated sciences, many collections, like Vavilov's, have been used as research material, to explore evolutionary processes and taxonomic relationships, identify centres of origin or diversity, discover distribution patterns of general variability or of particular characteristics or genes, and to relate these to the environments in which they occur. Studies of this kind range from multi-character analyses of world collections to detailed studies of more limited proportions of geography or subject. As a result of the intensified interest in genetic resources, the interest in the study of germplasm collections has also intensified. A discussion of strengths and weaknesses of existing collections is in place, partly to draw attention to shortcomings which might be avoided in those now being established or extended.

The use of germplasm collections as research material is based on the premise that accessions representing a country or locality had evolved in response to the selection pressures prevalent in the particular environment.

In many collections the material consists of genuine landraces or selections from them; in others this is only partially the case, hence discrimination is needed between material that in this discussion is termed indigenous or authentic, and material which is derived from introduction or from hybridization with introductions. Discrimination is not always applied, or even possible, and this and other limitations of collections call for serious consideration. Some of the shortcomings which limit the use of germplasm collections as materials for evolutionary or biogeographical research, are discussed in Box 4.7.

From the observations in Box 4.7 it is evident that *germplasm collections which are used as research material must be subjected to critical examination.* Few entire collections will come up to the high standard set by Holcomb *et al.* (1977) that 'only an ideally representative and well-documented collection would be useful for evolutionary and ecological studies of variation'. Many, however, contain substantial numbers of accessions which can be identified as valid. Material of uncertain or inappropriate status or origin should not be included in surveys or analyses. So-called world collections often have defects of the kinds mentioned, because as a rule they derive from a multiplicity of sources. They contain well-authenticated accessions alongside others entered with little discrimination for origin or status.

Authenticity rather than numbers must be the main criterion for research material. In the absence of a critical examination one may have more reason for concern at seemingly excessive than at apparently inadequate numbers. In the durum wheat collection used by Jain *et al.* (1975), for example, the small number (7) from Sicily, on which the authors comment, is more likely to be genuinely Sicilian than the 836 from Turkey are likely to represent genuinely Turkish cultivars. Similarly, there never have been 24 cultivars of safflower cultivated in Australia, and one may even question the number of 135 from USA, both derived from the USDA world collection of safflower (*Carthamus tinctorius* L.). The inclusion of accessions in country or regional statistics should rest on the assurance that they originated, or at least are, or have been, cultivated there, i.e. have demonstrated adaptedness to the environment.

A collection which is highly representative of the germplasm of a crop is that of safflower, referred to above, thanks to 'the dedication and foresight of Professor P. F. Knowles and the USDA', whose timely collecting effort only just preceded the main impact of the introduction of improved varieties (Ashri 1973). The collection was intensively studied for evolutionary divergence between countries (and regions), for variability in different countries, and for geographical sources of agronomically desirable traits.

BOX 4.7. LIMITATION OF GERMPLASM COLLECTIONS

First, in some major collections, especially of the temperate cereals, landraces are poorly and unevenly represented, Vavilov's own collections being notable exceptions. The biological status of accessions, whether landrace, selection, hybrid variety, introduction, etc., is not often adequately documented.

Second, sampling procedures during collecting and maintenance are varied and often poorly documented. They may range from a random sample, or a random plus a biased sample (representative of 'distinctive' types), to a biased sample, often without a record of actual procedures. But even if a population was adequately sampled in the field, its diversity may be reduced by selection of one or few 'representative' individuals, the remainder being discarded. Genetic integrity may be further infringed by natural hybridization, natural selection, or genetic drift occurring in the course of reproduction for maintenance, which until the advent of cold-storage facilities was carried out at least every 3–5 years.

Last, but not least, not many collections have precise information on the geographical origin of all their accessions. This restricts evolutionary and ecological interpretation of variation patterns, the planning of repeat collecting, and the establishment of core collections (see section 4.5.2).

To what extent a germplasm collection represents the indigenous genetic variation of a region depends to a large degree on the impact that plant introduction and breeding had made by the time of collecting. Sixty years ago, at the time of Vavilov's collecting expeditions, landraces prevailed throughout the geographical centres of diversity. This was no longer the case when many subsequent collections were assembled, at least not for those crops which were subject to early breeding efforts, such as the temperate cereals. There is a 'vintage' factor in the composition and status of collections: older collections should provide more valid information on evolutionary patterns, recent ones on areas where genetic diversity still persists. This must be watched especially when different collecting 'vintages' are combined in one collection.

Species differ in the extent to which they have been modified by plant breeding, and this is reflected in germplasm collections. Crop species range from those only in the initial stage of intensive plant improvement, like safflower, chickpea or pigeonpea, to species like wheat or barley which have been transformed by introduction and intensive breeding on a near-global scale.

Ashri (1973, 1975) used 20 morphological characters to assess variability and divergence in different parts of the world. Some traits were uniform throughout, others showed variability throughout, others had localized variability. On the basis of these and earlier observations, Knowles (1969) and Ashri (1973, 1975) concluded that in cultivated safflower fairly distinct types can be recognized which are found in what Knowles (1969) calls 'centres of similarity'. Ashri comments that the variability within centres is at least in part due to the data being assembled for political rather than agro-ecological regions. But this may not be the only relevant factor. Indeed, variation within countries may be more significant and of greater scientific interest. Knowles (1969) reported on different regional types in India, and Ashri comments that 'it would be worthwhile to record the detailed collection locations and analyze the data for divergence'. We tend to agree. Analyses based on detailed information on collection sites and their ecological conditions could be more revealing of adaptive evolutionary processes, and more effective in identifying sites of valuable genetic resources, than large-scale analyses of world collections.

A large and representative collection of groundnut (*Arachis hypogaea*) has provided evidence of coevolution of the host and two parasites. Established at ICRISAT, the International Crops Research Institute for the Semi Arid Tropics, the collection has been extensively evaluated for characters of agronomic and/or biological interest (Moss *et al.* 1989). One of the principal objectives is to discover sources of resistance to the many diseases and insect pests which attack the groundnut plant in different parts of its global distribution. Of particular interest is the identification of the geographical origins of resistances, which would facilitate studies of the evolution of host–parasite interactions.

This appears to have been achieved for two damaging fungal diseases of groundnut: rust, caused by *Puccinia arachidis*, and late leaf spot, caused by *Phaeoisariopsis personata*. Among the 10,000 accessions examined for resistance to the two parasites 42 were resistant to rust alone, five to late leaf spot alone, and 39 to both. The majority (87%) of the resistant genotypes belong to the ssp. *fastigiata*, 13% to ssp. *hypogaea*; the two subspecies are cross-compatible. Of the resistant genotypes, 74% originated in Peru, and another 10% elsewhere in South America, with links to Peru. Of the 182 accessions from Peru which were examined, 47% were resistant to both or either leaf disease (Subrahmanyam *et al.* 1989). On the basis of these results, the authors conclude that resistance to rust and to late leaf spot of groundnut originated in Peru, and pre-eminently in the Tarapoto region, where groundnut had been cultivated since about 2000 BC. Since there is evidence that ssp. *fastigiata* has its origin in Peru, the authors conclude that the host and both its major leaf diseases, rust and late leaf spot, coevolved there.

Some of the many studies of multigene diversity in germplasm collections, using isozyme analysis, have already been mentioned in section 3.2.2, as have been surveys for resistance to pathogens. A different approach is represented by the attempt by Peeters and Martinelli (1989) 'to classify and order' the variability represented in germplasm collections by the use of hierarchical cluster analysis. The objective was to examine the extent to which relatedness could be established by the use of this statistical technique; and further, to examine the potential for predicting the extent of segregation from the similarity rating of the parents. The material consisted of 247 accessions of barley derived from 52 countries, from which the five most highly related, five least related, and five random pairs were used. Entries were described by at least 20 characters, and observations were conducted over a period of three years. The results show that similarity within country gene pools was greater than between them, and segregations in crosses between country gene pools were more extensive than those within. These results indicate that collections can be partitioned by simple classifications such as the country of origin, as a basis for establishing a core collection; and, in the absence of records of origin, on the basis of hierarchical cluster analysis.

4.5.4 Genetic stock collections

Genetic stocks are difficult to define since the term has been applied to widely differing types of biological systems, from culture collections of microorganisms to single-gene variants of plants or animals, or even RFLP markers. Within the frame of this book, a narrower definition is appropriate. Though being genetic resources, genetic stocks differ from those dealt with so far, being aids to research rather than objects of, or source materials for research. Further, as a rule genetic stocks are defined or definable in genetic or cytogenetic terms, or they are ancestral or other crop-related species used in genetic or evolutionary research. We exclude a range of materials that are sometimes included, such as advanced breeding lines in plant breeding material, which are definable only in terms of economic performance, whereas genetic stocks as a rule are of no direct economic value. This discussion is further restricted to stocks of plant species used in agriculture, though the principles are derived from some of the oldest culture collections, such as *Drosophila* and *Neurospora*.

Most genetic stock collections grew out of research by individual scientists whose names became associated with their crops, for example, E. R. Sears with wheat, C. M. Rick with tomato. Genetic stocks were developed as research tools in genetic analysis and/or as reference materials in related fields such as plant physiology, plant pathology and plant breeding. The growing number and complexity of genetic stocks, combined with the growing demand for access by other scientists, led to the establishment of genetic stock collections.

In a comprehensive review of the conservation of genetic stock collections of species important in agriculture, D. R. Marshall (personal communication) lists the following categories of genetic stocks:

(a) Single gene or single trait variants
  (i) Morphological or physiological variants
  Examples: fruit colour in tomatoes, patterns of pathogenicity in wheat rusts
  (ii) Electrophoretically detectable protein variants
  Examples: seed storage proteins
  (iii) Restriction fragment length polymorphisms (RFLPs)
  Stocks with known genetic markers; RFLP probes

(b) Cytogenetic stocks
  (i) Variants for chromosome structure
  Deletion, duplication, inversion, translocation
  (ii) Variants of chromosome number
  Haploids, polyploids
  Nullisomics, monosomics, trisomics, etc.
  (iii) Alien addition/substitution lines

There are genetic stock collections of most of the major crop species in which a good deal of genetic research has been carried out, including maize, wheat, barley, soybean, tomato, pea, sorghum and others. Their role, hence their composition, varies between species. Accordingly, the requirements for expertise in maintaining stocks differ. They may include checking of linkage relations between mutant genes, of the chromosome constitution of aneuploid chromosome variants, and many more. As a rule, genetic stock centres maintain a database and act as information centres on genetic stocks and their applications.

In the past, one of the foremost achievements resulting from the use of genetic stocks has been in gene mapping of chromosomes. Molecular techniques (RFLP mapping) have enormously enhanced the availability of gene mapping and its application in research and in breeding practice.

4.5.5 DNA banks as germplasm collections

Samples of DNA extracts stored as a 'gene library' or DNA bank provide a new option for an assemblage of plant germplasm (Peacock 1989, Adams *et al.* 1994). Broadly, the samples in such collections are of three kinds: (i) total genomic DNA, (ii) DNA libraries, or (iii) individual cloned DNA fragments, including probes for RFLPs, satellites, etc. Samples of genomic DNA are made with DNA isolated directly from plant tissue. Preparing samples as DNA libraries requires the further step of fragmenting the DNA with a restriction enzyme and packaging the diverse mixture of fragments into cloning vectors. The aim in constructing each library is to retain each fragment from the original DNA extract, so that the whole genome is represented in the mixture. Unlike a DNA library, the third kind of sample is fixed and genetically pure as each vector molecule in the sample is host to the same plant DNA fragment.

Stored DNA samples are ideal for two contrasting purposes. The samples are convenient experimental material. They are ready for immediate shipment without quarantine problems and ready for further molecular analysis and manipulation. At the other extreme, DNA samples are ideal as a 'time capsule' approach to conservation. They are literally frozen genetic resources, potentially the most stable form of preserved germplasm, and one that does not require recurrent regeneration to retain its future utility indefinitely.

The two purposes differ in their linkage to utility of the samples. In the first case, immediate use is the reason for banking samples, and becomes the obvious criterion for judging sampling priorities and the value of any bank. In contrast, for the 'time capsule' approach, the current use of the samples is not an issue; the actual use is for the future to face. This leads to the unanswerable question of whether a DNA time capsule is worthwhile.

The three kinds of samples listed above differ in their suitability for replication *in vitro*. When any sample of total genomic DNA is depleted, it is best replenished by re-isolation from whole plants. A sample in the form of a DNA library can be amplified in a bacterial host, but it is subject to increasing bias in the fragments that survive. Only isolated cloned fragments can be replicated at will in a controlled fashion.

As germplasm collections, DNA banks share several of the problems of seed banks, but in a more acute form. Proper, reliable documentation and labelling of samples is crucial to their use. Whereas seed or vegetative accessions may still be of use despite partial or erroneous data associated with them, DNA samples are dependent on accurate information. Further, the despatch, receipt and use of samples require a higher level of technical resources than do seed samples. In evaluation, the recognition of phenotypes (such as disease resistance, stress tolerance, etc.) is restricted to those characters encoded by known genes. Problems of ownership and control of the samples are likely to be more complex with DNA banks than seed banks, because of the additional equity in technology that the bank contributes.

Major factors limit the role of DNA banks in the conservation of biodiversity. Despite the development of 'immobilization' techniques, and the power and promise of PCR amplification (Adams 1993), total genomic DNA samples are essentially non-renewable. Therefore DNA banking offers no solution for conserving endangered species. Samples are liable to be expended, raising policy issues about access and amounts distributed. In addition, DNA storage only allows for recovery of single genes, not of genomes (Peacock 1989). The advantage of conservation as seeds is that all the genes in the genome will function. For most characters, coordinated gene function *in vivo* is needed to identify a desired phenotype.

The formation of an international network of DNA repositories (DNA Bank-Net) for the storage of genomic DNA has recently been vigorously promoted (Adams 1993). Such a network has a role to play in making biological resources more widely available for human benefit through increased research and use. However, it is strictly an adjunct and in no sense an alternative methodology in conservation. Thus DNA storage cannot replace field gene banks (section 4.4.10) despite it being much cheaper than the latter. 'Conservation' of DNA samples does not relieve us of the burden of providing for the biological conservation of plant species.

4.6 ORGANIZATION AND MANAGEMENT

It remains to outline the institutional context in which genetic resources are assembled, evaluated and conserved. Plants have been, and will continue to be collected, used, stored, and in many instances discarded by individuals and institutions. Yet in the precarious position in which the world's genetic resources are placed, the idea has gained ground that an organized effort is needed to salvage as much as possible of what is left in the fields and in the wild and to ensure that the existing collections are preserved in properly organized and equipped gene banks or genetic resources centres; and further, that germplasm collections are characterized, recorded and made available for use. Clearly, this is a task for the world community as a whole, since all would profit from endeavours to salvage, study, use and preserve the Earth's biological heritage: the developing countries which are hosts to as yet unexploited reserves, and the developed countries which have contributed most of the science and technology to make both preservation and exploitation possible.

The organization which has emerged inevitably depends on the interaction of national and international institutions. This is the case at the level of the genetic resources centres (GRCs) or gene banks, which include both national and international institutions. It is also the case at the level of networks, whether crop-specific (the 'crop networks'), or all-inclusive (the 'global network') in which national and international centres associate in world-wide collaboration.

4.6.1 International beginnings

The idea of a genetic resources centre (GRC) emerged from the vast and still unparalleled collections assembled by N. I. Vavilov and his colleagues, the tangible result of their collecting activities in many parts of the world. In the 1920s and 1930s the USSR Institute of Plant Industry became a highly developed and successful GRC, embodying world-wide exploration, extensive study and characterization, and long-term conservation. It provided the inspiration for corresponding activities in USA, Great Britain, Japan, Australia and elsewhere, though most of these were mission-orientated 'plant introduction services' whose objective was to acquire materials currently required by agronomists or plant breeders rather than to obtain a representation of the existing genetic diversity. However, in European countries and in the United States, sizeable collections were assembled for all the major and many of the minor agricultural crops.

The broader concepts of genetic resources research, clearly visualized by Vavilov but in jeopardy in the USSR during the 25 years of Lysenko's dominance, began to be recognized at a conference convened by FAO in 1961 (Whyte & Julén 1963). These were further developed at a conference organized by FAO and the International Biological Programme (IBP) in 1967 (Frankel & Bennett 1970b), where, incidentally, the term 'genetic resources' was introduced.

The decade from 1965 to 1974 saw the emergence of what has been called the genetic resources movement. The initiatives came largely from the FAO Panel of Experts on Plant Exploration and Introduction. Tangible contributions were two volumes which explored and clarified the scientific and technical issues involved in the various aspects of genetic resources work (Frankel & Bennett 1970b; Frankel & Hawkes 1975). In the political arena there was the impact the Panel or its members made on FAO itself, on the

United Nations Conference on the Human Environment, Stockholm, 1972, on administrators in various countries, and, last but not least, on the scientific public. Members of the Panel played a prominent part in a meeting convened at Beltsville in 1972 by FAO and the Consultative Group on International Agricultural Research (CGIAR), which made recommendations for a global network of GRCs and for its coordination. This ultimately led to the establishment in 1974 of the International Board for Plant Genetic Resources (IBPGR) by CGIAR. The events leading up to these international activities were described by Harlan (1975b). For an historical account of the genetic resources movement from 1961 to 1974, see Frankel (1985, 1986a, b, 1987). The developments following the establishment of IBPGR are documented in its Annual Reports.

4.6.2 Gene banks: procedures and responsibilities

Gene banks have two kinds of function: to assemble materials and to make them available and accessible to potential users; and to participate in their preservation for the future, which is the subject of the earlier sections of this chapter. Here we deal with the procedures designed to provide information for the utilization of genetic resources.

Conservation of genetic resources of domesticated plants – or animals – has little purpose except for actual or potential utilization in agronomy, forestry, or plant or animal breeding, or in research in genetics, physiology, evolution, microbiology, biochemistry, nutrition, or processing technologies. The utilization of a germplasm collection for any purpose whatever necessitates that entries are recorded in a data bank, and that they are described and characterized with regard to the projected form of utilization. Since accessions are prevailingly used in biological or agricultural research or its application in plant breeding or agronomy, procedures for characterization and documentation developed by IBPGR have wide application.

Characterization and evaluation

The procedures consist of five operations (IBPGR 1991a):

1. Establishing the origin (*passport data*). These are *accession data* if an accession has been received from a breeder, another institution, etc., if possible including site of origin, ancestor or pedigree; or *collection data* if an accession is derived from a field collection, providing a collecting site description.
2. *Characterization.* Recording of characters, prevailingly of the mature

plant, which are highly heritable, can be easily identified, and are expressed in all environments.

3. *Preliminary evaluation* includes recording of plant development and plant characters of the growing plant.
4. *Further evaluation* may include reactions to physical stress and to pathogens and predators, and further biological characterization.
5. *Management* data, including the handling of seed, its distribution, regeneration, etc.

IBPGR has codified descriptors and descriptor states for each of these operations and for all the major and many of the minor crops. These were prepared by groups of crop experts and published in a series of booklets (IBPGR 1991a, b are recent additions), which are periodically revised.

Databases and information management

With the growth in number, size and diversity of germplasm collections in the last two decades, the increasing amount of evaluation data and the near universal availability of computers, there has been a transformation in the volume and diversity of information that is of direct relevance for the use and management of genetic resources. Basically, there are two kinds of databases: those concerned with the plants and their characteristics, and those concerned with aspects of management of genetic resources.

Crop-specific databases may relate to the genetic diversity of a crop species, to accessions of a crop species in a germplasm collection, or to a group (or network) of germplasm collections of a species. An example of a database designed to represent the genetic information contained in the germplasm of one crop species is the pea model for documentation (Blixt 1982). The structure of the database is generated by existing information on the genetics of the species. The model can be applied to any species for which corresponding information is available.

A database of *in vitro* stored crop germplasm has been established by IBPGR. Based on surveys conducted every two years, it lists information on the material, its characterization, the field of interest and the culture method used (Withers *et al.* 1990).

Catalogues of individual germplasm collections are being replaced by databases in electronic form which may be complemented by passport, characterization and evaluation data. The current emphasis on crop networks gives added support for the establishment of crop databases as a source of information for plant breeders and other users; an example is the database for *Beta* (Frese 1992). With national institutions assuming responsibility for the preservation of crop genetic resources (see section 4.6.3 below), crop databases will be the only sources of information which would make the contents of collections widely accessible. In addition, they would be helpful in management measures through identifying gaps or duplications, or in setting up core collections.

The size and complexity of present-day gene banks impose a need for information systems covering the various processes of management in the day-to-day running of the gene bank itself, and its interactions with other gene banks, with plant breeders and with other recipients of accessions (see Konopka & Hansen 1985).

4.6.3 Genetic Resources Centres. The global network

The FAO Panel of Experts proposed the Genetic Resources Centre (GRC) as the operational unit (FAO 1973; Frankel 1975). It was to consist of two component parts.

1. a 'base collection', for long-term conservation of seeds and other biological materials (see section 4.4.2), and
2. an 'active collection' for the multiplication and regeneration of seeds and their supply to users, and for the evaluation and documentation of accessions in the germplasm collection.

These could be in one institution (as at IRRI: see Figure 4.3); or a base collection could be on its own (like the US National Seed Storage Laboratory at Fort Collins), though functionally linked with active collections. Active collections should be dispersed throughout the distribution of a crop species. Commonly they would be part of, or associated with, institutions with extensive programmes in plant breeding and plant protection.

The general concept was adopted by IBPGR on its inception in 1974, but emphasis shifted from the integrated GRC to its base collection component, i.e. to the low-temperature seed storages. It is not clear why IBPGR chose the base collections to fill the central place, rather than GRCs or active collections, which would have been the logical choice in view of their links with plant breeders and other users.

Over the years the 'global network of base collections' grew to about 50 institutions which had accepted IBPGR's invitation to participate, with more than 100 species being represented. Institutions which joined the network included some of the international centres for agricultural research (IARCs) and some national centres. They acknowledged their responsibility to IBPGR to safeguard the collections in their charge and, through active

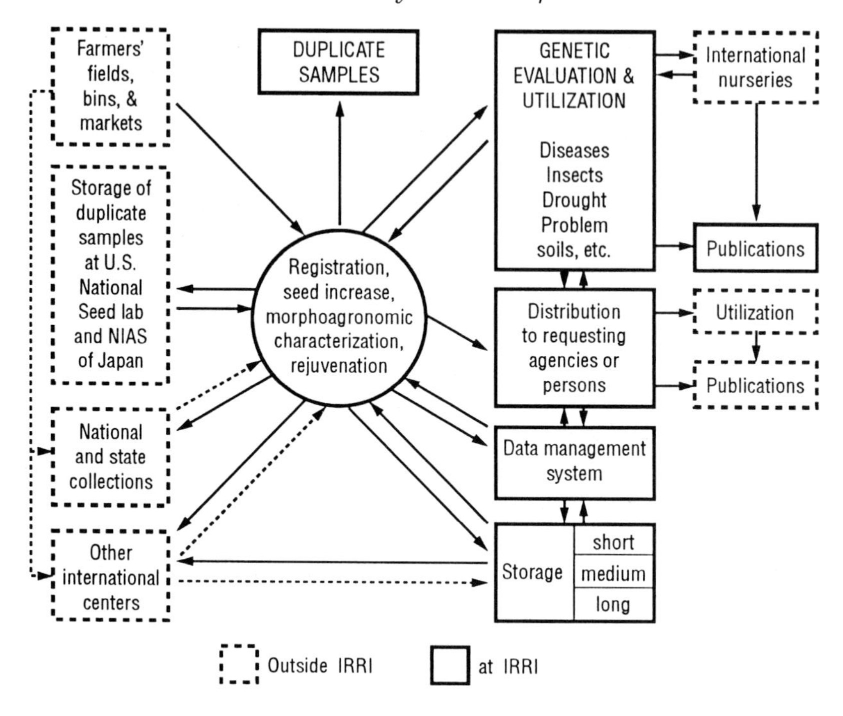

Figure 4.3. A Genetic Resources Centre. Flow chart of the different IRRI operations in acquiring and increasing seed, and cataloguing, preserving, evaluating, and utilizing rice germplasm, with the collaboration of other agricultural research centres (reprinted with permission from T. T. Chang, Genome *31: 825–31. Copyright 1989, Natural Research Council, Canada).*

collections, to make them available to bona fide users. Conditions of storage and management had been defined and the operations examined by IBPGR staff.

In recent years moves have been initiated by FAO to enter into formal agreements with the governments and institutions responsible for base collections, and to take over the coordinating role previously carried out by IBPGR.

4.6.4 National institutions and international networks

A critical attitude towards IBPGR, combined with political claims for national ownership of genetic resources, generated new ideas on the structure of national and international institutions. Changes were proposed involving the replacement of the global network of base collections by a series of international 'crop networks', each responsible for an individual crop. These would include all involved in a crop, from collectors and curators to agronomists and plant breeders (Marshall 1990; van Sloten 1990). The networks were to be formed by national collections which were to become the main operators of the system, responsible for all phases of genetic resources operations, from establishing and maintaining collections to long-term conservation and documentation. International links would be developed on a crop-related basis.

These ideas were adopted and developed in the strategy of the organization which took over from IBPGR early in 1993, the International Plant Genetic Resources Institute (IPGRI 1993), 'IPGRI's strategy is built on the premise that national programmes should be the foundation of any global genetic resources conservation effort.' Hence its first objective is 'to assist countries to assess and meet their needs for plant genetic resources conservation'. But since 'no country can rely solely on the genetic resources within its borders, it is imperative that all countries cooperate and that barriers to the exchange of genetic resources be minimized.' The second objective is, therefore, the encouragement of international cooperation through the formation of crop networks. It is not clear how these are to operate with regard to the establishment of data banks, supply of materials among and outside the network membership, standards of maintenance, etc. What these standards are to be is not clear; indeed, the organization and management of long-term conservation is left open, suggesting that it is to be left to national institutions.

The role of IPGRI emerges essentially as that of a consultancy which will 'in partnership with other organizations, undertake research and training, and provide scientific and technical advice and information.' (IPGRI 1993). No longer does it appear to have the responsibility for international initiatives to survey and assemble, evaluate and preserve and to make universally available the world's genetic resources, which was the mission of IBPGR at its inception. It appears that such responsibilities, as far as they are still conceived, have devolved on national institutions.

5

Plant species conservation and population biology

5.1 INTRODUCTION – THE POPULATION AS THE UNIT OF SPECIES CONSERVATION

In this book, we strongly emphasize long-term conservation, and the ecosystem as its vehicle (Chapters 8 to 10), while recognizing the roles of gene (Chapters 2 to 4) and of species conservation. Conservation biology at the species level has predominantly focused on animal species, particularly the large mammals (see, for example, Primack 1993). The dominance mirrors the disproportionately high level of the funds for conservation that these species receive. However, 'charismatic' animals are inadequate models for the conservation of plant species. Many of the crucial aspects of large animal conservation are behavioural, such as territoriality which affects home-range requirements, or courtship and mating schemes which affect the avoidance of inbreeding. In contrast, plants are relatively immobile and possess diverse and flexible breeding systems, which profoundly influence the science of their conservation.

This chapter therefore introduces the main theoretical aspects of population biology that bear on plant species conservation. It leads to discussion of conservation *in situ*, in natural populations of various kinds of plant species (Chapter 6); and *ex situ* in botanic gardens (Chapter 7).

Plant species exist in a series of more or less isolated populations between many of which at least some level of seed or pollen migration takes place. Virtually all members of any one population share a number of features. Unless precluded by virtually complete uniparental reproduction (autogamy or apomixis), they are likely to share a recent common ancestor. They encroach on one another competitively, they support the same local population of symbionts, herbivores and parasites, and through mating they share a common evolutionary future. Hence the population is the basic unit of conservation at the level of the individual species. Our discussion will deal first with population dynamics and the concept of the minimum viable population or MVP. The MVP has emerged as the basis for conserving individual species (Soulé 1987). Then we treat the microevolutionary forces – genetic drift, mating system, gene flow and selection – that shape the future of small populations, because population size is the central issue in the conservation biology of populations.

5.2 CONCEPT OF MINIMUM VIABLE POPULATION

The minimum viable population is an estimate of the minimum number of individuals of a species necessary to form a viable population. Box 5.1 lists a number of descriptions or definitions of the concept. From these definitions, it follows that any estimate of the MVP for a species is not a unique figure, but has several restrictions.

BOX 5.1. DEFINITIONS AND CONCEPTS OF THE MINIMUM VIABLE POPULATION

Frankel (1974) – the (area and) population size required for the survival of a species; the size that is likely to yield the required level of variation to afford the flexibility for evolutionary persistence.

Shaffer (1981) – 'A minimum viable population for any given species in any given habitat of the smallest isolated population having a 99% chance of remaining extant for 1000 years despite the foreseeable effects of demographic, environmental, and genetic stochasticity and natural catastrophes.'

Soulé (1987a) – '. . . the minimum conditions for the long-term persistence and adaptation of a species or population in a given place . . . without significant demographic or genetic manipulation'.

Ewens *et al.* (1987) – The demographic concept, 'the size of the population that guarantees 95% probability of survival for *y* years'. The genetic concept, 'the population size for which the rate of loss of genetic variation takes a given accepted value'.

Nunney and Campbell (1993) – The size 'that will ensure (at some statistical level) its persistence for a specified time.' In 'density dependent stochastic models, MVP is the carrying capacity that fulfils the stated persistence requirement . . . It is the expected size of the population in the absence of stochastic events.'

1. The estimate of a MVP has an ecological context and applies to a particular *habitat*.
2. The estimate refers to a specific *time span* and a maximum tolerable *probability of extinction*. The future of a population is subject to uncertainty and the probability of its survival decreases with time. Thus the persistence of a population for a particular period is a probabilistic concept. The selected time span and persistence probability interact; for example, the size required to be 90% sure that a population persists for 100 years might be the same as the size needed for a lower survival probability (e.g. 50%), but persisting for a longer period (500 years).
3. The estimate has a genetic context, because the environment of the population will inevitably change, requiring adaptive change in the population. It must therefore possess the necessary *genetic variation* to meet future environmental trends.
4. When the estimate is derived from genetic considerations, it is usually in terms of an '*effective population size*' or N_e. As discussed below, the actual size equivalent to such an effective size may be five to ten times larger than N_e (Nunney & Campbell 1993).
5. The estimate has a *metapopulation* context. It will require clarity as to whether it applies to the species as a whole, or the many individual, partially connected populations dotted across the landscape.

In directing attention to the MVP threshold, Shaffer (1981) suggested a maximum probability of 0.01 for extinction after 1000 years. Although he noted that these two critical values might be set over a range, his choice of example values is notably and perhaps unrealistically stringent. It may be appropriate for charismatic animal species but less so for a plant population. It is sobering to imagine how many contemporary populations of any plant species could be expected to survive intact after a millennium, even if habitat destruction were completely halted from today. Can we realistically hope for strategies that would maintain 99 in every 100 populations after a millennium? Few contemporary populations would meet Shaffer's criteria and thus be defined as exceeding minimum viable size. Indeed many more are likely to survive in a partial manner, because a sample of their genes will have found their way into other populations, or into the derivatives of an interspecific hybrid, or will form a new species.

In like manner, contemporary populations are the living descendants of only some fraction of the populations alive 1000 years ago. It is therefore important to regard the values of acceptable assurance probability and

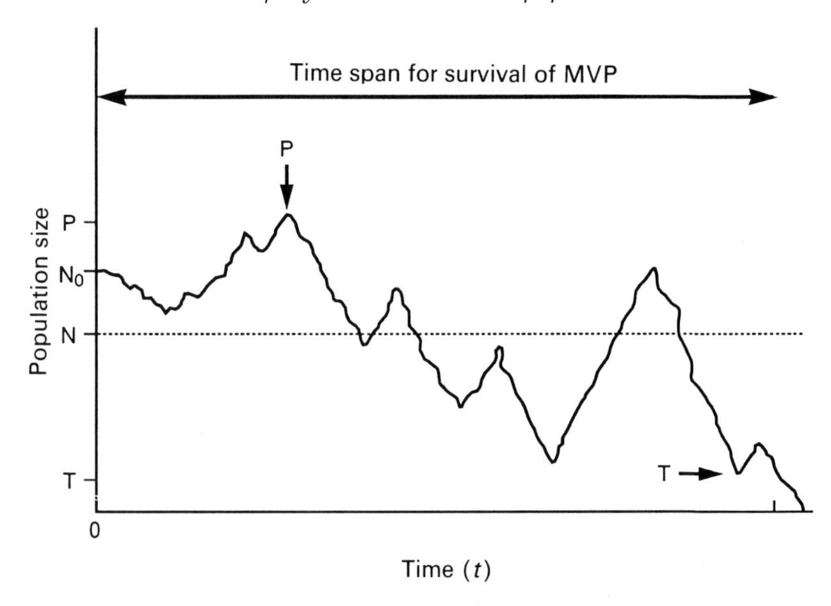

Figure 5.1. Hypothetical size for a population in future generations. The population is assumed to endure beyond the limit set for a MVP measured from today.

duration as arbitrary. They are likely to differ widely among different kinds of species. Thus for individual populations of a cosmopolitan and numerous species, the acceptable maximum value for survival probability should be much lower than that for the solitary population of a highly endangered species.

Further conceptual difficulties arise because of the ongoing stochastic nature of the process. Suppose that a population occupies a fixed habitat that is not subject to long-term steady deterioration. The population still must cope with random environmental changes and the MVP estimates the minimum size needed to cover this future environmental uncertainty. Suppose the current observed population size is $N_{(0)}$. In response to uncertainties, the population size will fluctuate with time or generations as a stochastic process with an absorbing barrier (Figure 5.1). At some points in this process the population will crash to low numbers. At the minimum trough size, *T* (other than at extinction), it is barely able to recover. During the process, the population will also pass some maximum peak value (*P*). Each such realization of the process, like the one Figure 5.1 depicts, would result in a distribution of population sizes observed during this time span, and this distribution would have an observed mean *N*. In Figure 5.1, the actual time to extinction happens to exceed the persistence time nominated for the MVP to achieve.

With the above scenario in mind, the question is – what is our conception of a MVP in relation to the observed variables: the current size [$N_{(0)}$], the future peak value [*P*], the average size [*N*] or the minimum trough value [*T*]? The current size is the correct answer, but the other values may loom larger for the conservation biologist in the field. Nunney and Campbell (1993) have discussed the MVP in terms of the carrying capacity needed to fulfil the stated persistence requirement, where the carrying capacity is a basic parameter of density-dependent models of population growth. 'Thus, MVP is larger than the mean population size, since it is the expected size of the population in the absence of stochastic events.' The important point to draw from their interpretation is that for the majority of the time, a population that is today considered to be a MVP, will in the future display sizes that are much less than 'the size of a MVP'. Further, at most points in the future but prior to extinction, it would register sizes from which the observer would conclude that it was no longer a MVP because it would be unlikely to survive the required time span as measured from those points. On these grounds, it is perhaps easier to think of a MVP in terms of habitat area or space needed to support the MVP (Soulé 1987a).

Boyce (1992) contends that population biologists must beware of the expectations from society that they can readily forecast the numbers needed to ensure long-term viability for a species. Rather than using MVP for such forecasts, he suggests that the importance of the concept of MVP and the role of population viability analysis lies in developing models and strategies for contemporary species management. We now turn more explicitly to discussing the factors that enter into such analyses.

5.3 POPULATION VIABILITY (OR VULNERABILITY) ANALYSIS

Population viability analysis (PVA) is the process of evaluating data and demographic models to estimate the likelihood that a population will persist for an arbitrary period of time (Boyce 1992). In principle the analysis would yield values of extinction probabilities, for a given population size, after any period has transpired. Hence PVA provides an estimate of the MVP for a given duration. Such analyses have been conducted for many animal species, particularly in the United States, pursuant to the US Endangered Species Act of 1973. Far fewer analyses have been made for plants.

Table 5.1. *Threats to Australian endangered plant species*

The number of endangered Australian plant species (out of 179 higher plant taxa) made extinct, or affected by various potential causes of extinction (from Leigh & Briggs 1992)

Threat	Listed extinct species: presumed cause	Endangered species	
		Past threat	Current threat
Very low numbers (< 10)		10	85
Roadworks	1	8	57
Weed competition	4	12	57
Grazing	34	51	55
Agriculture	44	112	50
Industrial or urban development	3	20	21
Fire frequency		10	17
Collecting		6	17
Mining	1	3	11
Forestry		10	10
Recreation			8
Dieback (*Phytophthora* species)			7
Clearing		2	5
Railway maintenance		2	4
Salinity			4
Insect attack			3
Quarrying			3
Trampling by pigs and buffalo		1	3
Drainage		1	2
Flooding		4	2
Rubbish dumping			2
Vehicle damage		1	2
Erosion	1		1
Altered artesian hydrostatic pressure		1	
Rock falls			1
Not specified or known	1	17	7

Two kinds of components pose threats to the future of a population. The first are systematic pressures such as habitat destruction, harvesting, changes in the physical or biotic environment (such as climate change, or invasion by highly competitive environmental weeds or increased grazing by feral animals). Table 5.1 gives a list of such factors identified as threats to the continued existence of Australia's endangered plant species. The list is almost surely incomplete, but it shows the range of pressures, directly or indirectly due to human activities, with which plant populations must contend. The appropriate responses to reduce such threats are usually evident without detailed modelling, the only question being whether the necessary steps can be taken to reverse them.

The second kind of threat arises from uncertainties for which there is no managerial remedy other than increasing the population size to reduce their impact. Hence this group of threats are the crucial variables in determining the MVP. Shaffer (1981) has divided these random perturbations into four types:

- Demographic stochasticity
- Environmental stochasticity
- Natural catastrophes
- Genetic stochasticity

He later replaced the word 'stochasticity' with 'uncertainty' (Shaffer 1987), which we will follow here.

Demographic uncertainty arises from random variation in the survival and reproductive success of the individuals in a finite population. This uncertainty particularly pertains to small populations, where differences among individuals in the number of their progeny matter. For *Astrocaryum mexicanum*, Menges (1991) computed that a population size of at least 50 is needed so that the probability the population survives such variation is greater than 95%. Faced only with demographic uncertainties, the expected persistence time of a population is a *rapidly exponentially increasing* function of population size. In these circumstances, the expected durability of a very small population increases slowly at first with increasing numbers, but ever more markedly as numbers continue to rise. In plants, the number of individuals needed to meet other forms of uncertainty (particularly variation in environmental factors) usually exceeds the numbers required to cover demographic uncertainty (Menges 1991).

Environmental uncertainty arises from random, or at least partly unpredictable temporal variation in the habitat variables (e.g. rainfall) and in the biotic environment (competitors, symbionts, herbivores, pathogens). In contrast to demographic uncertainty, the expected persistence time for a population faced with environmental uncertainties is a *steadily increasing, near-linear* function of the population's size.

Natural catastrophes include fires, floods, droughts and extreme winters. This category refers to events that result in a drastic reduction in population size, if not total destruction. The impact of any such calamity is relatively insensitive to the size of the population. Conversely, the expected persistence time of a population faced with a random, periodic series of catastrophes increases only in proportion to the *logarithm* of population size – progressively larger and larger populations yield diminishing dividends in terms of increased persistence times (Shaffer 1987).

Genetic uncertainties occur as random changes in the genetic structure of populations due to inbreeding, random drift, founder effect, or the breakdown of species reproductive isolation. The genetic changes include alterations in allele content and frequency, in levels of heterozygosity, or the break-up of coadapted complexes. Such changes are likely to reduce the survival and reproductive probabilities of an individual. In the case of changes in heterozygosity, the pattern of response is negative exponential like that for demographic uncertainty. However, for most purposes the general relation between population size and persistence time due to genetic uncertainties is approximately *logarithmic*, like that for catastrophes.

The first published example of population viability analysis in a plant species is Menges' (1990) analysis for populations of the endangered perennial herb, *Pedicularis furbishiae*. This species consists of only 28 colonies on river-banks within a 200 km stretch of the St John River in Maine, USA. Four components made up the analysis. (i) Over 6000 plants in about 20 populations were marked, mapped and followed for four years to estimate the probabilities of transition from one stage of the life history to the next, and to estimate the reproductive effort as seedling recruitment. The average values in a population were used to estimate the observed rate of increase (dominant eigenvalue of the Leslie matrix) for that population. (ii) Several ecological variables were assessed for each population (e.g. vegetation cover, substrate type, degree of disturbance). This tested the variables with greatest effect on the stage-specific transition probabilities. (iii) The annual variation in the transition estimates was used in simulation models to test the effect of environmental uncertainty on population viability. (iv) Estimates of the probabilities of local population extinction and re-establishment were made because in its riverine habitat, each population of the species is virtually certain to die out as a result of varying hydrology. In such a series of interconnected populations (a metapopulation), the key question is the persistence of the whole system, rather than of any individual population.

The results of the PVA were as follows. First, the viability estimates for individual populations ranged from 0.64 (rapid decrease) to 1.81 (rapid increase) with considerable year and site effects (i.e. environmental uncertainty). The demographic statistics suggested that most populations individually would be ephemeral. Second, seedlings were much more likely to advance to later stages when cover was low than in more competitive situations. Juvenile plants were more likely to advance when herbaceous species rather than woody plants dominated the community. Third, there was a high probability of catastrophic extinction of each individual population due to bank erosion or ice scouring in its riverine habitat, or extinction due to the competition from shrubs that establish in later succession. The analysis showed that the conservation of *Pedicularis furbishiae* is dependent on metapopulation dynamics and a balance between local rates of colonization and extinction.

This PVA is a detailed ecological study of population dynamics as they shape the likely persistence of the species in the future. It highlighted ecological issues of persistence, such as the importance of sensitive phases of the life cycle. However, the study did not include a genetic component, which would have greatly increased its complexity. Further, such PVAs will only be possible for a limited number of species, and a very restricted replication of populations within a species.

In summary, the important features of plants that affect a viability analysis of their populations are listed in Box 5.2. These features distinguish PVA for plant species from those for higher animals. They can be grouped into genetic variables, features of plant growth and life cycles, and ecological features depending on the biotic and physical environment of the individual plant. We now turn to the genetic processes in populations that have a major bearing on PVA.

5.4 POPULATION GENETIC PROCESSES

Since the publication of Frankel and Soulé (1981) much has been written about genetics and microevolutionary processes in relation to conservation biology. Important contributions include those edited by Schoenwald-Cox *et al.* (1983), Soulé (1987a), Falk and Holsinger (1991), Fiedler and Jain (1992) and Loechshke *et al.* (1994). In this treatment, we outline the major population genetic processes from the standpoint of plant conservation biology. This will form the necessary basis for discussing the *in situ* conservation of various kinds of plant species in Chapter 6. The reader wishing a more extensive treatment of population genetics is referred to texts by Crow and Kimura (1970), Hedrick (1983) and Hartl and Clark (1989).

5.4.1 Genetic drift

Reductions in population size, especially substantial reductions, initiate the process of random genetic drift – a key process in conservation genetics, the

BOX 5.2. DISTINCTIVE FEATURES OF PLANTS THAT AFFECT POPULATION VIABILITY ANALYSIS (ADAPTED FROM MENGES 1991)

Feature	Causal mechanism
Genetic	
Small neighbourhood size	Plant immobility, limited pollen & seed dispersal
Microsite specialization	Environment a fine-scale mosaic
Diversity in breeding system	Flexible mating systems
Potential for selfing	Capacity for autogamy
Inbreeding genetic uncertainty	Low within-population variation
Habit or life history	
Phenotypic plasticity	Indeterminate growth by reiteration of modules
Clonal reproduction	Indeterminacy
Cryptic phases in life cycle	Seed bank; dormant organs
Ecological	
Resource generality	Niche dimensions hard to quantify
Disturbance response	Sessile nature limits escape from unfavourable environment
Clumped spatial distribution	Spatial aggregation, limited dispersal
Size better than age as predictor	Competitive size advantage
Specific mutualisms	Plant-pollinator, disperser, symbiont

'genetics of scarcity'. In this process, a restricted and variant sample of the genes present in the parental population survives into the next generation. The random changes in allelic frequencies that occur due to sampling error, including the loss of alleles, are called *random genetic drift*. The magnitude of random drift is directly measured by the *effective size of a population* (N_e). This is defined as the size of an ideal population whose genetic makeup is affected by random drift to the same degree as is the genetic makeup of the real population under study.

The *ideal* or model *population* is one in which all plants are equally likely to be the parent of any progeny. Individuals are diploid, sexually reproducing and bisexual, producing both male and female gametes simultaneously. Gametes combine at random and self-fertilization occurs at the rate of N_e^{-1}. The number of progeny per parent is a Poisson random variable. The generations are disjunct and no selection, migration or mutation takes place. Such a model is the yardstick for predicting the effects of genetic drift.

When a large population is reduced in size such as during a bottleneck, genetic drift becomes important because of the following effects.

1. *Loss of alleles*

The expected loss depends on the distribution of allele frequencies. Allele frequency distributions range from being uniform or 'even', with alleles equally frequent, to highly skewed, with few frequent alleles and many rare ones (Chapter 2). For loci with similar allelic richness (numbers of alleles), the loss of alleles resulting from random genetic drift is much less when distributions are even than is the loss from skewed distributions.

A convenient intermediate model is the equilibrium distribution of allelic frequencies that Kimura and Crow's (1964) model of an infinite number of neutral alleles in a finite population (with parameter θ) generates. For this distribution, an approximate logarithmic relation exists between the actual number of alleles at a locus (i.e. allelic richness, n_a) retained in a sample or bottleneck of size S *gametes* ($=S/2$ diploid outbred individuals):

$$n_a \approx \theta \log_e[(S+\theta)/\theta] + 0.6$$

where $0.1 < \theta = 4N_e u$ (u = mutation rate, N_e = effective population size) and $10 < S < N_e$ (Brown 1989b). Table 5.2 contains values of allelic richness after sampling from large populations with differing levels of polymorphism. Thus if a population of 500 individuals is reduced to a bottleneck of 25 individuals ($S = 50$ gametes), and $\theta = 2.0$, about 40% (11.8 to 7.1) of the alleles are lost.

When the shift to a much reduced population size is prolonged, genetic drift continues. If no selection opposes the loss of alleles, or no further mutation occurs, complete uniformity will ultimately be reached. In the neutral allele model, mutation acts to generate new variation and the population reaches a new equilibrium. Let us assume that the mutation rate remains the same before and after the population reduction. We can use the figures in Table 5.2 to assess the final loss of allelic richness when the new equilibrium is reached. If the population of $N_e = 500$ with $\theta = 2.0$ is reduced to a new size of $N_e = 50$, with no change in mutation rate, the new value of θ will be 0.2. The equilibrium allelic richness is 1.8, representing a drastic loss of 84% of the alleles.

2. *Erosion of heterozygosity, or genetic variance*

The reduction in heterozygosity (ΔH) due to genetic drift is an inverse function of size of bottleneck.

$$\Delta H = -H/S$$

Table 5.2. *Allelic richness (i.e. number of alleles per locus, n_a), and heterozygosity (or gene diversity, H) in populations of various effective sizes* (N_e diploid individuals) *in equilibrium for the neutral allele model (parameter* $\theta = 4N_e \times$ mutation rate), *and in samples or after bottlenecks of size S gametes*

	Allelic richness (n_a)					Gene diversity (H)		
	Population (N_e)			Bottleneck (S)			Bottleneck (S)	
θ	50	500	5000	25	50	$\theta/(1+\theta)$	25	50
0.2	1.8	2.3	2.7	1.6	1.7	0.17	0.16	0.16
0.4	2.6	3.6	4.5	2.3	2.5	0.29	0.27	0.28
0.5	3.0	4.1	5.3	2.7	2.9	0.33	0.32	0.33
0.6	3.3	4.7	6.1	2.9	3.3	0.38	0.36	0.37
0.8	4.0	6.0	8.1	3.4	3.9	0.44	0.43	0.44
1.0	4.6	6.9	9.2	3.9	4.5	0.50	0.48	0.49
2.0	7.2	11.8	16.4	5.8	7.1	0.67	0.64	0.65
3.0	9.3	16.2	23.1	7.2	9.2	0.75	0.72	0.74
4.0	11.1	20.3	29.5	8.3	11.0	0.80	0.77	0.78
5.0	12.6	24.1	35.6	9.1	12.6	0.83	0.80	0.82
10.0	17.8	40.8	63.8	10.8	18.5	0.91	0.87	0.89

Note:
Note that N_e of 50 diploid individuals implies 100 gametes.

Values in Table 5.2 illustrate the effect on heterozygosity (H) where $H = \theta/(1-\theta)$ in the neutral allele model. With $S = 50$ there is only a 2% loss in heterozygosity which is much less than the bottleneck effect on allelic richness. Once again the effect becomes more noticeable in the long term. For the population with $N_e = 500$, $\theta = 2.0$, changing to $N_e = 50$, $\theta = 0.2$, there is ultimately a 75% loss of heterozygosity.

3. *Reduction in the proportion of alleles under selective control*

Nunney and Campbell (1993) point out that as population size contracts, a greater fraction of the allelic variation in the population behaves as if it were selectively neutral. This is because drift overrides selection in controlling the fate of an allele when the selection coefficient s is less than the inverse of N_e, whereas, when s exceeds N_e^{-1}, selection dominates. Thus when N_e is large, weak selection either for or against an allele suffices to retain it or remove it from the population. As N_e reduces, the critical value of s for selection to dominate increases.

4. *Reduction in the diversity of allelic combinations*

A similar argument applies to the effect of contracting population size on the number of combinations of alleles at different loci. The reduction in size reduces the number of combinations and increases the amount of randomly generated linkage disequilibrium between loci. The correlation (r) in allelic state between the selectively neutral alleles at two loci with recombination fraction c is an inverse function of recombination and the effective population size (Hill & Robertson 1968):

$$r^2 \approx 1/(1+4\,c\,N_e)$$

Effective size and census size of a population

Since the effective population size is the crucial variable in gauging the impact of drift on population genetic structure, what is the relationship between the effective size (N_e) and the census size (N) of a plant population? This relationship is fundamental to conservation planning because a large difference between N and N_e may mislead managers as to the status of a species. In essence, the relationship depends primarily on the variation among individuals in their lifetime reproductive success. Thus it brings together both population ecological and population genetic variables.

Variation among individuals in a plant population for their life history parameters can be extreme. For example, individuals in a population of the perennial palm *Astrocaryum mexicanum* show a tenfold variation in net height gain (a measure of growth rate variability that affects survivorship), and differ by up to 25-fold in their fruit production (Sarukhán *et al.* 1984).

The concept of effective population size refers to an idealized population of size N_e as defined above. Real populations of plants depart from this idealized behaviour in several ways, that usually operate to reduce the effective size of the population below its actual size in terms of the effects of random genetic drift (Crawford 1984). As noted, the effective size of a population is the number of individuals in the idealized population having the same magnitude of random genetic drift as the actual population. The effects of random drift can be measured in several ways, each giving rise to a type of effective size: (i) its effect on average inbreeding coefficient, giving an *inbreeding* effective size; (ii) its effect on variance of allele frequencies, giving a *variance* effective size; (iii) its effect on rate of loss of heterozygosity, giving an *eigenvalue* effective size; (iv) its effect on rate of loss of segregating loci, giving what Wade and McCauley (1988) term an *extinction* effective size; and (v) Ewens (1990) has defined a *mutation* effective size as having the same stationary probability that two random genes are identical when subject to a mutation–drift balance.

For many populations, the values for these kinds of effective sizes will be very similar. Most of the standard discussions of genetic drift are in terms of inbreeding effective size. We will use this to examine several ways in which real plant populations depart from the simple theoretical idealized population. However, let us note that from the standpoint of conservation, the loss of allelic richness is arguably the most important and dramatic effect of genetic drift. It would seem desirable therefore to develop an appropriate concept of effective size with respect to allelic richness (richness effective size). Thus, the *richness effective size* is defined as the size of an idealized population whose allelic richness for selectively neutral polymorphism matches that in the actual population.

1. *Fluctuation in population size between generations*

A population is likely to vary in size with time (e.g. Figure 5.1). If N_i is the size at time i during a sequence of t generations, the effective size (N_e) is the harmonic mean, and not the arithmetic mean of the sizes during this period:

$$N_e^{-1} = t^{-1} \sum_{i=1,t} N_i^{-1}$$

Such a mean value is strongly dependent on the smallest N_i in the sequence. Hence any generation of extremely small population size, or *bottleneck*, is very important in determining the amount of genetic drift in the whole sequence. Alternatively, any severe reduction in size requires a protracted period of extremely large sizes in subsequent generations to compensate and mitigate the effects of random genetic drift.

2. *Unequal sex ratio*

Separation of the sexes in to different individuals (dioecy) is an uncommon mating system among plant species. However, the potential for a biased sex ratio is not confined to dioecious species. In hermaphroditic species, sex bias may result from differential timing of male and female functions, or the phenomenon of environmentally induced, or developmental switching of gender (as, for example, happens in the oil palm, *Elaeis guinensis*: see Lloyd & Bawa 1984). Indeed many factors can cause a severe discrepancy in the number of individuals that are effective in contributing pollen as compared with the number that contribute ovules. In any one reproductive episode, if the number of functional males in the population is N_m and that of females is N_f, the effective size is

$$N_e = 4\, N_f N_m / [N_f + N_m]$$

Separation of the sexes becomes restrictive on effective population size when the sex ratio is severely imbalanced. In the extreme, suppose that all the effective pollen originates from only one individual. In this case the value of N_e can never exceed 4, irrespective of the number of functional female plants.

3. *Fertility variation*

A prominent source of discrepancy between actual and effective population size in plants is variation among plants in the number of progeny, or of gametes per plant in the population. Suppose that ν_k is the average number of progeny per individual in a population of N diploids and that the variance of the number is var(k). The effective population size is approximately (Hedrick 1983):

$$N_e = (N\nu_k - 1)/[\nu_k - 1 + \text{var}(k)/\nu_k]$$

If the number of progeny per plant (k) is a Poisson variable, var(k) $= \nu_k$ so that N_e approximates to N as expected. In most cases this variance exceeds that of the Poisson. Heywood (1986) collated data on means and variances in fecundity in 34 species of annual plant, and found that the ratio N_e/N ranged from 0.10 to 0.67. In 10% of cases variation in fecundity increased the amount of genetic drift by an order of magnitude over that expected from simple random gamete sampling ($N_e/N < 0.2$).

Likewise, variation in male reproductive success among individuals reduces the effective number of a population. For example, Schoen and Stewart (1987) found significant variation in male fertility among 33 distinct clones in a seed orchard of *Picea glauca*. The male gametes for more than half of the seed came from fewer than 20% of the clones. The variation in male fertility in this seed orchard had a marked effect on its effective population size, reducing it to about 20% of the census size.

We note that when the population size is stable ($\nu_k = 2$), and fecundity is invariant [var(k) $= 0$], then $N_e = 2N - 1$. This shows that controlling the variance in both male and female success will double the effective size of a population. This is an important point in considering methods for the regeneration of genetic resources (Chapter 4; see also Breese 1989).

4. *Age structure and overlapping generations*

Natural populations of perennial plants normally consist of a range of individuals of various ages and stages of maturity, with overlapping generations and age-specific mortalities and fecundities. Such a complex structure generally lowers N_e below the total census number. If the age structure is stable and the population size is not changing rapidly, the effective size is the same as that for a population with discrete generations having the same variance in lifetime progeny numbers and the same number of reproductive individuals entering the population each generation (Crow & Denniston 1988).

One particularly important feature of the life cycle of plants, that actually *increases* N_e above the value expected from the current visible population size, is the dormant seed, bulb or tuber bank (Ellstrand & Elam 1993). Dormancy, whether innate or induced, is a strategic alternative to dispersal (Harper 1977). It gives the possibility of avoiding unfavourable seasons and waiting for a suitable environment to reappear. Particularly for annuals, if the natural seed bank is of appreciable size and the dormancy period before germination a random variable, then additional cohorts are added to the total population: the seen and the unseen. This natural seed bank importantly acts to increase effective population size through increasing generation length and overlap. This reduces the effects of genetic drift. Genetic variation between young and old cohorts in the seed bank has been found in *Carex bigelowii* (Vavrek *et al.* 1991) and *Luzula parviflora* (Bennington *et al.* 1991). (The important role of artificial seed banks to assuage the loss of genetic diversity in cultivated plants is a direct analogy: Chapter 4).

5. *Neighbourhood size*

For populations that are continuously or unevenly spread out in space, yet within which isolation by distance occurs, Wright (1946) introduced the concept of neighbourhood size to evaluate their effective population sizes. Crawford (1984) reviewed and developed the formulation for when pollen and seed dispersal are taken into account. In this case, the observable variables are δ = the number of breeding individuals per unit area, t = outcrossing rate, σ_p^2 = the one-way variance of dispersal distance between the site of the pollen parent and that of zygosis, and σ_s^2 = the one-way variance of seed dispersal distance; and the effective population size is

$$N_e = 2(1+t)\pi\delta[t\sigma_p^2/2+\sigma_s^2]$$

This formula assumes that dispersal distances are normally distributed random variables. In reality, distance distributions in plants tend to be leptokurtic, which will reduce the values for the neighbourhood size. Crawford (1984) computed values ranging from one to a 100 square metres for the neighbourhood areas of 11 herbaceous species.

6. *Inbreeding*

The interaction of inbreeding and genetic drift is not discussed extensively in standard texts. Because many plant species are inbreeding, we will consider it further here; particularly the question of the effective population size of an inbreeding population. Inbreeding in a population acts to reduce the effective population size. Where F denotes the inbreeding coefficient, the standard formulation (Crawford 1984) to express this tendency is

$$N_e = N/(1+F)$$

This formulation is adequate under mild forms of inbreeding such as mild biparental inbreeding, i.e. mating between neighbouring plants that are more related than random pairs because of limited dispersal. However, ambiguities arise when inbreeding is due to a regular system of mating that splits the population into separate subpopulations (such as self-fertilization, sib-mating). For example, the effective size of a fully selfing population can be either half the census size, or unity, depending on the context (see Box 5.3).

Under predominant selfing (assuming the frequency of outcrossing is much less than the inverse of the population size), the population has a strong tendency to fracture into essentially isolated lineages. Selection or sampling among such lineages increases the identity by descent steadily, even if the progeny sizes show a Poisson distribution. Hence both the variance effective size and the inbreeding effective size are reduced below $N/2$. Further, selfing and the lack of recombination allows correlations in genotypic state (fixation) across loci to build up (Weir & Cockerham 1973, 1989). This complicates the common assumption that the dynamics of a single locus across many replicate populations are equivalent to those of several loci within the one population.

Drift and mutation

As we have already noted in section 5.4.1, random genetic drift relentlessly reduces the genetic variation within a population. Unless some other force opposes this effect, the population will eventually lose all genetic variance. One obvious candidate for a balancing force is new mutation. It is therefore logical to ask what is the minimum population size at which the loss due to drift equals the gain in genetic variation through mutation. Franklin (1980) considered this equilibrium in the case of neutral variation for quantitative traits and computed that an effective population size of 500 was needed to maintain additive genetic variance. Such genetic variance would then be available for continuing evolutionary change. This figure became a benchmark for deducing the minimum size for continuing evolution and hence conservation in the long term (Frankel & Soulé 1981). Lande and Barrowclough (1987) analysed the case of traits under stabilizing selection. They

BOX 5.3. EFFECT OF PREDOMINANT SELFING ON EFFECTIVE POPULATION NUMBER

Consider two of the various concepts of effective population size, namely inbreeding and variance effective number. The above formula for N_e with inbreeding $[N_e = N/(1+F)]$ might appear to be adequate for the concept of variance of allele frequencies. Thus a population of size N that is at equilibrium under complete selfing (and fully homozygous with $F=1$, but not necessarily monomorphic) has a variance of allele frequencies equivalent to that in an idealized outbreeding diploid population half its size. However, if selfing in an otherwise idealized population proceeds for some time, an inexorable loss of lineages will occur, due to the random Poisson process of progeny distribution among lines. Eventually a state of complete allelic identity by descent will be reached where all existing copies of a gene trace back to one individual. In this state the variance of allele frequencies is zero and equivalent to that in a fixed idealized population of size 1. Thus the effective size under full selfing is eventually 1 (Crow & Kimura 1970). However, if we add mutation and consider the Kimura and Crow (1964) infinite, neutral alleles model, the loss of variation due to drift is opposed by gain from mutation. In this case the variance effective size of a selfing population is half its actual size in accord with the formula, and not unity.

Turning to the concept of inbreeding effective number, the recursions for the two cases (idealized and actual) between inbreeding coefficient at generation t, F_t in terms of its value at generation zero are as follows.

Drift in the panmictic, idealized population:

$$1 - F_t = [1 - 1/2N_e]^t \, [1 - F_o]$$

Complete selfing in the actual population:

$$1 - F_t = (1/2)^t \, [1 - F_o]$$

Clearly for equivalence to hold between the actual and idealized population in terms of changes to inbreeding coefficient, the effective population size of a selfing population would have to be 1 (Crow & Kimura 1970). However, when the recursions are expressed in terms of the coefficient of coancestry (between two genes in two different random individuals from the population) we obtain the value for the inbreeding effective number of N from the original formula (with $F=1$).

The ambiguity of inbreeding effects on effective population size highlights the circumstances when genetic variation within populations of selfing species is likely to be extremely low (fragmentation, long distance colonization, or between line selection), and when it is likely to be moderate (among population migration, or diversifying selection). The greater interpopulation variation in gene diversity for inbreeding plant species is well documented (Schoen & Brown 1991).

found that 'a population with $N_e \geq 500$ individuals can maintain nearly as much genetic variance in typical quantitative characters as an indefinitely large population'.

Hedrick and Millar (1992) compared the equilibrium approach with an alternative approach of computing how large a population is needed to maintain some fraction of the current genetic variance (e.g. 90%) for a certain period (200 years) in a species with given generation time. However, not only is the chosen period arbitrary, but this approach assumes that the current situation is a relevant standard for long-term survival. The presumed magnitude ($10^{-3} \times$ additive genetic variance), beneficial nature and steady rate of the mutational input, and its presumed independence of population size, are crucial assumptions in both schemes and open to question. In addition, Gilpin (1991) has shown how subdivided populations in a metapopulation framework induce 'patch coalescence' through local extinction of a patch and subsequent recolonization with very few migrants. If the source patch has low variability, the new colony will resemble it and will have replaced one of diverging variability. Such a repeated sequence of events steadily erodes the variability of a system even when each patch numbers around 500.

Retention and replenishment of allelic richness, rather than of additive genetic variance, gene diversity or heterozygosity, may provide an alternative formulation of the minimum size for continuing evolution. We have noted (section 5.4.1) that reduction in allelic richness is perhaps the most significant effect of drift in small populations from the standpoint of conservation. The intriguing theoretical question is how many alleles per locus are needed on average for long-term evolutionary survival and, given a generational input by mutation, what is the minimum population size to guarantee that average. The figures in Table 5.2 show the logarithmic relationship between allelic richness and population size in the neutral allele model. This justifies the assertion made in section 5.3 that genetic uncertainties have a logarithmic response. On the other hand, the number of new mutant alleles per generation ($N_e u = \theta/4$) is linearly related to population size.

5.4.2 Sexual reproduction

Plant species display a rich variety of breeding systems. These systems determine in what quantity, zygotic state and linked arrangements the genes of a population pass from one generation to the next. Indeed, mating is a crucial process in conservation biology. Its continued success governs which genes and which species survive and evolve in time. Further, since sexual reproduction involves an assemblage of individuals (except in completely autogamous species), the population emerges as a basic unit of conservation. A sudden decline in population size can profoundly effect the mating process. One immediate consequence of such a bottleneck is that fewer individuals are available to participate in mating. The next generation these few produce, whatever its size, will consist of individuals that are much more related than were individuals in previous generations. Mating among them will mean higher levels of inbreeding and the question becomes what is the effect of this on the next generation of progeny.

Darwin pointed to the many adaptations in flowering plants that ensure outcrossing and hypothesized a selective advantage to the avoidance of inbreeding. Morphological or developmental devices that promote outcrossing include dioecy, monoecy (separate male and female flowers on the same plant), protandry (anthesis preceding stigma receptivity within the flower), protogyny (anthesis following stigma receptivity) and herkogamy (spatial separation of anther and stigma in the flower). Incompatibility systems may be either homomorphic, with no morphological differences between mating types or heteromorphic, which includes various types of heterostyly. Homomorphic incompatibility may be either sporophytic, when the maternal genotype controls the incompatible pollen reaction; or gametophytic, when the pollen genotype conditions the reaction.

Despite these many mechanisms that promote outbreeding, the frequency and intensity of inbreeding in plants in general is far greater than in animal groups. Several factors are involved – the sessile habit of plants, their limited dispersal of seed and pollen, the bisexuality of most species and the capacity for self-fertilization in many of these, and often their small population sizes.

Inbreeding is measured by Wright's inbreeding coefficient F, here defined as the probability that the two alleles at a locus in an individual are identical by descent. The effect of inbreeding (increased F) is to reduce the frequency of heterozygotes. This is shown in a simple model with two alleles at a locus, where the frequencies of the genotypes at equilibrium are:

$$\mathrm{A_1A_1} : \mathrm{A_1A_2} : \mathrm{A_2A_2} = \mathrm{p}^2 + pqF : 2pq(1-F) : \mathrm{q}^2 + pqF$$

Inbreeding depression

The term *inbreeding depression* refers to any decrease in growth rate, survival or fertility that arises when the parents of an individual are more related than would be two randomly chosen members of some base or reference population. The harmful effects can be evident in any component of fitness

Table 5.3. *Fitness values for self seeds or progeny relative to the value for outcrossed seeds or progeny*

Number of species and plant group	Character	Mean	Source
12 Gymnosperm trees	% filled seed	0.43	1
7 Gymnosperm trees	germination rate	0.81	1
12 Gymnosperm trees	plant size	0.70	1
8 Angiosperm herbs	seed production	0.63	1
6 Angiosperm herbs	germination rate	0.88	1
4 Angiosperm herbs	plant size	0.73	1
6 Angiosperm herbs	fertility	0.58	1
6 Angiosperm herbs	inferred viability	0.31	2
5 Angiosperm herbs	fitness	0.46	3
2 *Lobelia* species	two-year fitness	0.41	4
2 *Begonia* species	total fitness	0.68	5
2 *Mimulus* species	inferred viability	0.26	6

Source: (1) Charlesworth & Charlesworth (1987); (2) Ritland (1989); (3) Barrett & Kohn (1991); (4) Johnston (1992); (5) Ågren & Schemske (1993); (6) Dole & Ritland (1993).

(reduced viability, vigour or fecundity) in a specific environment. As just noted, the main genetic outcome of inbreeding is reduced heterozygosity and increased homozygosity through increased identity by descent. Hence, the two major hypotheses to account for inbreeding depression are the overdominance hypothesis and the partial dominance hypothesis. Under the former hypothesis, inbreeding depression arises from the loss of beneficial heterozygosity, whereas under the latter, it eventuates from increased homozygosity for deleterious recessive alleles. Much research has aimed at discriminating between these two hypotheses. In a recent review, Charlesworth and Charlesworth (1987) noted that most of the careful experimentation has strongly supported the role of partial dominance, although some contribution from overdominance cannot be ruled out.

The magnitude of inbreeding depression can be very great. Inbreeding depression in plants is commonly measured by comparing the performance of progeny derived by self-pollination with that of progeny derived from deliberate cross-pollination. The sorts of values for the ratio of fitness components of selfs to those of outcrosses are given in Table 5.3. An alternative approach developed by Ritland (1989) is to use marker genes to determine the actual frequency of self-pollination in a population and relate this to changes in the observed genotype frequencies over two generations.

This indirect approach estimates the viability of progeny derived from selfing relative to that of progeny derived from outcrossing.

Several technical problems attend either the direct or the indirect estimates. According to Charlesworth and Charlesworth, many of these problems lead to underestimates of the fitness reduction that would accompany inbreeding in nature. First, many estimates are of single components of fitness, rather than the complete life cycle. Second, measurements of relative performance made in benign environments (such as in the greenhouse) are likely to underplay the amount of inbreeding depression.

In general such measurements have detected a disadvantage for the progeny of selfing compared with outbred progeny of some 50% in outbreeding species. This level is much more than that found in selfing species. Yet a significant amount of inbreeding depression is found in many inbreeding species (in the order of 25%). This observation raises some doubts about the meaning of these measurements from the standpoint of conservation. Clearly, for the inbreeders, the observed amount of 'inbreeding depression' so defined has not proved terminal. Indeed, the selfing habit brings compensations such as the ability to found new populations from just a single migrant. Further, fluctuating levels of inbreeding in time may allow a period for the 'purging of lethals', or the silencing in the genome of mutable gene systems, which might otherwise build up unchecked, as a hidden genetic load to affect later generations.

Therefore, for the conservation of outcrossers, the question is what reductions to the size of the population lead to an excessive inbreeding depression. The answer to this question formed what Frankel and Soulé (1981) called the basic rule of conservation genetics. The rule is due to Franklin (1980) who used evidence from animal breeding to argue that natural selection for performance and fertility can balance inbreeding depression if the change in inbreeding coefficient (F) is less than 1% per generation. This equates of an effective population size of 50 individuals. Such a figure is a useful guide as the irreducible minimum population size for the short-term preservation of fitness.

The same rationale is applicable to outbreeding plant species, but cannot be invoked for inbreeders. In such species, the answer to the question of irreducible minimum size in the short term should be fixed using population dynamic parameters and ecological considerations. In a perennial species, populations of a single individual may be sufficient, provided that the individual is virtually certain to replace itself within its life span. For example, *Gossypium sturtianum*, an inbreeding shrub of semi-arid Australia, exists in one-third of its populations in the Flinders Ranges as single isolated individuals in dry creek beds. In a favourable season, a fresh crop of seedlings appears nearby such a mature individual, providing the replacement adult. In other species with a lower probability of individual survival and successful reproduction, more individuals would be needed for short-term survival. If the chance of individual replacement is 0.75, at least four individuals are needed for $>99\%$ assurance of population survival.

A genetic answer to the question for such species should perhaps use the concept of allele richness rather than heterozygosity and inbreeding depression, as in the case of minimum samples for long-term conservation (see section 2.5). Arguing that common alleles merit priority, Marshall and Brown (1975) developed a rationale for the minimum sample size for genetic resources based on the number of individuals required to assure retention of alleles above a specific frequency (0.05) with 95% probability. (Their figure of 59 independent gametes (=59 fully inbred individuals) is coincidentally similar to the Franklin number of 50.)

Recombination

Genetic recombination is an important process that deserves more attention than it gets in population conservation biology. In outbreeding species, sexual reproduction 'is not simply a device for transmitting sameness from one generation to the next. On the contrary ... the genetic processes that occur during reproduction provide the species with a shifting and dynamic variability system' (Carson 1983).

During the meiotic processes of crossing-over and segregation of chromosomes, a sample is drawn from the vast array of potential combinations of alleles that are present at loci throughout the genome. Further, each zygote results from the chance union of such gamete arrays at fertilization. Thus each reproductive cycle generates a new assemblage of diverse genotypes. The potential number of different phenotypes is so vast that only a tiny fraction are produced, and many of these have low relative fitness.

It is this continuing capacity to generate diversity that is crucial to the evolutionary success of sexual reproduction. Too often recombination is viewed in conservation theory merely as an unwanted disturbing force that breaks up favourable coadapted gene complexes. Recombination and directional selection are the primary means that plant breeders have used to improve crop plants steadily and successfully in the face of continually evolving pests and diseases. There is every reason to believe the same advantage is important in natural systems. The attitude we take to recombination determines tactics in the management of valued populations, as discussed in Chapter 6.

Favourable coadapted complexes occur on two levels (see also the discussion of variability structure in landrace populations of crop species: section 3.2.2), namely, within populations and among populations. The components within populations are polymorphic complexes held together either by chromosomal inversions or close proximity on the chromosome. In considering their sampling, Marshall and Brown (1975) argued that the complexes that are more crucial to the adaptedness of a particular population would be those in high local frequency and possess the dynamics of single alleles at a supergene. In other words, they would be complexes resistant to fracture by recombination. On the other hand, particular combinations of different genes conditioning, for example, resistance to different pathogens can, through recombination, be assembled on a 'needs' basis. Continuing evolution in pathogens necessitates the flexibility of response that sexual recombination provides (Hamilton *et al.* 1990).

Different complexes can arise in different populations, for example as ecotypes, each of which is adapted to a specific habitat. In this case, the complex is not necessarily polymorphic locally, nor held together by tight linkage. Recombination following hybridization between differentially adapted ecotypes is seen as a threat to the integrity of local gene pools. Yet it can be argued that recombination is a neutral force in this scenario. Complexes that are readily fractured in a hybrid swarm by recombination may, given the appropriate selection pressure, just as easily be reassembled or even improved upon by recombination. The lesson is that there is nothing inherent in the recombination process to justify extreme conservatism to issues such as 'fidelity to the site of origin', when deciding upon material for reintroduction or replenishment of endangered populations (section 6.2.3). The question of recombination between different gene pools arises again in the next section.

5.4.3 Migration and gene flow

Migration is the dispersal of gametes or individuals and their merging with a population different from the population of their origin. When the source and receiving populations differ in allele frequency, gene flow is said to occur. The amount of gene flow, or influx of new genes, is dependent on both the rate of migration and the degree of genetic divergence between the immigrants and the recipient populations. Although gene flow is strictly defined (e.g. by Slatkin 1987) as the *change* in gene frequency due to the movement of gametes or individuals from one place to another, it is commonly measured as the migration rate (m).

Higher plants have two distinct modes or opportunities for dispersal, a gametic mode by means of pollen, and a zygotic mode by seeds or vegetative propagules. The wide variety of pollination systems and seed dispersal mechanisms in plants presents a complexity in which the optimal levels of gene flow for conservation will be hard to define. Clearly, gene flow delivers to a population fresh genetic variation that provides scope for evolutionary change. Hybridization may also produce individuals with enhanced fitness through luxuriance associated with heterosis. On the negative side, immigrants may be less well adapted and recombination with the resident population might disturb favourable gene complexes (see section 5.4.2).

Another complication in the study of optimal levels of gene flow in conservation, is the usage of the term itself. At one extreme of the scale, gene flow refers to rare gene exchange between otherwise isolated and divergent populations. In this case, low migration rates may be at least tolerable whereas *high* rates could be *dysgenic* (see section 6.2.1, p.156). At the other extreme, some authors (e.g. Schaal 1980) have used the term for the local movement of seed, or of pollen from one plant to the next in the normal process of mating. In Schaal's experiment, gene flow was measured on the scale of metres within a population of *Lupinus texensis*. In contrast, the pollination system of a sparsely distributed species of tropical forest tree, such as *Pithecellobium pedicellare* with two individuals per hectare, can ensure near-complete outcrossing ($t = 0.95$) despite the great distances between individuals (O'Malley & Bawa 1987). In both these cases, *high* rates of pollen dispersal are presumably *optimal* for conservation. To avoid confusion, we will reserve the term 'gene flow' to connote movement of genes between populations (Brown 1989c; Ellstrand & Elam 1993).

The population structure of most plant species falls between the extremes of one single panmictic unit, or the simple island model of many isolated populations. Commonly their populations are connected through migration to varying degrees to form a metapopulation; and each population is subdivided into partially isolated subpopulations. In such a system, migration and gene flow will have many effects, ranging from the beneficial (recolonization of extinct local patches) to the potentially adverse (retarding population divergence).

Measurement of intraspecific gene flow

An exact description of gene flow in a species strictly includes the varying rates of migration among metapopulation units that have diverged to different degrees for different loci. In practice, gene flow is measured in a single parameter of effective migration (m) in a model system, analogous to the concept of effective population size using a model idealized population. The model system is that of divergence in the infinite neutral alleles model for n island populations (Latter 1973). The measure of divergence is Nei's G_{ST}. The estimator of the average number of successful immigrant per generation is

$$mN_e = \{(1/G_{ST}) - 1\}/\{2n/(n-1)\}^2$$

Slatkin and Barton (1989) have shown this to be a relatively reliable method for estimating historic levels of migration. Genetic markers such as allozymes or molecular differences provide an estimate of G_{ST}. Table 2.1 summarizes the average values of G_{ST} based on allozymes and typical for various kinds of plant species.

Figure 5.2 plots the relation between the observed genetic divergence and migration rate, for the case of a pair of populations ($n = 2$) and for a large number of populations. It is evident that problems arise at either end of the scale of G_{ST}. For low population divergence, a small error in the measurement of G_{ST} will produce a large change in the estimated migration pressure. In other words, there is little difference between $mN_e = 1$ or $mN_e = 5$ or more,

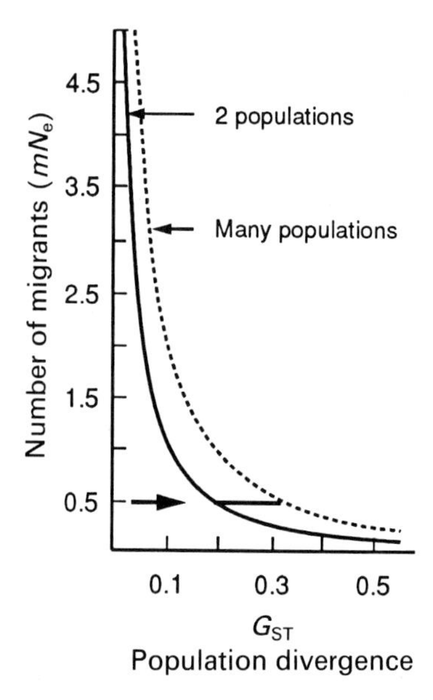

Figure 5.2. Theoretical relationship between the G_{ST} measure of population divergence and the number of migrants each population receives (mN_e) in the island model of neutral variation with either two, or many populations.

from a genetic standpoint. From an ecological standpoint, however, this may not be true. Alternatively, high values of G_{ST} are relatively weak in detecting changes in the level of migration. In this case, the model assumption of detectability of each new mutation is critical. The failure to resolve with electrophoresis all differences between allozymes is a handicap for this technique in estimating migration among very divergent populations.

Gene flow in conservation biology

As noted by Ellstrand and Elam (1993), small populations are expected to receive gene flow at a higher rate than are large populations. Large populations are more likely to be donors than are small ones, even when the latter are in close proximity. Both trends mean that gene flow is an important issue in conservation biology. The question is, how can the conservationist judge when gene flow is a major concern.

Since gene flow may be either beneficial (e.g. in replenishing genetic variation after a bottleneck) or detrimental (e.g. in adding maladapted genes), Ellstrand and Elam (1993) suggest that most concern should arise when current levels depart substantially from historic levels. Historic levels can be estimated indirectly using genetic markers as given above (Slatkin 1987). Ideally, these can be compared with direct indicators of present levels (e.g. pollinator prevalence or recent changes to the degree of fragmentation of populations). For a species with historically high levels of gene flow ($mN_e > 0.5$), a sharp drop might be remedied with transplants, or transport of seed or pollen. Conversely, a radical increase in gene flow in a species with historically low levels ($mN_e < 0.5$) may require a buffer zone to absorb the mobile propagules or pollen, or their dispersal agents; or for removal of weedy populations (Ellstrand & Elam 1993).

Interspecific gene flow

Interspecific gene flow through hybridization and introgression presents a dilemma to plant conservationists. On the one hand, many regard it as a threat to species diversity, particularly through the swamping of rare species. For example, Ellstrand and Elam (1993) estimate that interspecific mating is possible in about 20% of Californian plant species of concern, and in 10% of the protected species in the British flora. If the hybrid progeny are vigorous and fertile, they may eliminate a rare parental species, whereas any hybrid inviability or weakness lessens the reproductive effort of the parent. Either outcome from an interspecific cross is unwelcome to conservationists. On the other hand, species hybridization is a common and central process in plant evolution. Indeed, the origin of a great many plant species involves interspecific hybridization (Grant 1981). Further, the more adaptive genes of an extinct ancestor will survive in the new species.

Therefore the case for intervention in hybridization between two naturally co-occurring species is relatively weak. Wilson (1992) cites Gentry's suggestion that tropical plant species exchange genes less frequently than those in temperate zones – hence species hybridization may be less of a problem in the more diverse ecosystems. Interspecific gene flow is more of concern when it arises from human disturbance, primarily associated with agriculture or forestry (Brown 1992). An example from California is the rare *Juglans hindsii*, which is at risk from hybridization with the cultivated walnut (McGranahan *et al.* 1988). Interspecific gene flow from a crop or plantation species into natural populations of its wild relative is a common phenomenon and requires assessment as to its impact (see section 6.2.4). The advent of genetic engineering has added a new dimension to this question (Raybould & Gray 1993). Some agencies now require an assessment of risk as to whether the release into agriculture of a cultivar with novel genes poses any threat to natural biodiversity, should such genes leak into wild species by hybridization (Seidler & Levin 1994).

5.4.4 Diversifying selection pressures

As we noted in Chapter 2, plant species differ greatly in the extent to which their populations have diverged genetically from one another. The several populations of a species may differ in their allelic content, allelic frequencies, multilocus associations and in the amounts of genetic diversity each contains. So far we have considered three main forces that are involved in this divergence, namely random genetic drift, reproductive mode and gene flow. But the forces of most consequence for conservation biology are the local selection pressures that vary from one population to the next (see Huenneke 1991 for review). These forces of natural multi-niche selection belong to the mode of selection that Dobzhansky (1970) termed diversifying selection.

At the phenotype level, there is abundant evidence from transplant experiments for the existence of such forces as adaptive ecotypic variation in plants (see section 2.3.1). At the allozyme level the divergence is less evident, but certain types of molecular data (rDNA genes, chloroplast DNA) show appreciable population divergence (Brown & Schoen 1992). Components of the physical environment and the biotic environment that differ in different places, or at different times, make up the diversifying selection pressures that can bring about population genetic divergence.

In many cases a population performs best in the environment to which it is most recently adapted (Bradshaw 1984). Invasive species are, however, a very notable and important exception to this trend. Some introduced exotic species have become serious threats to biodiversity (see section 10.4.2, p.250), particularly when they move over long distances and leave behind their density-dependent regulating agents such as disease, predators and competitors. Two examples of serious environmental weeds threatening whole plant communities are bitou bush and rubber vine. Bitou bush (*Chrysamnthemoides monilifera*) is an invasive weed of coastal sand dunes and heath in Australia, originating from southern coasts of South Africa. Rubber vine (*Cryptostegia grandiflora*), a native of Madagascar, has been introduced to most tropical and subtropical countries as a potential source of rubber, or as an ornamental. In its adventive range, it is an aggressive woody climbing liana capable of smothering tall trees and forming impenetrable thickets. In both these examples, several insects and fungi, potential agents of biological control, occur in the native ranges (Parsons & Cuthbertson 1992).

The amount of adaptive population divergence is a fundamental parameter in the conservation biology of populations. Clearly in species made up of populations that are highly specialized in their present habitat, the transplanting of organisms from one site to another is more likely to fail than in species whose populations are broadly adapted. The concepts of general adaptability and specific adaptation are very familiar to plant breeders. Conservationists undertaking more active forms of population management may need to borrow these concepts.

As well as the fitness of immigrants or transplants, the amount of divergence is critical for another reason – the fitness of hybrid progeny, the products of gene flow. The question is whether these hybrids will display 'outbreeding depression'.

Outbreeding depression

Outbreeding depression refers to the reduction in fitness that occurs following intraspecific hybridization between individuals from spatially separated populations. If local populations are differentially adapted to their habitat, gene flow between them is likely to produce progeny that are less well adapted compared with the residents. In plants, local selection pressures, sessility, small effective sizes, limited dispersal and inbreeding favour the development of subpopulation structure on various scales. Do plant species in fact show significant 'outbreeding depression'? This is an important question for plant conservationists to resolve. If substantial outbreeding depression were a general phenomenon in plants, it would justify restriction of replenishment materials to extreme fidelity to the site concerned.

At present there is insufficient evidence to reach a firm conclusion on this question (see Barrett & Kohn 1991 for review). Price and Wasser (1979) developed an experimental approach to the problem. They proposed that since matings between neighbouring plants are likely to be matings between relatives that produce inferior inbred progeny, and matings between widely displaced populations are also likely to produce poorly adapted progeny, there should be a distance between mates for 'optimal outcrossing'. Their studies in *Delphinium nelsonii* and *Ipomopsis aggregata* found that crosses between plants at some optimal distance (about 10 m) set more seeds than did crosses between adjacent or between much more distant plants. However, later parallel studies by others, reviewed by Barrett and Kohn (1991), have obtained a variety of relationships between proximity and fitness of progeny. Further, it is important to consider the variance in fitness that may develop in an introgressed swarm. Some individual progeny may well show 'outbreeding depression', and indeed the mean fitness of the hybrid derivatives may lie below that of the parents. However, several lineages could have enhanced fitness relative to the parental sources. The situation should be viewed as dynamic and not static. Natural selection is expected to seize on this increased variance in fitness and possibly in the space of a few generations, drive the population towards improved adaptation.

5.5 CONCLUSION – IDENTIFYING RISKS TO POPULATIONS

Population dynamic and genetic processes form the key to understanding the prerequisites for species conservation. This is because the survival of plant species in time entails both reproduction in populations and evolutionary change. The challenge for population biology in conservation is to identify the crucial demographic and genetic risk factors that populations must face (Jain 1994). In this chapter, we have discussed the principal concepts – the minimum viable population, actual and effective population size, inbreeding depression, optimal gene flow and outbreeding depression and the metapopulation structure of plant populations. All these are problematic to differing extents, in concept, in measurement, or in application to conservation strategies. Nevertheless these concepts are basic to the development of strategies for the conservation of a range of plant species, as discussed in the next chapter.

6

The conservation in situ *of useful or endangered wild species*

6.1 INTRODUCTION – SPECIES THAT MATTER

This chapter considers the objectives and methods of programmes for *in situ* conservation at the level of a single plant species. At first sight, such endeavours appear to be a contradiction in terms. Strictly speaking, *in situ* conservation is conservation of the whole ecosystem, that is, conservation of the community in its natural location without special focus upon any particular species. Indeed, for any plant species to survive in nature unaided by humans, it must do so within a community of interacting organisms. Some of these interactions (e.g. with pollinators, seed dispersers, microbial symbionts) are crucial to its survival. Thus it is impossible to have meaningful species conservation *in situ* without ecosystem conservation, and any prescription for a species must recognize this. On the other hand, the converse is not necessarily true – it is possible to have ecosystem conservation while some species disappear.

However, whenever the community includes particular wild species of concern to humans, a general ecosystem approach to *in situ* conservation could prove inadequate. Single species become prominent in conservation planning for a variety of reasons, as follows.

1. When the plants are *directly harvested* or grazed in the wild. This includes forest trees, and many medicinal plants, spices, ornamentals, food and forage plants.
2. When populations are used as sources of *propagating material* for planting elsewhere, and as sources of genetic variation for breeding programmes. This includes forage plants, and wild relatives of field and horticultural crops.
3. When a species is crucial for the well-being of an ecosystem. This

includes dominant or *keystone species* and food plants of animals of major concern.

4. When a species is designated as *endangered*, particularly when it is chosen as the object of a recovery plan (see also Chapter 7).

Several examples are discussed in the following sections. We begin with the conservation of forest tree species as a paradigm of *in situ* conservation at the species level. We discuss issues of purpose, programme initiation and maintenance. Issues like the optimal selection and size of populations, the involvement of local people and the use of resources are shared with the other kinds of targets. Basic to them all is the question of intervention, with which the chapter concludes.

6.2 SPECIES TARGETS FOR CONSERVATION *IN SITU*

6.2.1 Forest trees

Prominent among wild plant species that call for individual conservation planning are the tree species used in forestry. They are outstanding, first, economically, as the source of many products essential to society such as industrial raw materials, timber, fuel, food and fodder. Therefore they are prone to decimation from harvesting. Second, they are prominent environmentally, as forests help to stabilize the environment by fixing carbon dioxide, by preventing soil erosion and by lowering the water table. Third, trees are dominant ecologically as they determine a wealth of interactions with other life forms in the community. Since forest tree species inevitably sustain reduction in numbers from human exploitation, safeguarding their genetic resources is the key to their conservation. It has long been recognized that conservation *in situ* is the primary method for this purpose (Frankel & Bennett 1970b).

Genetic resources of forest trees

The genetic resources of forest trees are in most instances located in primaeval, or 'old growth' forests. These are steadily shrinking – at the annual rate of 11 million hectares from the world's total forest and woodlands of 4500 million hectares (National Research Council 1991). About two-thirds of this loss is from the closed forests of the moist tropics. However, there is still scope for conserving a broad range of genetic diversity for most species, if possible in primaeval forests, but otherwise in secondary and remnant populations.

The major threats to forest genetic resources come from the increasing tempo of direct use, and from population pressure for more arable and grazing land (Richardson 1970; National Research Council 1991). This pressure has greatly intensified, especially in the tropics. In South-East Asia the once vast resources of tropical moist forests continue to shrink rapidly, due to the switch from selective to intensive logging, and the clearing of land for agriculture and plantation crops.

Today, foresters are increasingly renewing sources of timber by planting fast-growing pioneer species, many of which are exotics. It is somewhat ironical that the advance of plantation species should coincide with the disappearance of the genetic resources on which the improvement of their productivity, pest resistance or timber quality will depend.

In some forestry species in the temperate zone, genetic erosion is occurring just as it is in the tropics (Ledig 1988). For example, Douglas fir (*Pseudotsuga menziesii*) is an important plantation tree in its native range on the west coast of North America, and as an exotic elsewhere. Although several native populations in California are protected in reserves, pollen from surrounding plantations, and aerial seeding of non-Californian provenances, threaten to erode the integrity and diversity of the indigenous populations. The position is different in the Monterey pine (*Pinus radiata*), with only five small populations remaining in its native habitat, and one of these close to extinction (Eldridge 1990). There is sound cause to preserve the others *in situ*, as with 3 million hectares in plantation the species is the most widely planted exotic conifer world-wide (Moran & Bell 1987). One may, however, wonder whether the secondary gene pools in the vast areas planted in Chile, New Zealand and Australia are a more useful genetic resource for tree breeding than are the declining relic populations in California. Isozyme analysis of the Australian breeding populations suggested that they originated from two of the primary populations and together retain all their rare alleles (Moran & Bell 1987). The three southern populations, however, are likely sources of new genetic material.

Thus indigenous forests have a dual function as genetic resources. First, with regard to the species that are used in plantation forestry, indigenous forests are sources of genetic diversity for adaptation to new environments, for resistance to diseases and pests, for changes in silvicultural practices and for new products (Eldridge *et al.* 1993). Second, indigenous forests are reservoirs of other potentially useful species. For example, Guzman (1975) compiled a comprehensive list of woody and herbaceous forest species in the Philippines that are potential sources of timber (27 species) and food, including fruits (94), oil (13), beverages (12), vegetables (27) and spices (7).

Problems of tree species conservation in nature reserves

National parks and wilderness areas that are generally protected from exploitation serve as gene reserves for the tree species they contain. However, such areas are often far from adequate (National Research Council 1991). Their size and location are determined primarily on aesthetic and ecological grounds rather than with genetic considerations for a particular species in mind. In many developed countries, the larger parks are to be found in inhospitable areas with extreme environments and poor soils. Hence exclusive reliance on them for tree species conservation may lead to biased samples of gene pools. An example of such bias is found in *Eucalyptus globulus* in Victoria and Tasmania, where the more valuable stands growing in coastal areas and Bass Strait islands are greatly depleted whereas the less desirable upland stands are secure in National Parks (Eldridge 1990; Eldridge *et al.* 1993). Further, less than a third of the world's biosphere reserves have been inventoried for the species they contain and the status of their populations. Finally, a non-targeted approach to management can leave a reserve vulnerable to species loss, contamination from gene flow and invasion of inferior provenances.

As many authors have noted (see Ledig 1988; Millar & Westfall 1992), a systematic approach to conservation *in situ*, with deliberate choices of conserved areas, should replace the haphazard sampling in nature reserves. Hence conservation of economic forest tree species necessarily entails a strategy for forestry reserves in addition to general nature reserves.

Over the last decade the Forestry Division of the Food and Agriculture Organization has played a leading advocacy role, coordinating schemes for the conservation *in situ* of forest genetic resources (FAO 1989). The Forestry Division has argued for the development of networks of conserved sites. The growth of such networks requires integrated databases that record species occurrence and ecological and floristic diversity. In addition to this general role, the Forestry Division has initiated some specific pilot projects such as conserving *Baikiaea plurijuga*, the Zambesi teak in Zambia, or the drought-resistant species *Prosopis cineraria* in Yemen (Palmberg-Lerche 1992).

From the standpoint of genetic resources of particular tree species, the main goal of *in situ* conservation is to manage the target species in its original habitat so as to minimize the loss of genetic diversity present within and among populations and organized in coadapted complexes. The challenge is to maintain variation within a mosaic of economically and socially acceptable options of land use. Options range from strictly protected areas, buffer zones, managed forests, nature reserves with multiple uses, to agro-ecosystems.

At this point it is important to recognize that many forest geneticists use the term '*in situ* conservation' in a broader context than do conservation biologists. In forestry

> *In situ* conservation means the conservation of genetic resources of a target species 'on site', within the natural or original ecosystem in which they occur, *or on the site formerly occupied by that ecosystem* [our emphasis]. Although most frequently applied to populations regenerated naturally, *in situ* conservation may include artificial regeneration whenever planting or sowing is done without conscious selection and in the same area where the seed or other reproductive materials were randomly collected.'
>
> *(Palmberg-Lerche 1993)*

Such a conception of *in situ* conservation combines the *reservation* of wild areas, and the *restoration* or regeneration of forests that have been partially or completely logged. The main proviso is that the restoration should attempt to reconstitute the original state by using unselected, local seed. The definition focuses on conserving the genes of the target species, but pays little attention to the myriad of interactions making up the forest community.

This scheme for *in situ* conservation of forest genetic resources has two aims: to maintain economic production, and to replant with local sources of seed. In having these aims it resembles schemes for the *in situ* conservation of landraces of crops (see section 4.2). However, it differs from crop landraces in using random or *unselected* seed, i.e. in proscribing either deliberate selection for phenotypic attractiveness ('plus trees'), or inadvertent selection due to variation in seed fecundity. In contrast, the *in situ* conservation of landraces of many crops (e.g. maize) envisages them as subject to traditional selection, both deliberate and unwitting.

For plantation forests, restoration that attempts to conserve genetic resources *in situ* still must have economic and sustainable production as its primary objective. Conservation of local gene pools is a secondary objective. However, can the same population serve two conflicting purposes? It is arguable that other forestry practices (such as replanting with exotic species, or alien, selected provenances, or clonal monocultures) are preferable on economic grounds. In some cases it may be better to divide the area into a productive part in which likely yield rather than fidelity to the site determines the replanting stock, and a reserve within a buffer zone to remain as a mature uncut forest. Such a reserved community would also conserve the intricate interactions that a reconstructed forest is unlikely to restore.

Initiating a programme of conservation *in situ*

Initiation of *in situ* conservation involves the selection of species and sites, decisions on the management strategy, and the development of infrastructural support for the project. Fully informed choices depend on scientific, biological, economic, sociological and legal information: perhaps this is why only a limited number of modest projects are in place thus far (FAO 1991).

The target species

The first step is to identify the target species and the purposes of the programme. The outcome will depend on the value of the species (both nationally and internationally), its scarcity and vulnerability, and, because of provenance variation, on the significance of the candidate populations relative to the species as a whole. For example, among the over 900 species of eucalypts, priority should go to the major plantation species around the world: the three east Australian coastal species *Eucalyptus grandis*, *E. globulus* and *E. tereticornis*, the continent-wide species *E. camaldulensis*, and the Indonesian species *E. urophylla* (K. Eldridge, personal communication). In addition, the riparian species *E. deglupta*, which occurs from Sulawesi and Mindanao east to New Guinea and New Britain, has great potential for planting in wet tropical lowlands (Eldridge *et al.* 1993). Exploitation of that species for timber, and alienation of its river-flat habitat for agriculture, has already led to such depletion of valuable provenances that the need is urgent for *in situ* conservation.

Choice of sites

The choice of sites is essentially a problem of germplasm sampling. How many sites are to be set aside for *in situ* conservation? How large should they be? How should they represent the available genetic spectrum, for example in gauging the relative merits of central and marginal populations?

Concerning the number of sites, at least two sites are needed for sampling the divergence between populations. This number should be doubled to meet the need for 'back-up' or insurance against major calamity. The occurrence of the species in more than one major ecological zone is reason to increase the number yet further. Evidence of zonation would emerge from studying the geographic distribution of the species, and the ecological range and community diversity that its distribution encompasses. Such an ecogeographic survey is a logical framework on which to divide the distribution into distinct zones to guide a structured sample. Other factors that bear on the number of sites are the existing land tenure, whether the species occurs in already established reserved areas, and the condition of the candidate populations.

Within an ecological zone, the primary criteria in deciding the number and disposition of the sites are local genetic diversity and ecological amplitude. Thus sites with the highest allelic richness are preferable, as are sites that would contribute to representing the geographic and ecological range of the species.

Size of unit

How many individuals are required, and hence how large an area is needed? Forest trees rank among the largest of plants, so that their space requirements per individual are maximal. Further, for rare species in tropical lowland wet forests the density of a species can fall well below one individual per hectare (Hubbell & Foster 1986). At 10 per square kilometre, a population of 1000 individuals would require a substantial area of 100 km^2. Hence a limit to the size of a population that may be conserved is more likely to apply in tree species than in other plants.

The theory discussed in section 5.4.1 suggested that the effective number of individuals needed for a viable population in the long term is at least 500. A multiplier of five- to tenfold converts this effective size to the actual census size. Many have suggested that a population viability analysis (PVA) can provide more precise and reliable answers to the numbers question. However, PVA is no panacea readily applicable whenever decisions on size are needed. It is difficult and time-consuming to estimate the population dynamical parameters needed for a PVA. The longevity of trees is an added source of difficulty for PVA, because of the hazard of extrapolating cohort survival probabilities estimated in a study that does not span a completed life cycle. The results of a PVA will differ for different species, and are likely to differ between populations of the same species. In the short term, the decisions for most species will have to be made without PVA, and can best be improved with thorough PVAs of a few examples of a range of life forms.

The purpose of PVA is to take account of the uncertain forces shaping a population's future (demographic uncertainty, etc.: section 5.3). However, for forest trees, stochastic events may be secondary compared with the forces under direct human control, namely management strategy, multiple uses, logging policy, pest control, etc. Further, rare catastrophic events in general play a large part in determining the final size that a PVA justifies (Ewens *et al.* 1987). But with valuable forest genetic resources at stake, it is sensible to assume that management will take direct action to mitigate catastrophes such as fire, disease epidemics and reproductive failure. Therefore populations smaller than those computed from naïve PVAs may indeed form viable units for conservation *in situ*, if backed up by *ex situ* collections and replenishment when needed.

Others have answered the numbers question by appealing to Kimura and Crow's equilibrium theory for the amount of heterozygosity expected for neutral genetic markers (Latter, in Frankel 1970b; Millar & Libby 1991). Depending on the mutation rate assumed for such loci, the effective population size needed to maintain current levels of heterozygosity ranges from values like 2230 to about one million. For several reasons, such computations lack precision in answering the question. First, the values are too sensitive to the assumed mutation rate. Second, stochastic equilibrium is assumed. Finally, rates of accumulation of molecular heterozygosity and the approach to equilibrium are very slow compared to population dynamical changes. Thus much of the observed heterozygosity could pre-date population divergence and probably pre-dates the divergence between related species (Gaut & Clegg 1993).

Linkage to ex situ *conservation and use*

The next major issue in the planning of *in situ* conservation of forest trees concerns the provision of material for use or conservation *ex situ*. Ideally seed should be sampled periodically. The frequency of resampling should approximate the longevity of stored seed because regeneration of expiring conserved samples is expensive for tree species. The main purposes of such samples are to serve as a source of germplasm to provide material for research and for seed orchards, and to form a back-up in case of a disaster that requires the need for restoration of the primary *in situ* population. Obviously sampling may have to be more frequent in populations that prove to be in heavy demand.

Once again the question of sampling strategy arises: in particular, the size and nature of such samples. Considering the logarithmic relation between sample size and allele retention (section 2.4.1), a starting point is to collect seed from the number of trees one order of magnitude less than the population size (e.g. from 100 in a population of 1000). The sampled trees should be spaced to cover the whole population but randomly chosen from among available neighbours within a clump. More than one fruit per individual should be collected as seed within a fruit may have correlated paternity. Since most tree species are outcrossing, the sampling of several fruits or cones from each tree will diversify the sources of pollen. In turn this will increase the effective size of the sample.

Local people and institutions

An issue of concern is the effect of conservation measures on the local people, particularly when some of the resource is to be harvested. The conservation of *Alnus jorullensis* in Peru affords a good example of the potentially complex interactions that can arise. Small remnant natural stands of this species are in danger from over-exploitation for timber, stakes and fuel. However, local support for their conservation followed from demonstrating that the remnants were a source of seed for plantings. As a border to terraces used for food crops, the species serves three purposes. It helps prevent soil erosion, fixes nitrogen for crop production and provides wood for the community. In turn, the plantings relieved pressure on the remnant stands from direct harvesting and led to better management (C. Palmberg-Lerche, unpublished data). Clearly in this as in many other cases of *in situ* conservation of domesticated and wild species, the strengthening of local institutions and the involvement of the community dependent on the resource will be crucial in determining the success of long-term conservation.

Management of conserved forests

Once a project is under way, the management strategy becomes crucial and must address a number of issues. These include the level and kinds of intervention, the intensity of harvesting, the source of germplasm and the practice of replenishment.

The major and minimum objective of the conservation of forest tree species *in situ* is to maintain a representative sample of their genetic diversity. This objective constrains the degree to which harvesting can take place. The continued functioning of a conserved ecosystem depends on a multitude of ecological interactions (section 8.3). Particularly important are the symbiotic relationships between trees and their pollinators, seed dispersers, and microbial and fungal organisms associated with their roots, and other interactions with animals that live in the ecosystem. The greater the level of harvesting, the greater the disruption to these interactions and the more knowledge about them is needed to ensure sound restoration practices (Roche & Dourojeanni 1992).

For most species, little direct intervention is feasible, other than regulating the removal of trees, preventive maintenance, fire control and monitoring disease and pest spread. More intensive measures should be confined to remedies to counter detrimental factors or dwindling populations. Examples are artificial regeneration using local seed, weed removal, or controlled burning to suppress competing species. Management of adjoining areas is an additional concern. For example, if neighbouring plantation forests are of other than local unselected stock, they could be a source of contaminating pollen (Millar & Libby 1991).

In reviewing the impact of humans on forest ecosystems, Ledig (1992) listed the following as major disturbances – deforestation, harvesting, fragmentation, demographic alteration (conversion of the forest to even-aged stands with shorter generation times), domestication, habitat alteration, environmental deterioration (due to pollutants), translocations of mammalian herbivores, browsers, competitors, pathogens, insect pests, and removing barriers to the movement of plants, animals and microorganisms. This provides a handy check-list for management strategies for conservation *in situ*. Ledig also points out that the more selective the logging, the less likely it is to affect the genetic structure and mean fitness of the remnant. Thus logging only the most attractive 10% of the population has much less effect on the progeny than does removing the top 80%. After intense selective harvesting, the genes in the remnant can recombine sexually to form the next generation with virtually the same fitness as the current one.

Following Scandinavian models, forest geneticists in California have developed a scheme for conservation *in situ* as a network of genetic resource management units or GRMUs (see Box 6.1). The designating of GRMUs resembles the choosing of core collections of *ex situ* germplasm collections of crops (see section 4.5.4). The GRMU scheme is a stricter procedure than are some allowed under the above definition of *in situ* conservation (section 6.2.1, p.151) because it recognizes the need to conserve interactions among species.

While much has been written about the desirability of conserving the genetic resources of forest trees *in situ*, the National Research Council (1991) concluded that there is still a long way to go before such conservation is widely in place. 'The potential conservation utility of these programs has not been realized and may not be for many years.' The challenge now is for urgent action, both in old growth forests as well as in secondary forests and forest fragments. Unfortunately the resources will not wait around for a change of practice.

6.2.2 Medicinal plants

Medicinal plants are another diverse category of plants directly used from the wild. In contrast to forestry, their use is usually not fatal to individual plants. For example, the plant parts from 54 species used in traditional Ethiopian medicine are leaves (35 species), roots (26), seed (13), bark (8), fruit (4) and whole plant (3) (Abebe & Hagos 1991). However, in many cases heavy or continued exploitation risks the regeneration of the natural source population (Cunningham 1990).

Perhaps even more than for forest trees, the principal threat to such

BOX 6.1. GENETIC RESOURCES MANAGEMENT UNITS OF FOREST TREE SPECIES

A GRMU has 'as its primary function, the protection of genetic resources, either of a single (nominated) species or an entire community, but (it) could be used for other economic benefits, such as grazing or timber harvest, as long as the other uses did not threaten the primary function' (Ledig 1988). In the choice of sites, the highest priority is to capture the core of variability within the species in areas that represent the variability of a region, and populations that include the most valued trees (Millar & Libby 1991).

Once candidate sites are chosen, each is assessed for associated native flora and fauna and the final selections are the sites that maximize protection of allied species. The choice should exclude those sites where the use of adjacent lands prejudices their genetic composition (through gene flow, habitat disturbance). A buffer zone should surround the GRMU, the amount of buffer depending on altered selection pressures from adjacent land usage (for example, limited selective harvest requires less buffer than if the adjacent land is clear-felled, or alienated from forestry). The ability of conifer pollen to disperse widely requires precaution. An alien species (and a non-invasive one) may be preferred as a buffer to a plantation of the target species but of non-local origin, as the exotic species would absorb and not generate contaminating pollen.

Millar and Westfall (1992) exemplify the procedure with a study of white fir (*Abies concolor*) in Eldorado National Forest, California. The first step is a genetic study of the extant distribution, using genetic markers, supplemented with morphological traits if possible. A multivariate analysis is then made, refined by knowledge of the ecological range of the species, so that the species area is stratified into a limited number of zones, between which genetic distances are large. This divergence is an attempt to measure the 'transfer risk' incurred if trees are moved from one site to another. 'The average transfer risk within zones is 5% and the average transfer risk among the extreme zones is 22%' (Millar & Westfall 1992). For each zone, two core GRMUs are chosen as a minimum to represent the average diversity within that zone (Millar & Libby 1991). The set of candidate GRMUs is compared to a map of pre-existing nature reserves where management is compatible with GRMU status, and for suitability of each existing population. If no area is in place, or cannot be set up, then this is noted as a gap in the conservation of the species, and other remedies (*ex situ* stands, etc.) sought. Once established, the network of GRMUs should be defended and monitored actively, studying species inventory, soils, ecological and community types.

species is not over-harvesting but the destruction and conversion of their habitats to other purposes. This is pre-eminently the case in the Amazonian tropical rain forests, which Schultes and Raffauf (1990) have dubbed 'The Healing Forest'. They compiled botanical and ethnopharmacological information on 1516 species which comprise only a tiny fraction of the total resource, and stressed the need to conserve both the plants themselves and the knowledge of their use possessed by indigenous peoples.

In several countries the bewildering wealth of material growing in a wide range of conditions can cause problems. For example, over 700 plant species in Sri Lanka have been recorded to have therapeutic value and many are used in traditional 'Ayurvedic' medicine (de Alwis in FAO 1989). Some of these medicinal plants are cultivated in home gardens, borders or disturbed forests. However, unless carefully surveyed to determine which species are at risk and areas set aside as reserves *in situ*, the danger of extinction through oversight exists. Further, 'since adequate hospitals and Western-trained doctors are not found in much of the tropics, the destruction of the rainforest will also destroy the primary health care network involving plants and traditional healers' (Balick 1990).

The conservation needs of medicinal plants were the subject of a recent international meeting organized by WHO, IUCN and WWF (Akerele *et al.* 1991). Medicinal plants are significant to both developing and developed countries. Estimates indicate that over 75% of the world's rural people rely on traditional herbal medicine. About half of the world's medicinal compounds are still derived or obtained from plants (Hamann 1991). Many of the most important drugs of recent times were first isolated from plants, including the curare alkaloids; penicillin and other antibiotics; antihypertensive alkaloids like reserpine; and both cortisone and contraceptive steroids that are derivatives of diosgenin.

There is clearly a great range of higher plants (Heywood 1991 estimates some 25,000 species) from which to draw, and a great repository of traditional knowledge in the various cultures of people using medicinal plants. A major challenge is to identify the species that merit priority for conservation. The degree of 'risk of extinction totally or locally' (as suggested by Heywood 1991) is one among many criteria to consider, others being value, actual and potential usage, cultural importance, and uniqueness.

A major aspect of the conservation of medicinal plants is their considerable social and economic value (Farnsworth 1988). To highlight the monetary loss from the extinction of medicinal species, Principe (1991) estimated that the potential annual market value in OECD countries of the species likely to vanish before the year 2050 is $60 million. This figure is about 0.15% of the amount spent on plant-based drugs. It represents a benefit foregone rather than an actual loss. It is, however, only a market value, and excludes other components of the total economic worth of the drugs, such as the cost to society of not having them and the benefits of good health. The total economic value could be five- to 50-fold higher (Principe 1991).

McNeely and Thorsell (1991) point out that reserved areas are significant sources of plants of real or suspected medicinal value, and that the entry of people into such areas to harvest species of use raises the problem of poaching. The authors advocate that the best way to enhance the role of protected areas in the conservation of valued plants, is to develop plans to enable the local people to benefit from the resources in a sustainable way, rather than to exclude them with sanctions, fines or imprisonment. They present guidelines for the selection, planning and management of protected areas. The basis is a survey of the useful species present in a region. Once species targets are decided, harvest techniques, limits and the mechanisms for sharing of benefits with the local people should be planned. As well, the maintenance of the resource may entail varying degrees of intervention. Species of subclimax communities may require management, such as grazing, burning, logging to open the canopy, thinning, controlling detrimental species or enrichment planting. Thus the aim of their conservation *in situ* is the sustainable supply of medicinal plants.

Indications that species are worthy of attention come from past and present usage, from taxonomic, genetic, biochemical, or other associations with species now used, or from new requirements by industry. An example of a systematic search among indigenous wild species was the Australian Phytochemical Survey between 1945 and 1970 (Price *et al.* 1993). Chemists in CSIRO and the universities conducted this collaborative study of alkaloids and anti-tumour constituents of indigenous species in Australia and Papua New Guinea. Some 2500 Australian species and 2250 from New Guinea were surveyed for compounds of pharmacological and other scientific interest. While the project has not placed a commercial product on the market, it enormously increased the knowledge of the secondary chemistry for much of the flora (Collins *et al.* 1990).

Several researchers have considered random versus other strategies of phytochemical surveying. These same strategic considerations are highly relevant to conservation priorities. Balick (1990) reported preliminary tests of plants from central America in which an *in vitro* anti-HIV screen detected as active 25% of species classified as 'powerful plants' by herbal healers, whereas only 6% of randomly collected species were active. In addition to ethnobotanical data, Gentry (1993) emphasized ecological clues and taxonomic considerations in sampling. He argued that the search should concentrate on taxa involved in 'coevolutionary arms races' with specific groups of insects, for example the families Passifloraceae, Aristolochiaceae, Apocynaceae, Asclepiadaceae and Solanaceae. Predictions are that climbing plants such as lianas are likely sources, because they rely for defence on specific, highly active, low molecular weight compounds, rather than tannins and lignins. The prominence of Bignonicaeae, Dioscoreaceae and Smilacaceae as sources of useful species supports a correlation between biodynamic properties and the scandent habit. Intricate interspecific interactions and scandent plants are prolific in tropical forests.

When a wild species is identified as being of commercial value, it may be exploited in its native habitat, or it may be taken into cultivation, which is likely to result in domestication. Either case will necessitate protection and conservation of the habitat: in the first case, to preserve the natural resource for continuing production; in the second, to conserve the genetic resources for maintenance and improvement of productivity. As for potentially useful species, their preservation forms a major reason for the conservation of ecosystems. This is particularly the case for tropical forests which are the most promising and diverse reservoirs of new useful plant species.

6.2.3 Forage plants

Many wild species are used in pasture and range lands for livestock production. Some of these have become domesticated, usually in countries other than their own: as, for example, Mediterranean, African and South American pasture grasses, and legumes in North America and Australia. Large *ex situ* collections of such species have been established. Over the last few decades, many species of subtropical and tropical forage legumes, most of which had not previously been cultivated, were collected and introduced to Australia (McIvor & Bray 1983). Some of these introductions, especially species of *Stylosanthes*, have succeeded and are now adopted by other tropical countries.

Collections continue to be made in the regions of origin to explore previously uncollected areas, or to obtain types adapted to new ecological conditions or for specific purposes. For example, despite several decades of introductions and breeding of *Trifolium subterraneum* in Australia, a first collecting trip to Sardinia in 1977 resulted in at least three new cultivars.

Just when and why introduction, rather than plant breeding, should be the main source of new cultivars in many forage species is an interesting question (Cameron 1983) and of direct relevance to the need for *in situ* conservation of pasture plants in their natural range. New forage cultivars must show persistence, a range of tolerances, and an ability to compete with previously released cultivars and volunteer strains in the field. Naturally derived genotypes are very likely to possess these features. In the case of *T. subterraneum*, Nichols and Francis (1993) suggest that the Sardinian source has climatic and edaphic similarities with the target pastoral areas, and has been subject to continuous, heavy grazing by sheep.

Since populations from the native range of forage species are an important genetic resource, there is need for concern about their conservation and continuing adaptation *in situ*. Many species are widely distributed, or locally common and well adapted to disturbed situations as befits a resilient forage plant. For example, *Stylosanthes scabra* and *S. humilis* occur in central America over 40 degrees of latitude (Williams *et al*. 1984). Harlan (1983) has noted that the clearing of tropical rain forest for grazing has opened up opportunities for forage species to colonize. However, overgrazing and misuse of the land can rapidly impoverish vegetation and lead to soil erosion. Vorano (in FAO 1989) describes a project of integrated resource management aimed at *in situ* conservation of forage plants in Argentina's Chaco region that addresses these concerns.

On the whole, the native sources of forage are not under acute threat except in specific areas where rough grazing land is turned over to agriculture, forestry or urban and industrial development. This is the case with rye-grass (*Lolium perenne*), with ancient stands in pastures and meadows in various parts of Europe (Ford-Lloyd & Jackson 1986). For years these populations have served as the reservoirs from which modern highly productive strains have been developed by plant breeders at the former Welsh Plant Breeding Station, Aberystwyth, and elsewhere. Now many of these old pastures either have been ploughed up or are threatened.

A new development is the reserving of landscape as national parks with multiple uses, such as the Sierra de la Cabrera in Portugal and the Serra da Estrela in Spain (Reid, personal communication). If traditional practices are encouraged in such parks, rather than irrigation, manuring, overgrazing or mowing regimes, species diversity is maintained (Garcia 1992). In addition, productive ecotypes of relatively neglected forage species like *Lotus corniculatus* and diploid medics can thrive. Similar areas with some relief from recent overgrazing are needed in countries like Turkey, Algeria and Morocco that are rich in species with a long history of supporting livestock.

6.2.4 Wild relatives of crop plants

The wild species related to crop plants differ widely in their distribution and abundance (Zeven & de Wet 1982). Some are widespread, like *Hordeum vulgare* ssp. *spontaneum*, the wild progenitor of barley, which occurs in disturbed sites from Morocco and Greece east to Pakistan, and in primary habitats in the Fertile Crescent (Harlan & Zohary 1966; Bothmer *et al.* 1991). The wild oat species *Avena fatua* has a global distribution and is one of the world's worst weeds (Holm *et al.* 1977). Other wild relatives like *Zea diploperennis* are known from only a few populations (Benz 1988). Some are stable components of mature communities, as for example are the common perennial species of *Glycine* found in open eucalypt woodland across Australia (Brown *et al.* 1984). Most Australian national parks and reserved areas thus contain populations of one or more *Glycine* species. In contrast, other wild relatives prefer disturbed agricultural habitats and are unlikely to be in reserves (e.g. weed races of *Solanum sparsipilum* (Hawkes 1990), *Helianthus annuus, Pennisetum americanum, Glycine soya,* etc.).

Conservation *in situ* rather than *ex situ* is preferred for wild relatives (Frankel 1970b; Ingram & Williams 1984). The main reasons are that *in situ* conservation can include a greater number and wider diversity of genotypes and species; that it allows continuing evolution of adaptation; and that *ex situ* conservation can be more difficult or expensive (Marshall 1989). Most importantly, *in situ* conservation is the method that preserves biological information on genetic diversity in context. Not only does it conserve the genetic diversity relevant to intraspecific and interspecific interactions among organisms and their associated pests and beneficial species, it is also present in populations that are or have been host to the relevant biotypes of the pathogen or symbiont. For example, natural populations of *Linum marginale* are polymorphic for resistance to rust and host to a diversity of pathotypes reacting differentially to these resistances (Jarosz & Burdon 1991). Certain natural populations of wild tomato species contain sources of resistance to drought or salt stress, excess moisture or predatory insects; resistances that are evident from their ecology (Rick 1973).

Hence population genetic divergence, or the extent to which the populations of a wild relative differ genetically from one another, is a crucial variable in planning for their *in situ* conservation. Such divergence may be present already between the different populations of the species *in space*, or it could arise in the future between generations of the one population *in time*. Unfortunately, in contrast to the many studies of genetic divergence between populations in space, there is little information on temporal change in natural populations and few studies of temporal variation in allele frequencies. Yet continuing evolution is a major rationale for *in situ* conservation of wild relatives.

What characters should be used to measure the genetic divergence of populations in space and in time? The choice includes neutral genetic markers (such as protein and DNA-sequence polymorphism), selected variation (such as disease, pest and stress resistance), ecotypic differences and morphological variation. Of these, characters that are under the control of major genes are preferable as they are much more amenable to transfer to crops than are characters due to polygenes. The expression of such complexly inherited characters will generally not remain intact when crossed into a plant breeder's elite parental germplasm.

In measuring divergence, not only the genes present, but also the range of environments of the various populations needs to be assessed. Aspects of the native environment such as climate, soil, and differences in pathogenicity of resident pathogens, provide a framework for hypotheses about variation in the wild relative. For the measurement of environments, the question arises of what scale deserves priority – microenvironmental variation or macro? Subtle variation within a local habitat such as minor differences in light intensity, while fascinating to the ecological geneticist, is unlikely to be of use to the crop breeder. Such genes are so sensitive in their adaptation (assuming they exist) that they will have too little adaptability to be worth transferring to related crops.

In contrast, macroenvironmental variation merits strong emphasis. The *in situ* conservation of wild relatives requires the preservation of representative samples of infraspecific variability in populations throughout their natural geographic range (Marshall 1989). Just as discussed for forest trees (section 6.2.1, p.153), ecogeographic surveys are needed to determine the portion of species gene pool already present in nature reserves and hence the optimal location of additional reserves.

The mating system plays a crucial role in the microevolution of natural populations and should be monitored. From allozyme surveys of plant species (section 2.1.2), it is clear that the mating system is a major determinant of patterns of population divergence. The fact that the close relatives of wheat, barley and oats are predominantly yet variably self-pollinated is basic to their conservation and their use. Such a mating system allows the maintenance of adapted genotypes, the rapid divergence of populations, and the occasional bursts of new recombinant genotypes (Allard 1975).

The research that has been done on the composite crosses of barley at

Davis, California and elsewhere (Allard 1988, 1990) offers a model for studying variation in wild relatives conserved *in situ*. Since samples of the various generations of these bulk hybrid populations have been kept, it has been possible to trace changes in several kinds of characters. The composites have shown how disease resistance changes, and how selfing leads to the rapid though incomplete loss of variability and the build-up of allelic correlations. Composites in which male sterility genes have been included have shown that elevated levels of outcrossing retain genetic variance (Jain & Suneson 1966).

Wild relatives are of major importance as genetic resources for the improvement of their related crops (section 2.3). Already proven as sources of disease resistance (Harlan 1976; Stalker 1980; Lenné & Wood 1991), wild relatives are likely to be the main sources of new resistances needed to cope with future evolution in the crop pathogens. This will especially be the case if host populations are maintained *in situ* to allow coevolution to occur. This proposal raises several crucial and essentially open questions about host–pathogen interactions considered in Box 6.2. These questions are relevant to the design of research projects for monitoring microevolution in populations conserved *in situ*.

In general, the coevolutionary dynamics of the total system involves four partners, namely crop and crop relative and the pathogen populations of each species. An example of such an interaction involves barley as the crop species, *Hordeum vulgare* ssp. *spontaneum* its wild ancestral species, and *Rhynchosporium secalis*, the causal agent of scald in both species, as the pathogen (Figure 6.1). Five genetic interactions are important to microevolutionary changes in the whole system:

1. Coevolutionary dynamics in the wild host and its resident pathogen population.
2. Evolutionary changes in the pathogen population infecting the crop species in response to changes in the barley cultivars planted.
3. The introgression of wild resistance genes into crop cultivars.
4. The invasion of pathogen biotypes from the wild alternative host onto the crop.
5. The response reaction shown by new wild resistance genes to the pathogen population of the crop.

Ideally, a project for *in situ* conservation of a wild crop relative whose aim is to generate useful resistance genes should address all these processes.

Conservation *in situ* of wild relatives frequently has major importance beyond the country native for the species. Prominent examples are the major industrial crops rubber and coffee, where world production relies heavily on plantations in countries other than the original home of the crop. In the case of natural rubber (*Hevea brasiliensis*), 'the economic future of this valuable industrial crop may well depend on the future survival of wild *Hevea* gene pool resources in Amazonian rainforests' (Oldfield 1989). The wild relatives *H. benthamiana* and *H. pauciflora* in Brazil are a crucial source of resistance to South American leaf blight, caused by the fungus *Microcyclus ulei*. The need for conservation *in situ* of the genetic resources of coffee (*Coffea arabica*) in Ethiopia arises from deforestation and dwindling of the original stands, the planting of introduced cultivars and the recalcitrance of seed to long-term storage (Ameha 1991).

The following two examples illustrate the kinds of projects for *in situ* conservation of wild relatives of crops. One aims to conserve a rare, wild species related to maize in Mexico, and the second intends to provide for continuing evolution in a wild wheat in Israel.

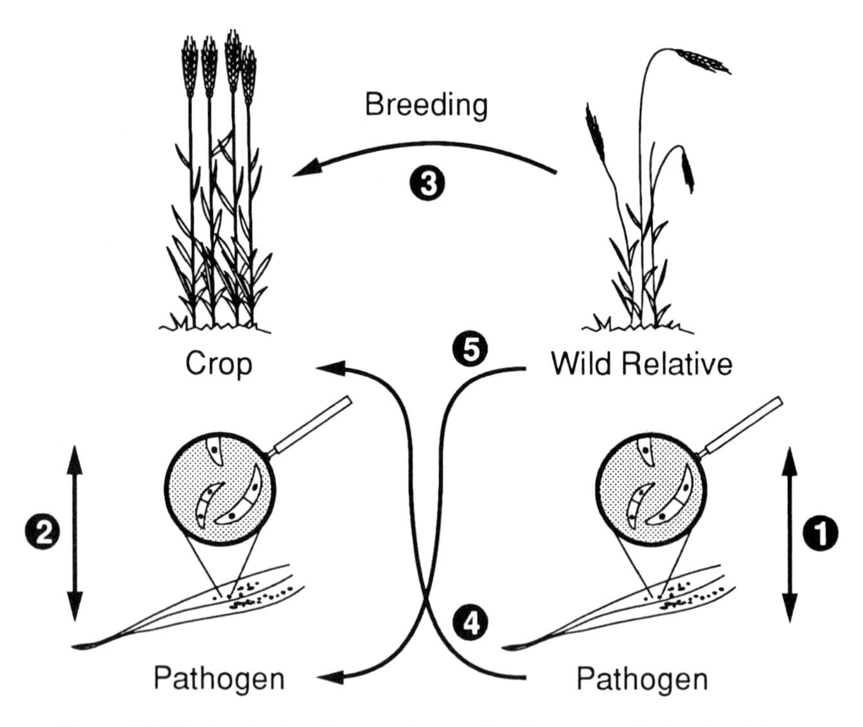

Figure 6.1. The interactions between host and pathogen populations involving a crop species, its wild relative and a pathogen common to both hosts (see text).

BOX 6.2. RESEARCH QUESTIONS FOR *IN SITU* POPULATIONS OF WILD RELATIVES AS A SOURCE OF USEFUL RESISTANCE GENES

The following are open questions concerned with evolution *in situ*, and relate to the linkages between the four players in the system depicted in Figure 6.1.

1. What evidence exists that the resistance structure of host wild populations evolves in time to meet new threats, and at what pace? Little is known about the origin of new resistances and temporal changes in host populations of wild relatives (but see Burdon 1994; Burdon & Thompson 1995). In crop populations, different outcomes are on record: for example, the successive generations of barley composite crosses II (CC II) evolved toward higher resistance, whereas those of CC V did not (Allard 1990), despite being grown in the same locality.

2. When might evolutionary changes in a population of a wild relative pose a threat to nearby fields of the related crop? Such a threat may arise if the wild alternative host allows the evolution of a diversity of pathogen genotypes. Two examples taken together show that this may or may not happen.

The first case is the tetraploid wild barley *Hordeum leporinum* which is a widespread agrestal weed and, in Australia, is a significant alternative host and source of inoculum of the scald pathogen (*Rhynchosporium secalis*). Isolates of this pathogen collected from wild barley differ widely in pathogenicity (J. S. Brown 1990). Wild barley populations carry many different resistances (A. M. Jarosz and J. J. Burdon, unpublished data) and act as a host that generates pathogen diversity. This case serves as a warning of the potential dangers of wild relatives growing in areas where the domesticated species is planted.

In the second case, the pattern of host–pathogen genetic relations has been monitored over time in a series of natural populations, situated in the Kosciusko National Park, NSW, Australia. The host species is the native wild relative of flax, *Linum marginale* and the pathogen is *Melampsora lini*, the causal agent of flax rust. One population of wild flax with a greater diversity of resistance phenotypes is host predominantly to a particularly virulent race of flax rust (Jarosz & Burdon 1991). However, the more susceptible and less diverse host populations nearby are infected with less virulent rust races. In this case apparently, the more virulent race from more resistant population has failed to displace them.

3. Will the expected, newly evolved resistances to the pathogen populations resident on native wild hosts, confer resistance to pathogen populations found in a cropping zone remote from where the wild relative grows? For scald, genes that confer resistance to Australian isolates of the

BOX 6.2. (*cont.*)

pathogen are common in populations of wild barley, *Hordeum vulgare* ssp. *spontaneum*, particularly from the more mesic parts of its range (Figure 6.2, Abbott *et al.* 1992). In many instances, wild resistance genes have proved useful for crop breeding in countries other than that where the resistance was collected. However, this history cannot be taken to prove the point that alien resistances will always be useful. It may be the outcome of biased or pragmatic screening of germplasm collections, rather than a systematic study of resistance to both local and exotic isolates of the pathogen. We cannot therefore assume that new resistances evolved *in situ* will necessarily confer resistance to both local and exotic pathotypes.

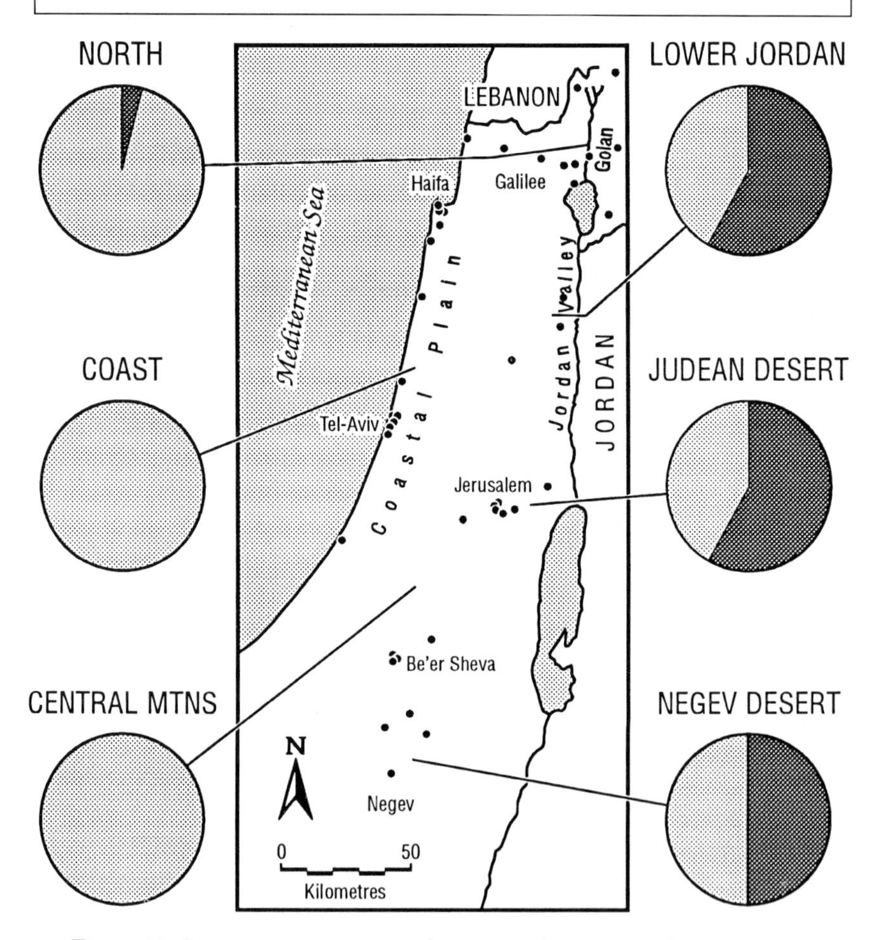

Figure 6.2. Geographic distribution of resistance (light shading) and susceptibility (dark shading) of Hordeum spontaneum *to* Rhynchosporium secalis *in Israel.*

Zea diploperennis in the Sierra de Manantlan Biosphere Reserve, Mexico

Following the discovery of the endemic perennial *Zea diploperennis* (Iltis *et al.* 1979), the idea of conserving this wild relative of maize in its natural habitat led to the declaration of the Sierra de Manantlan Biosphere Reserve, in Jalisco, western Mexico in 1987. Populations of the wild annual relative, *Z. mays* ssp. *parviglumis*, and the Tabloncillo and Reventador races of maize traditional for this area, are further targets for conservation.

Benz (1988) has pointed out that the situation of *Z. diploperennis* is common for wild relatives of many crops, namely, they tend not to occupy pristine environments like nature reserves, but have evolved jointly with the cultigen and occupy habitats disturbed or maintained by humans. To conserve such species *in situ* will require conserving the whole agro-ecosystem, including the spontaneous relatives as well as the traditional crops. Limits on external inputs, such as exotic improved germplasm and chemicals will need to be set, so as not to endanger the wild relative. Yet the author argues that ways can be found to provide opportunities for the cultivators involved in managing the system. Benz is optimistic that *Z. diploperennis* and the three other taxa can be conserved *in situ*, without excluding human activity from the reserve. Indeed, research has shown that populations of *Z. diploperennis* virtually require cultivation and grazing in adjacent fields to prosper.

Triticum turgidum var. *dicoccoides* at Ammiad, Israel

The wild tetraploid wheat, *Triticum turgidum* var. *dicoccoides*, is the progenitor of most cultivated tetraploid wheats. Its present distribution and abundance are limited, and genetic resources in primary habitats warrant conservation. In 1984 a site east of Galilee, in Israel, was selected for intensive genetic, ecological and pathological studies (Anikster & Noy-Meir 1991). Permanent transects were set in place, and a detailed assay of microgeographic patterns of variation for marker genes, demographic and morphological traits, and environmental variables was carried out. The genetic variables are to be monitored through time.

Horovitz and Feldman (1991), in drawing together the current findings of this cooperative project, stressed the importance of topographic diversity in the selection of sites for conservation. So far the data confirm that such selection maximizes the benefits of the *in situ* approach by conserving many genotypes. Preliminary guidelines for establishing *in situ* reserves (Hawkes 1991), based on the Ammiad experience, are similar to those for GRMUs in forestry (Box 6.1), except that they specify that sites for wild relatives should be large enough to contain at least 1000 individuals of each species.

6.2.5 Endangered species

Endangered species call for individual attention not because of a recognized use, but because they are deemed to be under threat of imminent extinction. As will be discussed in the next chapter, endangered species are the focus of *ex situ* conservation in botanic gardens; the present section deals with the field aspects of their conservation *in situ*. Species are in this predicament for a variety of reasons (Table 5.1), but most often due to human activities. It is usually a difficult and expensive task to reverse the situation. We would thus expect that the *in situ* conservation of endangered species will require a higher level of human intervention than the conservation of forest trees, forage plants, or wild relatives of crops.

Broadly we can recognize two phases in safeguarding endangered species *in situ*. First is the recovery phase, and second the maintenance phase. The former is the more intensive effort, requiring a definite plan. A *Recovery Plan* is a detailed strategy to conserve an endangered species (Cropper 1993). Its essentials are a review of botanical and ecological knowledge of the species, the current and likely threats to its populations and the proposed research and management for existing and any new populations of that species. Its ultimate objective is to establish self-sustaining populations in nature.

World-wide, the number of endangered plant species is daunting, particularly in the tropics (see section 7.1). Lists from temperate areas in developed countries number in the hundreds. Such numbers raise the question of criteria to determine priority ranking of species to be conserved. Holsinger and Gottlieb (1991) reviewed the process of listing species as endangered, threatened or rare and questioned the nomination of individual cases based solely on assessment of scarcity of populations and their vulnerability to loss. 'Given that 700 plant taxa are said to be at risk of extinction in the United States in the next decade, we believe it is essential to develop biologically based criteria to identify the ones whose loss would be felt most keenly, so that their survival can be guaranteed.' It seems inevitable to accept the notion that the case for specific action in some plant species will be more compelling than for others, and that criteria are required to identify them. Box 6.3 contains a list of criteria for the comparative assessment of recovery plans, adapted from those of the Endangered Species Unit of the Australian National Conservation Agency and from Holsinger and Gottlieb (1991).

BOX 6.3. CRITERIA FOR COMPARATIVE ASSESSMENT OF RECOVERY PLANS

The main criteria for assessing priorities for species recovery are:

1. The degree of threat to survival in nature.
This includes the smallness and scarcity of the populations of the species. However, the ranking of species primarily of the grounds of likelihood of extinction is unsatisfactory, because much effort could be wasted on a hopeless cause.

2. The ecological importance of the species to the ecosystem.
This covers the cases of a dominant species, a keystone species or a unique resource for other species. Such cases are likely to be unusual among endangered species, but would apply to the dominant tree of an endangered ecosystem, for example *Eucalyptus albens* (Prober & Brown 1994).

3. The genetic (taxonomic) distinctiveness of the species.
This is one of the more significant factors to consider. Monotypic genera or families deserve the highest priority, and endangered species are of more direct concern than endangered subspecific taxa (Holtsinger & Gottlieb 1991). It is a useful criterion and relatively simple, despite the debate over whether the pattern of genetic divergence should take precedence over the degree of divergence (see Box 7.3).

4. Any value (biological, economic, social, cultural, scientific) of the species to humanity.
This includes endangered species belonging to categories already considered in this chapter: namely, forest trees, medicinal plants, forages, or wild crop relatives.

5. The potential for the species to recover.
This raises the question of likely population dynamics once remedial action is taken.

The recovery phase

In dealing with endangered plants, the central issue is obviously *numbers*. For example, the average endangered species in New South Wales, Australia, consists of about three populations with a modal size of about 20 individuals per population (from data in Brown & Briggs 1991), although these statistics vary among species over a wide range. (Holsinger and Gottlieb (1991) portray the typical rare endangered species as comprising fewer than five populations and with fewer than 5000 individuals.) To increase numbers above such precarious levels will require steps to restore the habitat and counter the perceived threats. It may prove sufficient to achieve this by land reservation. Usually, however, forms of habitat maintenance like fencing, and protection from fire, invasive plants or diseases, will be needed. The effectiveness of such steps will depend on an analysis of population dynamics and detection of the major risk factors (Pavlik *et al.* 1993). The recovery phase amounts to an experimental test of hypotheses about population dynamics.

A further possible step in the recovery is either to replenish existing populations, or to introduce material into sites where the species is known to have occurred, or to translocate individuals to a new suitable habitat within its historic range. Here botanic gardens have a major role to play in developing the material for planting and the horticultural expertise to handle it (see section 7.3.4). A major issue is the nature of planting material. Most practitioners advocate supplying propagating material that originates only from that site for reasons of microhabitat adaptation. However, if only a few individuals are now left, genetic erosion will have been severe. The recovery effort is more likely to succeed if it uses a genetically heterogeneous sample such as a composite cross of several sites (Barrett & Kohn 1991; Pavlik *et al.* 1993). This could provide vigour from heterosis, and segregating variability for local selection and the avoidance of inbreeding depression in subsequent generations. DeMauro (1993) found that mixing populations of the self-incompatible lakeside daisy *Hymenoxys acaulis* enhanced outcrossing potential because it corrected local deficiencies of cross-compatible genotypes.

Some authors (e.g. Rieseberg 1991) have noted that extremely wide crosses might be subject to outbreeding depression and should be avoided. On the whole, however, such depression has been much more difficult to demonstrate in plants than has inbreeding depression (section 5.4.4). Clearly, very wide crosses such as between divergent ecotypes should not be used.

The continuing maintenance phase

The periodic monitoring of conserved populations is a major commitment of effort, yet it is necessary for success. Patterns of population recruitment and mortality require attention, as well as threats to the quality of the habitat (diseases, herbivores, grazers and weeds). Cropper (1993) has produced a useful detailed monograph on the management of endangered populations, with ten cases of endangered Australian plants as examples.

Falk (1992) among others has suggested that monitoring of genetic variation provides helpful data on the state of the population, particularly its breeding system. However, as Holsinger and Gottlieb (1991) note, 'the active management of the genetic structure of populations of an endangered plant species will require an enormous investment of time, money and expertise. Only a few of the most important species will warrant such heroic efforts.'

Much can be learned from case studies, as Cropper (1993) shows. Through clearing for agriculture, the tree species *Eucalyptus rhondantha* is now confined to six sites in Western Australia (Sampson *et al.* 1990). Detailed monitoring and genetic studies have provided prescriptions for its conservation, including land acquisition for reserves, replenishment of populations to a minimum of 25 individuals, control of fire, back-up seed storage, and the avoidance of inbreeding. However, the long-term aim must be towards a lesser dependence of any species on active human intervention, or else it will fall victim to the first change of heart or inevitable budget cut.

Pavlik *et al.* (1993) have succeeded in founding a new population of the endangered heterostylous herb *Amsinckia grandiflora* within its historic range. They were fortunate in having germplasm that stemmed from collections made in the 1960s by Dr R. Ornduff. The other ingredients for success were a careful choice of microsites, the use of isozyme screening to compose the founding population with maximum allelic diversity, the control of competing non-native grasses using fire or a specific herbicide, and careful demographic monitoring to identify critical life-history attributes and prospects for sustained, long-term recovery. The authors stress the need for a well-designed experimental approach.

6.3 CONCLUSION – MANAGEMENT AND CONSERVATION *IN SITU*

In situ conservation of endangered species encounters three major dilemmas or areas of controversy. These are the durability and long-term significance of the effort, the problem of choosing priorities for conservation, and the desirability of intervening in natural systems. This last problem brings us to the issue of human intervention in the management of natural systems for conservation *in situ*.

A general question arises when conserving *in situ* the various kinds of species discussed in this chapter: what place is there for genetic change? Should we encourage it, allow it to take its course, or adopt measures to slow it down?

This question is fundamental to our approach to conservation management (Frankel 1983). The issue is whether and when to intervene in natural processes. The options can be represented in a three dimensional model of interacting gradients (Figure 6.3), where the axes or variables are the fundamental forces determining the rate of microevolutionary change. The managed population occupies some point in space, determined by the current level of the three variables.

The first dimension is the *population structure* axis. This axis has received much attention in the conservation literature as it deals with the design of reserved areas (see Chapter 9). It includes the question of one large as against several small subpopulations. Perhaps its most important parameter is the migration rate (m) among subpopulations, because migration affects the rate of evolution and population divergence. We have already noted that foresters are concerned about unwanted gene flow into a conserved site and take steps to prevent it. On the other hand, a population of an endangered species may benefit from new genetic variation from outside because locally it has passed through a bottleneck. The other major parameter of population structure is local population size (N_e). Again this may need deliberate altering from the *status quo*. For example, replenishment of an endangered population may be called for.

The second axis is the *recombination* axis. It has received far less attention

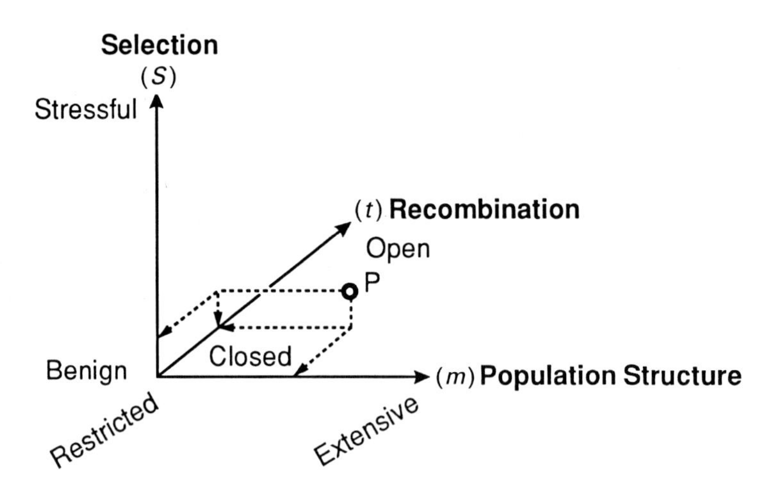

Figure 6.3. Model of interacting management options for in situ *conservation that affect the rate of genetic change. The three hypothetical axes are 'population structure' with migration ranging from restricted to extensive, 'recombination', ranging from closed to open and 'selection intensity' ranging from benign to stressful. The current position of the population is P.*

and is perhaps more controversial. Microevolutionary rates will be much influenced by the rate of recombination, which can range from closed to open. The linkage distance between genes is not amenable to alteration but the recombination rate can be varied through changes in the mating system. Closed mating systems, like complete autogamy with no outcrossing ($t = 0$), are very conservative in terms of the origin of new genotypes. Therefore one option for a manager is to alter the present recombination rate, for example, by using a male gametocide on a fraction of the population. Do we encourage, or rogue out 'hybrid swarms'? In the case of wild relatives, a good case may exist for enhancing recombination when the spontaneous level in a conserved population is exceedingly low.

The third axis is the *selection intensity* (s) axis. This axis concerns many options such as: should we encourage disease levels by spreading inoculum, or discourage disease with fungicide; should we reduce or ignore weeds? Even the controlled reduction of grazing may initiate evolutionary changes as plants spread from refuge microsites to colonize nutrient rich, open niches. We have already noted that modulating the frequency of fires was important: in addition to affecting the population structure of the target species, the occurrence of fires affects other species which could have indirect selective effects (section 10.4.2, p.246).

In Figure 6.3 the model of management options as a three dimensional space emphasizes that the outcome in terms of genetic change in a population will depend on interaction between the three axes. The current position of the population is the point P. In general, genetic change is the greater, the further P is from the origin. While the manager of a wild population may choose not to manipulate any of these axes, such a choice should be reached on scientific rather than dogmatic grounds. Each of the three axes needs to be assessed and the tempo and mode of genetic changes optimized. Overall, three criteria emerging from this chapter should be met. These are (i) population survival, (ii) maintenance of evolutionary potential in the form of genetic diversity, and (iii) in the case of wild relatives, development of new genotypes.

7

Ex situ conservation of threatened and endangered plants

7.1 RARE AND ENDANGERED PLANT SPECIES

Threatened species present scientific, economic and moral challenges: scientific, because their extinction would remove evolutionary links that contribute to an understanding of plant life; economic, because endangered species – or some of their genes – might prove useful in the future; and moral, because mankind is to blame for having caused, or contributed to their endangered state. These or similar views are held by many biologists, environmentalists and conservationists and by a growing number of writers and concerned members of the public. It is understandable that botanic gardens, the collectors, custodians and systematists of the plant kingdom, are seen and increasingly see themselves as the guardians of endangered plant species. 'The fact remains that plant species are becoming extinct in the wild, and some should be conserved in botanical gardens' (Ashton 1988). This remark, by the Director of the Arnold Arboretum of Harvard University, signifies in the simplest terms the widely shared concern of botanic gardens and arboreta for the conservation of endangered plant species.

A concern for the preservation of threatened plants is more recent than that for the survival of rare and endangered animal species. This is due partly to lesser awareness, partly to the fact that many animal species have been subject to selective reduction by humans. On the other hand plants, and of course many animals, are mainly endangered by deprivation of habitats through the destruction of natural ecosystems, rather than by human impact on particular species. A concern for threatened species was strongly expressed by the United Nations Conference on the Human Environment held at Stockholm in 1972. In 1974, the International Union for the Conservation of Nature and Natural Resources (IUCN) established the

Threatened Plants Committee (TPC) with the objective to enlist world-wide participation of botanists in gathering information on threatened species (Lucas 1976) and their location, and on means for their preservation. The TPC was succeeded by the IUCN Botanic Gardens Conservation Secretariat (see below). A great deal of information on the conservation of threatened plants in living collections and on the more general role of botanic gardens in nature conservation was brought together at a conference held at the Royal Botanic Gardens, Kew in 1975 (Simmons *et al.* 1976). It brought the need for the preservation of endangered species to the attention of botanic gardens world-wide.

How are threatened species defined? A definition is important in formulating conservation policies and strategies. It can also be basic for managerial or legal restraints on the use of resources and thus may assume political significance. The TPC proposed a modification, for use by botanists, of the categories introduced by IUCN for both animals and plants. The proposed terms were 'endangered', 'rare', and 'vulnerable'. Endangered species are at serious risk of disappearing from the wild state within a few decades if present land use and other causal factors continue to operate. A rare species is one 'that either occurs in widely separated, small subpopulations or is restricted to a single population' (Drury 1974). A more recent definition, in use in California, recognizes endangered, threatened and rare species, in descending order of presumed imminence of loss (Holsinger & Gottlieb 1991).

Though rarity is a likely risk factor in the survival of species, it is not the prime cause of the massive rate at which species currently are becoming endangered and extinct. The principal factor is the extensive damage to, or destruction of, the habitats of vast numbers of species, especially in the species-rich tropics. A few statistics quoted from Raven (1987) indicate the magnitude of current species extinction. Of the 250,000 vascular plant species, 170,000 are tropical or subtropical endemics. The rain forests, the richest tropical communities, are subject to destruction by exploitation and land clearing on an unprecedented scale. Raven (1987) suggests that the closed forests of western and northwestern Amazonia, the interior of Guyana, and of Zaïre, containing an estimated 40,000 species, may have the best chance of survival for the time being. Of the balance of 130,000 tropical species, some 60,000 will be at risk of extinction within the next half century. Of the 80,000 species of the temperate zone, about 8000 are threatened and several hundred endangered. Raven concludes, 'we are living at a time of utmost threat to the world's biota. In order to ameliorate this situation and secure the aesthetic, scientific and economic benefits of

BOX 7.1. THE NATURE OF RARITY

A fuller understanding comes from an analysis of a common feature of endangered species – rarity. Rabinovitz (1981) and Rabinovitz *et al.* (1986) applied three dimensions in such an analysis.

1. Geographic range – whether large or small.
2. Habitat specificity – whether broad or restricted.
3. Local population size – whether large somewhere, or always small.

The three characteristics are combined in a 2 × 2 × 2 table (Table 7.1). Only the top left-hand combination is really common, and all but the top two left ones are liable to be rare.

Table 7.1. *A typology of rare species based on three characteristics; geographic range, habitat specificity, and local population size*

Geographic range	Large		Small	
Habit specificity	Wide	Narrow	Wide	Narrow
Local population size				
Large, dominant somewhere	Locally abundant over a large range in several habitats 58	Locally abundant over a large range in a specific habitat 71	Locally abundant in several habitats but restricted geographically 6	Locally abundant in a specific habitat but restricted geographically 14
Small, non-dominant	Constantly sparse over a large range and in several habitats 2	Constantly sparse in a specific habitat but over a large range 6	Constantly sparse and geographically restricted to several habitats 0	Constantly sparse and geographically restricted in a specific habitat 3

Source: From Rabinovitz (1981); numbers from Rabinovitz *et al.* (1986).

The validity of the classification system was examined using the British flora as a database (Rabinovitz *et al.* 1986). Confinement to a specific area such as Britain has the advantage of data reliability, although it must be recognized that species may be rare in a restricted area though common elsewhere. Detailed ecological data were available for 177 species (the 'Biological Flora'). These species were assessed for their position in Table 7.1 by 15 independent judges. Of the 160 species for which there was a reasonable measure of consensus, 129 were common, with a broad geographic distribution and large population size, and 31 were rare; numbers are shown in Table 7.1. Lower numbers are associated with restricted geographic range and with small population size.

The authors note that habitat specificity does not enhance rarity; on the contrary, among the species with both large and small geographical range, those with a specific habitat outnumber the wide-ranging ones. This fact points to the importance of studying and conserving the habitats of rare

BOX 7.1. (*cont.*)

species. However, since the authors found the three ecological characteristics to be independent of one another – at least in the sample of the British flora used in this study – they conclude that conservationists need to pay attention to all three. Indeed, the analysis of the ecological elements of rarity should prove helpful in indicating appropriate sites for sampling of material and for *in situ* conservation.

these species for our children and grandchildren, we must find out as much as possible about the plants, animals and micro-organisms and move quickly and effectively to save as many of them as possible.' He stresses that special attention is to be given to species of scientific or economic importance, actual or potential.

Knowledge of organisms, their distribution, ecology and breeding system, is essential for conservation at the level of species and their genetic diversity. It is at the specific level that botanic gardens have a traditional role in knowledge gathering and, more recently, in conservation. In line with their important functions in education and information, botanic gardens are well placed to play a significant part in gathering and publicizing information on endangered and threatened plants.

7.2 THE ROLE OF BOTANIC GARDENS IN CONSERVATION

The most ancient and widely known function of botanic gardens is to assemble and maintain a diversity of plant species, in the open, in glasshouses and, for reference and study, in herbaria. They conduct or facilitate botanical research, especially in plant taxonomy, for which they are the traditional centres. They have a long history of concern for species of economic significance, dating back to the medical gardens of the fourteenth century. From the seventeenth century onwards, botanic gardens played an enormously useful role in the introduction and exchange of food and fibre plants between Europe, the Americas, Asia and Africa, an activity that only in the last hundred years was taken up by government agencies, universities and individuals. It is relevant to recall these important contributions to economic botany in order to see recent developments in botanic garden strategy (see below) in their historical perspective.

The 1975 Conference on Threatened Plants (Simmons *et al.* 1976), as we have seen, recognized the conservation of endangered species as a responsibility of the world's botanic gardens. The First International Botanic Gardens Conservation Congress, held at Las Palmas, Canary Islands, in 1985, took the planning for international cooperation a good deal further. It examined the role of botanic gardens in the context of a world conservation strategy and recommended the establishment by IUCN of the Botanic Gardens Conservation Secretariat (BGCS) 'to co-ordinate and monitor the plant conservation activities of the botanic gardens of the world.' (Jackson 1989).

7.2.1 International Botanic Gardens Conservation Strategy

The Second International Botanic Gardens Conservation Congress held at Ile de la Réunion in 1989, introduced a drastic change of direction. The major emphasis was shifted to germplasm conservation, with plants of economic significance replacing threatened and endangered species as the principal targets for botanic garden conservation efforts. The Botanic Gardens Conservation Secretariat was charged with the task to 'monitor and promote research into the conservation of useful plants, including germplasm of medicinal species, wild crop relatives, primitive land races of crops, and the so-called minor crops which are not covered by other germplasm networks or organisations.' (Jackson 1989). Propagation and re-introduction of rare and endangered species are to be 'actively encouraged', though apparently with lower urgency.

The new Botanic Gardens Conservation Strategy (WWF/IUCN/BGCS 1989) was elaborated by the IUCN Botanic Gardens Conservation Secretariat (BGCS). This wide-ranging document presents a great diversity of opportunities for botanic gardens to participate in a world-wide conservation effort. It recommends participation in *in situ* conservation, with mini-reserves of 5 to 50 hectares dedicated to the conservation of individual species with their genetic diversity. It stresses the many uses of *ex situ* conserved material in the economy, and as a source of rehabilitation in nature. Genetic diversity is emphasized as an essential condition for use and conservation. As indicated previously, strong emphasis is given to wild relatives and to primitive landraces of crops, and to 'minor' species that are cultivated though as yet not domesticated; these are to be prime targets for the botanic gardens conservation programme. At the same time a world-wide effort by the 250 larger botanic gardens should endeavour to preserve the 35,000 estimated endangered species, in addition to the 15,000 species of economic significance. This would result in an average responsibility for 200 species, each with a representative range of its genetic diversity.

Seeing that the majority of endangered species are tropical, whereas the botanic gardens are predominantly in temperate climates, much of the burden of preserving endangered species must fall on greenhouse and seed storage facilities. Logistic considerations point to the inevitability of selection for preservation, but the strategy does not suggest guiding principles for the determination of priorities.

It remains to be seen which aspects of the strategy are being adopted by botanic gardens individually and collectively, and what impact it will make on the survival of endangered species. Indeed, Hurka (1994) argues that botanic gardens are likely to make their most telling contribution to conservation through education and their influence on public opinion. He stresses the display of biodiversity, if possible in such a way that explains the functioning of ecosystems. Second, botanic gardens should give prominence to local flora.

7.2.2 A national centre for plant conservation

In distinction from the world-wide strategy promoted by BGCS (see the preceding section), a national centre in the United States places the main emphasis on the preservation of the endangered native flora. Both have in common a strong emphasis on the genetic diversity of species in need of protection or preservation. The Center for Plant Conservation, established in 1984, is now located in the Missouri Botanical Garden, St Louis, Missouri. Its aim is to establish a comprehensive national programme of plant conservation within existing institutions. Its resource is the National Collection of Endangered Plants of the United States, which is a living collection maintained at 21 botanic gardens, supplemented by a seed bank maintained within the National Germplasm System of the US Department of Agriculture. The Center aims to develop an information system and databases on the conservation status and biology of endangered species and the conservation technology applied in the United States.

The approach to conservation advocated by the Center is based on the 'integrated conservation model' (Falk 1990). It consists of three basic components. (i) The definition of the target level of biological organization – alleles, individuals, populations, taxa, ecosystems, all or any of which are of ecological and evolutionary significance. (ii) The identification of specific threats at a specific target level. At the species level, for example, there are 208 federally recognized threatened and endangered species, out of a total of some 5000 rare or threatened taxa in the United States. (iii) Consideration of the full range of conservation methods and resources. The emphasis is on integration of *in situ* and *ex situ* methods and on direct intervention. Biological management of nature reserves is to replace the traditional 'laissez-faire' principle. Botanic gardens, arboreta and seed banks are to play increasingly important roles at different target levels: seed banks, for example, at allelic levels.

Falk (1990) stresses the importance of comprehensive biological information as an essential requirement for effective conservation strategies, involving *in situ* and *ex situ* preservation as well as reintroduction efforts. In particular, population genetic information such as the partitioning of genetic variation or the effect of inbreeding in small populations, is strongly emphasized as a precondition for successful conservation. Basic to the conservation of endangered species is the availability of a comprehensive range of the genetic diversity of a species. This is regarded as essential at all target levels of conservation, from the ecotype to the ecosystem.

7.3 ISSUES IN WILD SPECIES CONSERVATION *EX SITU*

In the foregoing outline of conservation strategies a number of issues have been touched upon which call for a fuller discussion. Foremost among them are the purposes and priorities of the conservation effort. Next, the problem of selecting specific conservation targets among rare and endangered species. Third, the scope and purpose of genetic representation of target species; fourth, the reintroduction of endangered species into natural environments; and fifth, the role of databases in botanic gardens conservation.

7.3.1 Purposes and priorities for botanic garden conservation

As we have seen in sections 7.2.1 and 7.2.2 there are considerable differences among protagonists of botanical garden conservation. The resolutions from the latest International Congress give 'particular emphasis' to the *conservation of useful plants*, while the propagation and reintroduction of rare and endangered species are merely 'actively encouraged' (Jackson 1989). On the other hand, the national plant conservation effort in the United States – which has been adapted for use in other countries – places its full emphasis on the *conservation of rare and endangered species* in the national flora. As we have seen in section 7.1, both strategies draw on past attitudes and activities of botanic gardens for their historical background. They need, however, to be considered in the light of current conservation needs, of alternatives to botanic gardens, and of the available resources and other responsibilities of botanic gardens.

A 'north–south' distinction is apparent between the two strategies, with geographical, ecological and economic connotations (see, for example, Kloppenburg & Kleinman 1987 for the north–south distinction in indigenous genetic resources). The Botanic Garden Conservation Strategy (BGCS) has a global perspective, with emphasis on the tropics where a large proportion of species faces extinction; its economic targets are also predominantly tropical. On the other hand, the Integrated Conservation Model (ICM) reflects the more limited scale of extinctions in the northern hemisphere, where a comprehensive conservation strategy is more realistic than it would be on a global scale.

The concern for endangered species derives from the main task and purpose of botanic gardens, to assemble and cultivate a diversity of plant species for study and use; because they are of interest to many people; and from a long history of leadership in the field of plant taxonomy. By comparison, the preoccupation with useful plants is recent, stimulated by public concern for biodiversity, and more particularly for the genetic resources of species useful to humans. Box 7.2 deals with these recent developments.

We conclude that with the significant exception of medicinal plants a botanic garden conservation strategy concentrating on useful plants is open to question. For species which are actual or potential genetic resources for plant breeders, international and national germplasm centres provide increasingly comprehensive protection. Universities, experiment stations and individuals maintain collections of many more. Species in peasant cultivation are maintained *in situ*, and some have been collected and protected in regional institutions, including botanic gardens. A comprehensive strategy for the preservation of such species, other than on grounds of being endangered, is scarcely justified.

7.3.2 Selection of endangered species for *ex situ* conservation

Rare and endangered species, irrespective of their usefulness, have a prominent place in any botanic garden that takes an active interest in species conservation. While it is generally agreed that *ex situ* preservation of endangered species is secondary to their conservation within the natural communities of which they form a part (see section 6.2.5), botanic gardens can play an important role in complementing *in situ* conservation (see section

BOX 7.2. SHOULD USEFUL SPECIES HAVE PROMINENCE IN BOTANIC GARDEN STRATEGY?

As outlined in section 7.2.1, the conservation priorities proposed by BGCS include a wide range of cultivated and wild species, from field and garden crops to forest species, from landraces to wild crop relatives. Prominence is given to three loosely defined categories. The first are the wild relatives of crops discussed in Chapter 6. The second are 'minor crops' not represented in germplasm collections, including the crops of peasant cultivation – the 'life support' and 'under-utilized' species discussed in section 3.2.7. The third category are wild species which are utilized, or may be so in the future. The question is which of these categories are appropriate targets for botanic garden conservation. A further category, named by BGCS the 'primitive landraces', is clearly inappropriate, since landraces are the primary objectives of germplasm conservation by international and national germplasm centres.

In a review of the holdings of *wild crop relatives* in germplasm collections, Heywood (1992) claims that, with few exceptions, only a small fraction of the wild gene pools are represented. The question must be asked whether botanic gardens should be called upon to fill the gap, or whether alternative institutions are more appropriate. The answer depends on the purpose, whether conservation or use, and on existing expertise and facilities. Since use in research or breeding is the most common objective, proximity to an active research programme, ease of access, contact with users, and long-term security should be foremost considerations. In all these respects existing germplasm collections rather than botanic gardens are the more appropriate guardians for crop-related wild genetic resources. The growing interest among plant breeders in the use of 'alien' materials, enhanced by the increasing application of genetic engineering, has stimulated extensive collecting activities by IBPGR and the international and national centres.

The second and third category species named above are not generally subject to national or international conservation programmes. As we saw in section 3.2.7, the *crops of peasant cultivation*, including the so-called 'life-support' species, survive and produce in environments which tend to be physically, economically and socially disadvantaged. Some species are widely distributed, but many are adapted to specific conditions of water or nutrient supply, temperature, altitude, etc. Like the 'under-utilized' species – as we have argued in section 3.2.7 – the 'life-support' species have scant prospects of a substantial expansion. Selection by cultivators is likely to continue, with professional breeding remaining exceptional rather than common. Species and their diversity are preserved *in situ*, i.e. in cultivation (see section 4.2). The case for *ex situ* conservation by botanic gardens of substantial numbers of such species does not appear convin-

BOX 7.2. (*cont.*)

cing. However, species coming into prominence in cultivation or in research are likely to stimulate collecting and preservation by concerned institutions or individuals.

The third category of species proposed for botanic gardens conservation are *wild species actually or potentially useful to humans*. Potentially, their number is enormous; Heywood (1992) suggests at least 25,000–50,000 for the tropical regions. Their uses extend in many directions – from emergency foods in isolated locations or disaster areas, to horticulture, forestry and as industrial raw materials. Ecological considerations may also play a part in species selection – for example, species on saline soils, or steep hillsides – depending on particular interests. It is in such diverse areas that the current utilization of wild species is extensive and diverse (cf. Heywood 1992), and future prospects are without limit and unforeseeable. Exploration leading towards exploitation requires on one hand a definition of objectives, on the other a wide systematic knowledge of plants, their locations, lifeforms, ecology and breeding systems. As sources of information, as well as of plant materials, botanic gardens have made, and can continue to make valuable contributions. Since it is clearly impossible to encompass even a sizeable proportion of all species that might come into review, target selection is inevitable (see section 7.3.2 below).

Wild species used as *medicinal plants*, in their natural state or in manufacture, are of particular concern because of their importance in disease control on one hand, their precarious condition on the other. Many of these species are over-exploited in their habitats and are subject to local or widespread extinction. Genetic diversity, essential for selection and breeding, is being eroded. There is a need for preserving representative samples of those species that are most endangered. Since medicinal plants have no institutional guardians, this role may well be assumed by botanic gardens, to preserve and to supply materials for research, education and utilization, and to provide exhibits of direct human interest for the general public.

7.2.2). Generally, they can provide a sanctuary for species that are endangered in their habitats or are threatened with habitat loss. Botanic gardens can assemble and maintain collections representing the genetic diversity of species, especially since seed banks have become recognized components of many botanic gardens. Such collections are available for research on the ecology, physiology or genetics of threatened species and for their re-establishment in natural communities. They are also available for selecting species of potential horticultural merit. As Raven (1976) pointed out, 'the simplest, most effective, and least costly way in which botanical gardens can ensure the perpetuation of particular species is to introduce them successfully into the horticultural trade'. Raven cited *Ginkgo biloba*, *Franklinia alatamaha* and *Encephalartos woodii* as successful examples of the salvage of rare and endangered species through their introduction into commercial horticulture. *Swietenia humilis*, virtually wiped out in its habitat in the lowlands of Panama, is now a common street tree in the Canal Zone. The beautiful wildflowers of Western Australia, at one time threatened by the extensive trade in cut flowers, have been saved by the establishment of commercial plantings of the most popular, and hence most endangered species, combined with protection in the wild.

In most instances, some form of selection among endangered species is inevitable. The scope may depend on botanic gardens acting independently, or as members of a cooperative network which may be national, like that associated with the US Center for Plant Conservation (Thibodeau & Falk 1987), or international (see section 7.2.1). Irrespective of any overall plan, local interest and concern, as well as ease of maintenance, are likely to result in some priority being given to endangered species in the local or regional flora. Keystone species and other species of ecological significance or of research interest are likely to receive preferential consideration, as may species of cultural or historical significance. Priority on grounds of usefulness or economic value has been considered in Box 7.2.

The need for a systematic approach to the selection of species arises from the current emphasis on biodiversity and its preservation. This in turn resulted from the realization that in many parts of the world, diversity, both between and within species, is being infringed or lost, and that over large areas, especially in the tropics, the indigenous flora continues to be wiped out, resulting in large-scale species extinction.

Selective principles other than those emphasizing evolutionary distance may have a place in conservation strategy. In diverse fields of biology, closely related species provide the materials for research programmes where the loss of an endangered species would remove what might turn out to be an important 'missing link'. Clearly, no single criterion can satisfy the priorities for preservation that arise in various areas of science or technology, in agriculture or in manufacturing. It might be necessary to make several sets of choices for different interests.

7.3.3 Preserving genetic diversity

In the two models of species conservation outlined in sections 7.2.1 and 7.2.2, the preservation of genetic diversity forms an essential part of *ex situ*

BOX 7.3. SYSTEMS OF CLASSIFICATION FOR SPECIES SELECTION

To preserve the highest level of biodiversity within limits of available resources it is necessary to adopt a system of priorities based on general principles. Preference for endemic taxa is one possible system. Given (1981) suggested that taxa confined to a region and differing most from taxa occurring elsewhere be given first preference, which would mean that endemic families and genera would have priority over endemic species of genera represented elsewhere.

The cladistic method of classification is designed to assess, and to a degree, quantify the distinctness between taxa considered for preservation. It is based on phylogenetic relationships between species expressed as divergence since their most recent common ancestor. Common characters – which may be as diverse as morphological characters or chloroplast DNA restriction sites – are used to establish interspecies relationships. Vane-Wright *et al.* (1991) introduced a taxic diversity measure which defines the degree of genealogical differences among species. Taxonomic weighting is based on hierarchical relationships which are expressed in the form of a hierarchical tree. For each taxon, the number of nodes to the root of the tree is used in determining its distinctness. The inverse of the node count provides a priority measure for preservation.

Vane-Wright's method is based on the branching order, or topology, of the phylogenetic tree, but does not involve the length of branches which, as Crozier (1992) points out, could result in differences in the lengths of independent evolution being obscured. Crozier proposes the use of genetic distance to obtain an estimate of branch length; genetic distances can be derived from a variety of information such as allele frequencies or RFLP data.

preservation. For this there are a number of reasons. Genetic diversity raises the chance of *ex situ* survival. It provides material for exhibits and for research and education. It is essential for re-establishing or enriching wild populations, and for as yet unforeseen exploitation for human use. The question is how representative or how large the preserved material should be, seeing that within-species diversity is competitive with the number of species which can be preserved. This resolves itself into the number of sites and the sample size per site needed, to constitute the *ex situ* preservation of an endangered species.

The implicit purpose in preserving the genetic diversity of endangered species is to keep options open: for research, use or conservation, or for a return to natural or man-made ecosystems. *Ex situ* conservation not only complements *in situ* conservation (Maunder 1994), it makes a specific

BOX 7.4. SAMPLING FOR RETENTION OF GENETIC DIVERSITY

With the number of sites usually small, the size of the sample per site is the main variable determining the retention of genetic variation, which is expressed as the number of alleles per locus (the allelic richness). Brown and Briggs (1991) reason that in the preservation of endangered species the main concern should be with common alleles, i.e. alleles with a frequency >0.05–0.10, as it is in the preservation of crop species (Marshall & Brown 1975). The aim is to recover at least one copy of such alleles with a probability of 0.90 to 0.95, which requires a sample of 50–100 randomly selected plants. Brown and Briggs (1991) justify the emphasis on common rather than rare alleles by the small and often truncated populations of many endangered species (cf. Table 7.1, in Box 7.1), and by the emphasis on survival rather than on the use of specific alleles as in plant breeding.

Depending on plant size, manner of propagation – whether seeds, tissue culture, annuals, trees, etc. – the maintenance of populations of a size of 50–100 from each site may impose a burden which for any but the most easily preserved species could be considerable and restrictive. Brown and Briggs (1991) have attempted to define minimum sampling targets, applying the logarithmic view of the distribution of genetic diversity, which assumes that useful variation increases in proportion to the logarithm rather than the absolute size of the sample. The sampling strategy derived from this concept proposes a minimum of 15 individuals per species, or five populations of 10 individuals.

Such guidelines may be appropriate for large-scale, non-discriminatory programmes of species conservation of the kind envisaged in the Botanic Gardens Conservation Strategy (see section 7.2.1 above). More substantial and discriminating sampling patterns may be in place for the extensive genetic studies advocated by Falk (1990) as a preliminary to recovery and restitution, and for the process of reintroduction itself (see section 7.2.2 above).

contribution through being readily available and under direct control. But the preservation of genetic diversity, even on the modest scale envisaged in the preceding discussion, involves substantial costs in scientific effort, organization and resources. As emphasized in section 7.3.2, as a rule some form of selection of species takes place. It appears that more often than not, selection is motivated by actual or potential interest, concern or use. But in their absence, can the burden of preserving genetic diversity be justified?

7.3.4 Reintroducing rare and endangered species[1]

The establishment of a threatened species as a stable component of a plant community is widely regarded as the most desirable process of species preservation. It must, however, not be viewed in isolation. Reintroductions as a rule derive from materials assembled and cultivated *ex situ*, and may rely on such sources for occasional replenishment. Clearly, *ex situ* and *in situ* preservation are functionally linked.

There is a good deal of evidence (quoted by Holsinger and Gottlieb 1991) that in annuals and perennials the success rate of reintroductions is highest with material derived from the transplant site. This suggests that microhabitat adaptation is a distinct advantage. When there is no such close adaptive relationship, a composite cross of representative stocks may have the best prospect of success, with natural selection establishing locally adapted populations (Barrett & Kohn 1991). Generally, a broadly representative sample would be the most appropriate resource for future reintroductions. This should not be unduly hard to obtain since, as Brown and Briggs (1991) point out, endangered species tend to be restricted to a small number of sites.

Reintroduction demands a good deal of management, skill and resources for operations extending from collecting of planting materials (usually seeds) and, as a rule, raising of seedlings, to the choice and preparation of transplant sites, planting, protecting and further safeguarding of transplant populations. Information on causes of, or circumstances contributing to a decline of the original population would be of interest for attempts to counter adverse ecological factors, especially in the seedling or juvenile stage. These may include the presence of predators, parasites, competitors, nutritional imbalances or deficiencies, or the decline or absence of beneficial co-occurring species such as pollinators or seed dispersers, or of an adequate water supply.

Falk (1990) gives emphasis to the need for studies of genetic variation within and between populations. No doubt, information on the breeding system and on inbreeding in small populations is required for an effective reintroduction programme. But, as Holsinger and Gottlieb (1991) point out, 'obtaining a representative sample of their genetic diversity for off-site collections does not depend on knowledge of population structure.' Indeed, if the 680 species noted by the Center for Plant Conservation (see section 7.2.2 above) were to require extensive genetic analyses prior to attempts at

[1] For an extensive discussion of recovery and maintenance see section 6.2.5.

reintroduction, some might not survive to benefit from the information. If reintroduction of endangered species is to become widely practised, ecological and genetic common sense predicate the need for sizeable populations, genetic diversity, familiarity with the planting sites – and 'green fingers'.

Reintroductions, as Maunder (1992) points out, in the main have as their objective the re-establishment of endangered species, or of degraded communities; they frequently interact. Maunder (1992) observes that reintroduction is 'a relatively high-risk, high-cost activity'. Most reintroductions are as yet in the experimental stage, and 'evidence is still awaited on their long-term success'.

7.3.5 Documentation and information

A comprehensive database is an essential condition for international cooperation in identifying and preserving threatened plants. This was recognized by the 1975 Conference on the Conservation of Threatened Plants (Simmons *et al.* 1976) and taken up by the Threatened Plants Committee (TPC, later TPU) of IUCN. By the mid-1980s TPU had identified over 18,000 threatened plants world-wide, of which about 5000 were located in botanic garden living collections (IUCN/WWF 1988). By this time many botanic gardens had established or were planning computerized record systems. An international database linking national ones became an urgent need.

Since its establishment in 1985, the Botanic Gardens Conservation Secretariat (BGCS: see section 7.2.1) has endeavoured to expand a central information system for botanic garden collections, with emphasis on endangered species. In 1990 BGCS developed a readily accessible, pc-based system of its own. International data exchange was facilitated by introducing an international transfer format (ITF) for botanic gardens (IUCN/BGCS 1987) which was widely adopted by botanical gardens. The BGCS now maintains the database on *ex situ* conservation assembled by the World Conservation Monitoring Center (WCMC), and records of about 60,000, rare and threatened plants, representing 10,000 taxa, held in about 400 institutions world-wide (Jackson 1991).

7.4 CONCLUSIONS

As we have seen, *ex situ* conservation to a large degree is subsidiary and complementary to *in situ* conservation. *In situ* conservation has the potential for long-term preservation of *communities and populations*, under conditions of continuing adaptation. *Ex situ* conservation provides the freedom to select *individual species* for preservation. Priority may be given on geographical or ecological grounds, for educational, scientific or economic reasons (see section 7.2.1 and Box 7.2), or because a species is endangered or threatened in its natural environment. Two areas have emerged in which botanic gardens can make distinct contributions to the conservation of endangered species.

1. Assemble information on endangered species, their geographical distribution, diversity, degree of rarity, endangered state, and their representation in botanic gardens or other collections.
2. Assemble and maintain representative samples of the genetic diversity of endangered species, to safeguard their continuing survival, and to serve as a readily available resource for research and education, for reintroduction into natural communities and for distribution to industries.

It is worth recalling that in discharging these and related responsibilities, preservation in botanic gardens, by comparison with nature reserves, has distinct opportunities and advantages, as well as restrictions. Botanic gardens are capable of maintaining species derived from a diversity of climatic conditions, if need be in greenhouses. They supply water and nutrients, provide protection from parasites and they can control the breeding system. Many species can be maintained as seeds or in tissue culture, thereby increasing population size and reducing cost of maintenance. On the other hand, live cultivation of endangered species faces restrictions of space and finance, and hence for any but small or short-lived plants, limitations of population size. Yet cultivation is important as a means of informing the public of the threat to the survival of species around the globe.

The growing international links between botanic gardens offer opportunities for cooperation across climatic zones. Hawkes (1992) notes that the Copenhagen Botanic Garden has strong contacts in South-East Asia and Latin America. Seeds of tropical species stored in Copenhagen can be regenerated in botanic gardens in the tropics, thus saving expensive greenhouse space in the North. Indeed, one can imagine many advantages which could flow from North–South cooperative networks.

Botanic gardens offer a refuge for a greater diversity of species and with somewhat better security than is available by other means. Yet we must recognize the very real limitations imposed by available space, staff and facilities, by competing responsibilities and demands, and, last but not least, by shifts in policy due to staff changes, or to shifts in political control or public concern. In terms of the time scale of concern (see Chapter 1), preservation in botanic gardens might not rate beyond a generation or two. Yet international agreements may succeed in strengthening the security of tenure which individual institutions are able to provide. Mutual responsibilities impose on participants at least moral obligations which, in case of default, could be taken over by others. Ultimately the long-term validity of preservation in this form depends on the interest, perseverance and dedication of individuals, in addition to traditions and obligations. All preservation is ultimately as much a social as a biological issue.

8

Community structure and species interactions

8.1 INTRODUCTION – COMMUNITIES AS THE ONLY VIABLE STRATEGY FOR LONG-TERM CONSERVATION

Plant communities vary in complexity from seemingly simple desert communities dominated by just a handful of species to tropical rain forests that are home to many thousands of species of plants and animals, fungi and bacteria. However, they are much more than simple serendipitous assemblages of a range of free-living species. Plant communities are also rich repositories of parasitic[1], mutualistic and symbiotic associations of varying complexity, specificity and interdependence that together shape and affect the composition of the communities of which they are part.

This diversity of community associations, the greater diversity of species embedded in them and the still greater diversity of interspecific associations are the crux of the problem faced in the preservation of the Earth's biotic wealth. Indeed, for all practical intents and purposes, *in situ* conservation of these communities is the *only* way the large majority of such individual species can be retained for the foreseeable future.

The dilemma created by this wealth of diversity is, then, one of which communities to conserve and how this can best be achieved in the face of a burgeoning human population. Answers to these critical questions require an understanding of the dynamic nature of communities and the forces that generate them. In this chapter a limited range of these forces are briefly described to highlight the web-like nature of the interactions that tie the members of natural plant communities together. As we shall see, these are not trivial effects. The loss of a single 'keystone' species (section 10.4.1, p.244) may precipitate a cascade of changing patterns of interaction that

[1] Taken in its widest sense to include herbivory as well as more intimate associations.

ultimately affects the integrity of the whole community and the survival of its component species.

8.2 THE STRUCTURE OF COMMUNITIES – COMPETITION AND THE PHYSICAL ENVIRONMENT

What are the factors that shape the formation of plant communities? Do these operate uniquely in each particular site thus generating unique community assemblages? Or do ecosystems developing under similar sets of edaphic and climatic conditions converge to the same community structure?

Studies of the processes involved in the development of plant communities have broadly focused on the successional changes in species occupation that follow initial colonization of newly exposed areas of the physical environment. Such successional progressions may occur on areas with no previous vegetational history (primary succession), or on areas recently denuded of their overlying communities (secondary succession). Instances of primary succession represent only a very small fraction of all communities in the throes of progressive vegetational change. However, the often stark nature of the patterns that typically occur in such communities (for example, coastal sand dune and glacial moraine succession) and their seemingly predictable end-points, encourage a Clementsian view of all communities moving inexorably towards stable, site-specific climax assemblages.

If correct, such predictability would be of great appeal to many conservationists as it greatly simplifies many of the apparently intractable problems concerning what areas to conserve. In reality, however, no such steady state exists and ecologists have long recognized the dynamic and spatially heterogeneous nature of virtually all plant communities. Rather than being seen as discrete, predictable units of coevolved species, the vegetation of an area tends now to be regarded more as a continuum of species assemblages that grade into one another as individual species react differentially to changes along environmental gradients. In many instances these gradients are subtle and community-level differences are notable only over large distances. In others, changes are more abrupt and switches in species composition are more marked.

At its simplest, succession is the sequential process of replacement of early colonizing species by later arriving or developing ones. Typically the process is characterized by a general increase in the structural complexity of communities, a progressive change from annual to perennial habit and, in very many cases, an increase in species diversity. Over the years, there has been considerable interest in the mechanisms whereby these sequential

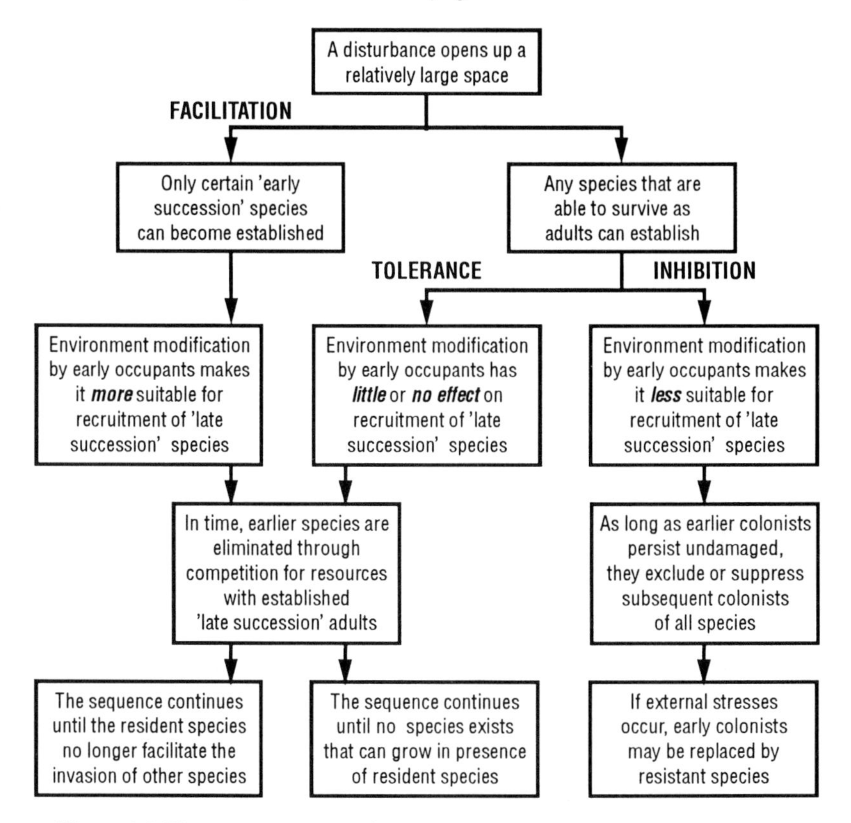

Figure 8.1. Three main models of community succession (adapted from Connell & Slatyer 1977, after Begon et al. *1986).*

changes occur. A wide range of models has been proposed encompassing three major themes (the facilitation, tolerance and inhibition models of Connell and Slatyer 1977; Figure 8.1) with an array of more specific variants (see Miles 1987).

The facilitation model particularly reflects the processes that occur during primary successional sequences of vegetation. Here early colonizing species alter otherwise inhospitable sites to such an extent as to make the entry of other species possible (for example, the stabilization of wind-blown sand by *Ammophila arenaria*). At the other extreme, the inhibition model reflects situations where development of a successional sequence is impeded indefinitely by the pre-emption of resources by earlier colonists. Further vegetational development will only occur when such species are damaged or killed by some local physical extreme or the action of herbivores or pathogens. The prolonged dominance of some old-field communities by rhizomatous *Solidago* species and their weakening by chysomelid beetle attack (McBrien *et al.* 1983) may well be an example of this long-term influence and its subsequent release. Finally, the tolerance model is an intermediate case between those of the facilitation and the inhibition schemes and is characterized by situations where species that appear later are simply those that arrived either at the very beginning or later and then grew slowly (Connell & Slatyer 1977). These species typically are tolerant of the reduced availability of particular resources such as light.

As Miles (1987) has pointed out, it seems unlikely that any successional sequence ever occurs in which only one of these models operates. Nor is there likely to be a set order to the species that follow one another in the developing community. Indeed, close examination of examples of both primary and secondary succession has detected complex networks of successional transitions. Londo (1974), working in dune slacks near Haarlem in the Netherlands, found a tendency for many assemblages to revert to dominance by species previously supplanted from the site. Similarly, Horn (1974) observed that the pattern of tree species occurring in a New Jersey forest formed on abandoned farmland reflected evidence of both the tolerance and the facilitation models. Furthermore, particularly in the case of secondary succession, the direction and rapidity of change will be heavily influenced by the range of species present on, or migrating to, the site early in the successional process (Miles 1987).

Even in relatively mature communities that are, seemingly, temporally far removed from the massive disturbances typically associated with the beginning of successional sequences, the vegetation is still constantly changing – only the scale is different. Rather than successional events occurring over hundreds, thousands, or tens of thousands of hectares, the vegetation constantly responds to much more minor disturbance ranging from localized windstorms, fires and other physical extremes to individual treefalls, digging by animals and a myriad of other factors that create highly localized openings in the community. In turn, each of these patches is subject to a process of colonization and succession. The importance of such localized disturbance events can hardly be over-emphasized (Grubb 1977; Sousa 1984) as they play a major role in contributing to the species richness of communities. Certainly, many species may maintain their place in communities by dint of their competitive prowess. However, for a large number of species this is not the case. These species rely heavily on being in the right place at the right time. For such 'fugitive' species, survival is achieved by a continuing process of movement of generations across the community landscape (Tilman 1994). As some populations are driven to local extinction by competitively more superior species, other populations become established on newly exposed disturbance sites thus maintaining the presence of the species in the community as a whole.

8.3 INTER-ORDER INTERACTIONS AND THE STRUCTURE OF COMMUNITIES

The interaction of plants with one another and with their physical environment produces a wide diversity of distinctly different communities. However, the diversity and complexity of the different plant communities present in the world today is not solely a product of these interactions. Indeed, in many ways these interactions provide the raw material upon which a further set of biotic forces exercises enormous leverage, altering the direction and rapidity of successional change, the relative dominance of species and the overall diversity of individual communities. The degree of recognition accorded these forces – herbivores, pathogens, mutualists and symbionts – has varied markedly but there is no doubt that their combined influence is immense and responsible for the detailed structure of most, if not all, plant communities.

8.3.1 Role of herbivores, pathogens and mutualists in shaping whole community structure

Vertebrate herbivores

A considerable body of evidence exists which demonstrates the fundamental importance of herbivores in shaping the composition, diversity and structural complexity of a wide range of individual plant communities. Not surprisingly, this role was first recognized for vertebrate herbivores that are easily seen, counted and excluded from experimental areas, with the most dramatic visual effects being apparent where fence lines separate two areas with a long history of different grazing regimes (Figure 8.2).

Inevitably, some of the best documented examples of the complex changes that occur in vertebrate grazing are found in agriculture, where controlled management of the timing, intensity and duration of grazing pressures has been used to manipulate pasture composition. Such manipulations aim to combine the animal's grazing preferences with different phenological patterns, growth rates and competitive abilities of plant species to generate communities dominated by particular species or combinations of species. In natural or less intensively managed communities vertebrate

Figure 8.2. Photograph of fence line effect showing differences in vegetation cover resulting from the presence (right-hand side) or absence (left-hand side) of grazing by sheep. Photograph reproduced by kind permission of Dr C. James.

grazing may also have profound effects on succession processes, although these effects may be less widely recognized as the immediate comparison afforded by an excluding fence line is absent. However, where suitable experimental comparisons are devised, the effects are manifest (see Crawley 1983 for an extensive review). A classic example is Tansley and Adamson's study (1925) of the floristically rich chalk grasslands of the English Downs. Comparison of areas protected from rabbits with those exposed to continuous grazing showed marked differences in species composition and diversity. Initially this community was composed of a diverse mixture of more than 25 grass and herbaceous species, none of which achieved significant dominance. However, once freed from the constraints imposed by grazing, the highly palatable grass *Zerna erecta* (*Bromus erectus*) rapidly increased in importance and within just six years the diversity and complexity of the community declined substantially. Continued exclusion of rabbit grazing over the subsequent 10 or more years saw the natural appearance of shrubby species as succession towards scrub and woodland communities intensified (Hope-Simpson 1940). Elsewhere in England, studies of other grassland (Watt 1957) and heath communities have shown rabbits to have similar diversifying effects.

Extermination or exclusion of non-native animals is a typical immediate response in the establishment of new nature reserves. However, as shown by the rabbit grazing example (the rabbit was introduced to England in Norman times), although such exclusions may result in dramatic community changes, these may not necessarily be the ones that would maximize floristic diversity. Furthermore, the effects of vertebrate herbivores are not necessarily consistent either across a range of species grazing the same community or across a range of plant communities grazed by the same herbivore. In the first instance, different herbivores frequently have different modes of feeding and distinctly different food preferences (Crawley 1983). Thus sheep feed differently from cattle, having different preferences – even distinguishing between different leaf mark morphs in white clover – and, as a consequence, may generate and maintain pasture communities of quite different botanical composition and physical structure from those induced by cattle.

In a similar way, even the same grazer may have contrasting effects in different communities where the relative competitive ability of particular groups of plant (grasses, forbs) and their palatability may differ markedly. Thus the effect of rabbits on English grassland communities is not predictive of the impact of rabbits on subalpine vegetation in the Koscuisko National Park, New South Wales, Australia. There a combination of exclusion and

BOX 8.1. GRAZING AND PLANT COMMUNITY COMPOSITION

A particularly good example of the effects of large grazing animals on the composition of plant communities is afforded by a series of trials involving sheep grazing of a permanent pasture dominated by *Agrostis* species together with *Festuca ovina, F. rubra, Holcus lanatus, Trifolium repens* and a range of other minor grass and forb species (Jones 1933). Depending on the timing and intensity of grazing the proportion of *T. repens* in the sward varied from 6 to 60% while that of *L. perenne* ranged from 5 to 35%. Under a hard spring grazing regime the sward was primarily a *T. repens–L. perenne* community (>80%). In contrast, a sequential pattern of over- and undergrazing led to a sward dominated by *Agrostis* species and a range of other grasses and forbs. The combined contribution of *T. repens* and *L. perenne* to this community was approximately 10%.

Maximal floristic richness, including sown species and many invading weeds, occurred when overgrazing in winter and spring was followed by undergrazing during early summer. As Harper (1971) pointed out, this treatment is most like that imposed in natural systems where overgrazing will occur during times of low plant growth.

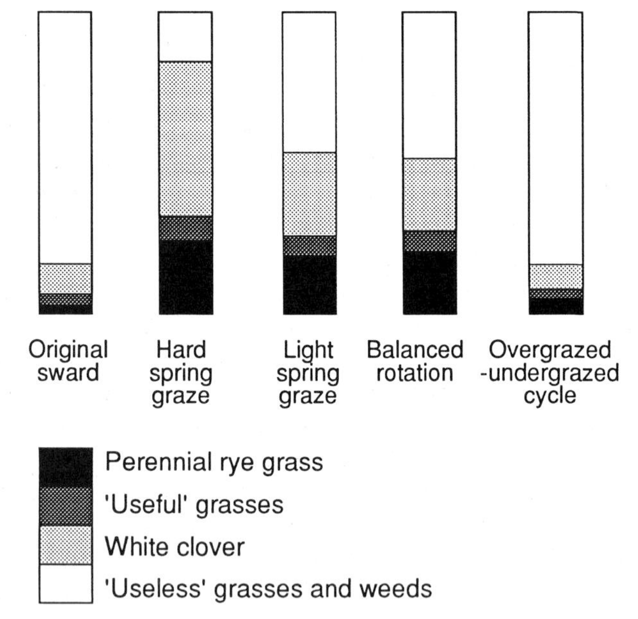

Figure 8.3. Frequency of different species occurring in permanent pasture subject to different grazing regimes (after Harper 1971).

enclosure treatments showed that rabbits caused a substantial decline in the percentage cover and diversity of a range of palatable forbs in herb-patch, frost hollow and tall shrub/grass communities (Leigh *et al.* 1987) – the exact opposite of the English experience!

Rather than simply halting the progress of successional processes, in some circumstances grazing may even throw these processes into reverse. This may occur when herbivore populations face starvation either through dramatic increases in population size (for example, following the relaxation of hunting pressures) or through catastrophic loss of food resources (for example, prolonged droughts or unexpectedly severe winters). In either of these situations, unusually heavy damage to tree or shrub species may lead to the re-establishment of earlier successional stages. The prime example of changes wrought in this way is provided by the activities of the African elephant which typically transforms closed woodland or thicket into open grassy savanna (Bond 1993).

Invertebrate herbivores

The successful control of a range of weedy plant species with insects (for example, *Opuntia* spp. in various parts of the world: Crawley 1989; *Hypericum perforatum* in California: Huffaker & Kennett 1959; *Salvinia molesta* in Australia: Room *et al.* 1981) attests to the potential effect that such insect grazers may have on plant populations and communities. However, in these situations biotic conditions are unusual in that both herbivore and host populations are largely free from natural enemies, and plant densities and populations are unnaturally large. As a consequence, such events give little insight into the *typical* rather than potential role of invertebrate grazers in determining plant community structure.

As with demonstrations of the role of vertebrate grazers, that of invertebrates is most convincingly shown in manipulative field experiments where judicious use of insecticides allows a chemical 'fence' to be erected for comparative purposes. By far the most extensive of such investigations is the long-term study being conducted at Silwood Park (Southwood *et al.* 1979; V. K. Brown 1990), where experimental plots have been established in three different stages of the secondary succession occurring on abandoned farmland. Each of these successional stages – i.e. ruderal sites created on bare ground; and early and mid-successional sites, three and 15 years post-cultivation respectively – have been variously treated with a foliar, systemic insecticide or left untreated. Over four years, marked changes occurred in the relative representation of annual and perennial forbs and of perennial grasses.

Total vegetative cover increased significantly in all successional stages, although this was delayed by a year in the mid-succession plot (V. K. Brown 1990). Within this overall change, annual forbs increased in the first year of succession in treated plots while the incidence of perennial forbs was reduced at all stages. The latter change apparently occurred as a result of competitive pressure from perennial grasses whose growth was greatly increased by insecticide treatment. The number of plant species present (species richness) was also affected by herbivory. In untreated plots, species richness was highest in the ruderal stage, declining in the early and mid-successional stages. Pesticide treatment enhanced this decline, with species richness rising in the ruderal stage but falling below that of the untreated control plots in later successional stages when enhanced growth of grasses tended to exclude other species (V. K. Brown 1990).

Control of subterranean herbivory also produced marked changes in the structure of the overlying plant community, although these differed from those produced by foliar feeders (Brown & Gange 1992). During the first four years of succession the main consequence of reductions in root-feeding insects was an increase in the relative cover of both annual and perennial forbs. The increased performance of forbs was also reflected in a consistent increase in the species richness of plots treated with the soil insecticide. This increase was substantially greater than that occurring on plots treated with a foliar insecticide only. Indeed, Brown & Gange (1992) recorded 17 plant species that were recorded only on the soil-treated plots (13 of these were forbs). In contrast, no species were restricted solely to the foliar treated plots.

The relative effect of foliar versus root herbivory on the cover abundance of component plant life forms in this early successional community clearly shows the differential effects of foliar and root feeding invertebrates. The greatest effect of foliar feeders was to restrict the growth of grasses while the impact of root feeders was concentrated on annual and perennial dicotyledonous herbs (Figure 8.4).

Periodic short-lived explosions in the size of populations of herbivorous insects are a well-known phenomenon in natural plant communities (Readshaw & Mazanec 1969; Kulman 1971). During such episodes host plants typically suffer considerable defoliation with a concomitant loss of competitive vigour. In one such situation involving an outbreak of three chrysomelid beetles (*Trirhabda* spp.) feeding on *Solidago canadensis*, foliar insecticides were used to assess the effects of herbivory on the composition of an abandoned meadow (McBrien *et al.* 1983). Although insecticide treatments did not start until after the main peak in beetle numbers had subsided, differences rapidly became apparent between treatment and control plots in the percentage cover of *S. canadensis*, four perennial grasses (*Poa pratensis*, *Phleum pratense*, *Agropyron repens* and *A. trachycaulum*) and a perennial herb (*Fragaria virginiana*).

In all plots, plants of *S. canadensis* were severely affected during the outbreak. Once spraying began, however, plants treated with insecticide were able to recover rapidly while unsprayed plants continued to suffer herbivory. Many of the latter plants died. The marked reduction in *S. canadensis* cover that resulted from herbivory was accompanied by substantial increases in the combined cover of the four grass species (up from approximately 20% to 70% over four years) and *F. virginiana*. As all these species are regarded as components of a somewhat earlier stage in the secondary succession that occurs on abandoned farmland, here insect herbivory has actually resulted in an apparent reversal of the succession process. However, McBrien *et al.* (1983) suggested that in the long run the overall rate of succession may in fact increase because of this interruption to the semi-stable period of *Solidago* dominance that may last for prolonged periods.

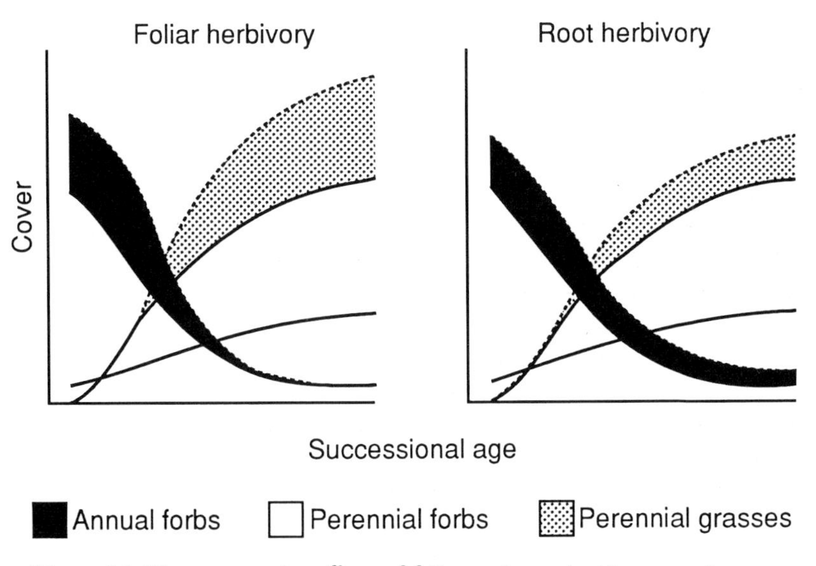

Figure 8.4. The comparative effects of foliar and root herbivory on the cover abundance of three broad groupings of plants in an early successional plant community (from V. K. Brown 1990).

Pathogens

As is the case with invertebrate herbivores, the most dramatic examples of pathogens affecting plant communities involve the epidemic spread of exotic pathogens through previously unexposed host populations. These changes have been particularly severe where the soil-borne pathogen *Phytophthora cinnamomi* has entered communities in Western Australia and Victoria. In both these areas a substantial proportion of the taxa of the dominant families in these communities (Myrtaceae, Proteaceae, Xanthorrhoeaceae) are extremely susceptible to this root-rotting pathogen and complex heath and eucalypt forest communities are rapidly being replaced by floristically depauperate ones dominated by sedges and other resistant species (Weste 1981; Shearer & Tippett 1989).

A complete contrast to this indiscriminate destruction is provided by *the* classic example of an exotic pathogen invading a natural plant community. In this instance, the pathogen *Cryphonectria parasitica* has been no less devastating to populations of its host, the American chestnut (*Castanea dentata*), but its overall effects on community structure have been far more subtle. In many situations, *C. dentata* has simply been replaced by one of several other co-dominant species (Woods & Shanks 1959; Stephenson 1986). However, where *C. dentata* was particularly common, the diversity of the tree and shrub layer has increased as the number, density and relative importance of species has changed. In moist situations this may be accompanied by an acceleration of the natural successional trend, while in drier ridge habitats the new community is more xeric than that which contained *C. dentata*.

Exotic diseases upset existing balances within communities and thus tend to produce easily discernible effects. On the other hand, the effects of pathogens that have been part of particular communities for millenia are generally less obvious. Where hosts in such situations are short-lived, documentation of the consequences of pathogen attack, like that of invertebrate herbivory, requires special steps to provide a necessary contrast. Such experiments have yet to be performed. However, there are circumstances where the required contrast may be provided naturally. This is especially the case for host–pathogen interactions that involve pathogens which spread relatively slowly and discretely from an initial focus, and which induce the death of relatively long-lived species that then persist as sentinels of past epidemics.

The slow spread of the root-rot fungus *Phellinus weirii* in conifer forests of the Pacific Northwest of the United States is a good example of this process.

These forests are dominated by *Tsuga mertensiana* but include a range of other species, the most prominent of which are *Abies monticola*, *Pinus contorta* and *P. monticola* (Cook *et al.* 1989; Dickman 1992). In the absence of *Phellinus weirii*, the local microclimate determines which of two successional pathways communities follow. A common *Pinus–Tsuga* early successional pathway is followed in mesic locations by an *Abies*-dominated climax forest or, in more xeric situations, by a *Tsuga*-dominated one.

The presence of disease caused by *P. weirii* may produce marked changes in these patterns. Although none of the conifer species in the forest are immune to attack there is considerable variability in their inherent susceptibility. *Tsuga mertensiana* is particularly susceptible whereas *A. amabilis* is considerably more resistant. This differential susceptibility, together with differences in the environmental requirements of *Abies* and *Tsuga*, may produce marked shifts in the structure of forests attacked by *P. weirii*. On moist sites, the diversity of the community within expanding infection foci remains the same as that in the surrounding unaffected forest. However, the successional sequence is accelerated as *Abies* is able to establish preferentially. In contrast, in drier situations *Abies* is unable to take advantage of the destruction of *Tsuga*. Instead, a renewed cycle of pine establishment occurs and species diversity increases (Cook *et al.* 1989).

Pathogenic organisms have recently been implicated as a driving force in the early stages of the most widely recognized example of primary succession – that occurring on coastal sand dunes. In much of northwestern Europe the grass *Ammophila arenaria* is the dominant stabilizer on primary dunes. At the leading edge of the dune, where fresh sand is continuously deposited, growth is vigorous. However, as sand drift diminishes, the growth and general vigour of *A. arenaria* also declines markedly. Comparisons of the occurrence of pathogenic soil organisms (fungi and nematodes) in fresh blown sand, and in sand sampled at three stages along a transect through this foredune succession, have found noticeable differences in the occurrence of these organisms.

In wind-blown sand few pathogenic organisms were evident; in mobile dunes harmful organisms were confined to the rhizosphere of plants; while in mature dunes they were widely distributed throughout the sand. Controlled experiments examining the performance of *A. arenaria* in sand taken from a number of these different sites clearly showed that the growth of the plants could be substantially increased through the application of either fungicides or nematicides (Van Der Putten *et al.* 1988, 1990). When these data are interpreted in light of the phenology and growth pattern of the horizontal and vertical rhizomes of *A. arenaria*, a strong case can be made for

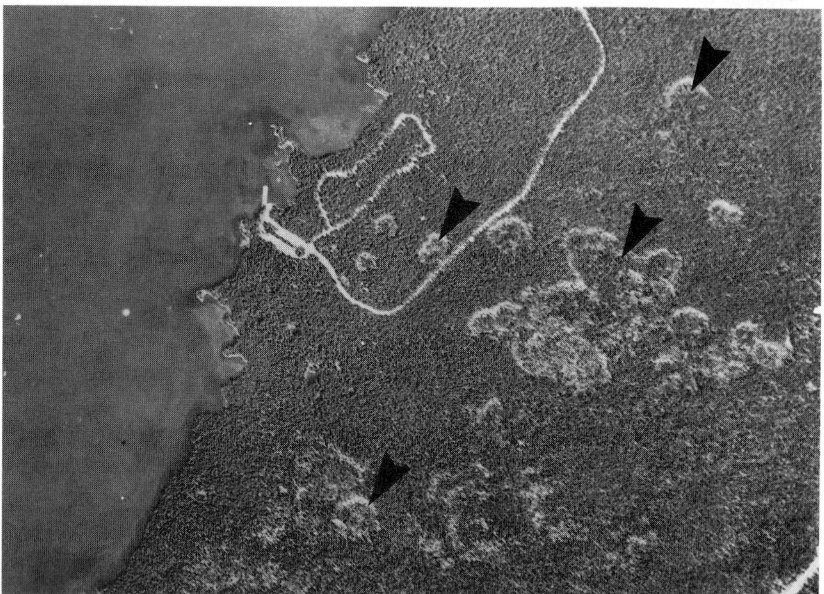

Figure 8.5. The effect of disease on whole plant communities. Upper photograph: Death of Banksia attenuata *in a* Phytophthora cinnamomi *disease centre (foreground) in the Stirling Range National Park, Western Australia. Photograph reproduced by kind permission of Dr B.L. Shearer. Lower photograph: Spreading disease foci (examples arrowed) caused by the root rot fungus* Phellinus weirii *in* Tsuga mertensiana *dominated conifer forests in the Cascade Mountains of Oregon, USA. Photograph reproduced by kind permission of USDA Forest Service and Professor E. Hansen. (Scale: c. 6 cm = 1 km.)*

the hypothesis that the decline of this grass on mature, stable dunes is at least partly determined by pathogen activity. Interestingly, it has also been reported that degeneration of *Hippophae rhamnoides*, one of the shrub species found in early dune slacks, has also been related to the occurrence of parasitic nematodes (Oremus & Otten 1981).

Mutualists and symbionts

Beyond the 'predatory' relationships delimited by herbivory and parasitism, plant communities typically display a wide range of mutualistic and symbiotic relationships. Frequently these are self-contained pair-wise combinations (for example, specialized pollinators of orchid species) that are likely to have little, if any, direct effect on overall community structure or function. In some cases, however, symbiotic or mutualistic associations may have wider ramifications affecting entire ecosystems. This may occur both in situations where the association is a highly focused pair-wise interaction and in more diffuse relationships where a single common partner interacts with several others. Here, a few of these associations are briefly discussed to indicate how they could affect the structure and diversity of plant assemblages.

Mycorrhizal, legume–rhizobia, and actinorrhizal interactions can be regarded as the archetypal symbiotic associations found in plant communities. Despite their undoubted importance in garnering nutrients for hosts, as facilitators in the nutrient dynamics of terrestrial ecosystems, and their widespread or, in the case of mycorrhizal associations, almost ubiquitous occurrence (Read 1991), evidence of their importance in shaping the structure and diversity of natural communities is relatively new.

The role of vesicular-arbuscular (VA) mycorrhizae in shaping community composition has been examined in experimental microcosms of 16 herbaceous species that typically occur in English calcareous grassland. These assemblages were grown in the presence or absence of VA mycorrhizae for one year. In the presence of mycorrhizal fungi, the yield of the dominant species (*Festuca ovina*) declined, while that of a range of subordinate species fared better. Indeed, several small-seeded species such as *Centaurium erythraea* suffered much greater mortality in the absence of the fungi (Grime *et al.* 1987). The net consequence of these effects was a significant increase in species diversity in the presence of mycorrhizae. These results are supported by recent attempts to quantify the effects of such infections on field communities by fungicidal application (Gange *et al.* 1990). In that instance, fungicide treatment lead to a reduction in the number of plant species recruited into the community.

BOX 8.2. THE EFFECT OF NITROGEN-FIXING ASSOCIATIONS IN ECOSYSTEMS

In undisturbed communities the effects of interspecific interactions are often difficult to ascertain. This is particularly the case in situations involving soil organisms where measures aimed at altering the number or density of one group (e.g. through pesticide application) may have unpredicted effects on other unrelated organisms.

Recently, however, the potential importance of nitrogen-fixing associations has been convincingly demonstrated in an example provided by the invasion of the exotic, weedy nitrogen-fixing tree *Myrica faya* into young volcanic sites in Hawaii (Vitousek *et al.* 1987).

Such sites naturally contain no native plants with nitrogen-fixing abilities and plant growth is severely limited by nitrogen availability. The invasion of *M. faya* and its actinorrhizal symbionts has increased the input of fixed nitrogen into these sites by more than 300%.

The longer term consequences of this massive injection of nitrogen into these systems has yet to be determined. However, one distinct possibility is that it will lead to increased growth rates and an acceleration of primary succession events.

Sources of nitrogen in open-canopied forest sites in Hawaii

	Nitrogen available (kg/ha/year)	
Source	No *Myrica*	With *Myrica*
Rainfall	5	5
Native N_2 fixation	0.6	0.7
N_2 fixation by *Myrica*	–	18

Source: Vitousek *et al.* (1987).

The widespread occurrence of VA mycorrhizae and their role in linking and facilitating nutrient transfer between different plant species in grassland communities is now well documented (Chiariello *et al.* 1982; Read 1991). Apparently VA mycorrhizae have a general role in contributing to the diversity of species found in nutrient-poor habitats (Read 1991).

Legumes are often seen as a key guild[2] of species that affect community development through their nitrogen-fixing association with *Rhizobium* (see, for example, Bowers & Sacchi 1991). However, the extent to which such species alter the nutrient status of natural soils sufficiently to influence subsequent community development is unclear (but see Box 8.2). In a similar way, species of *Alnus* and *Dryas* utilize their nitrogen-fixing association

[2] A group of species assuming the same functional role in a community or communities.

with *Frankia* to adopt an early pioneering strategy on glacial moraines. Again, however, it is unclear whether the subsequent substantial accretion of nitrogen in the soil (Harley 1970) simply accelerates successional changes or precipitates a sequence of plants totally different from that which would have occurred in its absence.

Flower pollinators are not a category of mutualists that is often seen as playing a pivotal role in the structuring of whole communities. After all, while insect pollinators may be essential for pollination in many plant species, any one of a generic group of pollinators will function more or less effectively. Thus in Australia the introduced honey bee (*Apis mellifera*) is equally effective as a vector of pollen of a wide range of native plant species (*Eucalyptus* spp.) as it is for agricultural crops (*Brassica napus, Malus* spp.) and weeds (*Echium plantagenium, Carduus* spp.) that were introduced from the same European–near Eastern region from which *A. mellifera* originated. Conversely, though, many plant species have evolved such specialized floral morphology and pollination mechanisms, ranging from elongated corolla tubes to the complex floral mimicry shown by several species of orchids, that the number of effective pollinator species may be limited. While these restrictions may threaten the long-term viability of individual plant species within a community, there is little evidence to suggest that the structure of the community itself may be at risk.

An exception to these potentially modest effects occurs if a significant fraction of the members of a plant community are dependent on one or, at the most, just a very few pollinator species. Plant communities occurring on oceanic islands that have depauperate pollinator faunas may well fall into this category. Indeed, in Samoa approximately 30% of all tree species occurring in the rain forest rely at least partially on just two species of flying fox for both pollination and subsequent seed dispersal (Cox *et al.* 1991a). The loss of such species through either hunting or natural disaster (e.g. cyclones) would probably lead to long-term changes in the reproductive success of their dependent plants, to changes in the degree of aggregation of parents and offspring, and ultimately, to the diversity and resilience of the forest as a whole (Cox *et al.* 1991b). Other island floras would appear to be equally vulnerable.

8.3.2 Role of soil microflora and fauna in shaping community structure

We are not concerned here with those relatively specialized soil dwelling organisms that form more or less intimate associations with the roots of host plants (e.g. mycorrhizal fungi, pathogens). Rather our interest centres on the myriad of soil-inhabiting algae, bacteria, fungi, protozoa, nematodes, microarthropods and other animals whose presence is all too often overlooked. Collectively these organisms are responsible for the decomposition of all organic matter and the recycling and/or fixing of a wide range of vital nutrients including nitrogen, phosphorus and sulphur.

The combined function of the soil's microflora and fauna as the disposers and recyclers of resources has enormous ramifications determining the continued functioning of all ecosystems. Without this invisible army of organisms all nutrients would be sequestered in living and dead plant and animal tissue and communities would be swamped in their own debris (Hawksworth 1992). However, although absolutely vital for ecosystem maintenance sufficient functional redundancy exists among these organisms such that the important decomposer and recycling functions are likely to continue despite any marked change in species composition. This does not mean that human activity cannot affect these processes to serious effect. For example, drastic changes in water levels may result in a virtual suspension of decomposition activities as dead plant material becomes permanently waterlogged and oxygen availability falls close to zero.

Of more immediate consequence is the role that many members of the soil's microflora and fauna have in developing and maintaining soil structure (Harley 1970; Hawksworth 1992). Organisms as diverse as algae and earthworms make important contributions to the physical properties of individual soils such as permeability (algal mats in desert soils) and aeration (earthworm activity) that may affect the species composition of the overlying vegetation. The degree to which such effects are duplicated by other soil organisms is largely unknown but would seem to raise the possibility that the loss of small groups of related species may well induce changes that would lead to long-term vegetational changes.

8.4 CONCLUSIONS

The plant communities we observe today are the product of the interaction of an extremely diverse range of physical and biotic forces. They are dynamic entities in which change is an integral, indeed essential part of the processes whereby the community is maintained. The widespread extent of these interactions and their dynamic nature are an essential foundation on which to build rational plans for the selection and management of community reserves.

9

Choosing plant community reserves

9.1 INTRODUCTION

Scale and pattern are features of immense importance in any understanding of the long-term demography and fate of individual species, and through this, whole ecosystems and communities. Few, if any, species are to be found spread evenly across their entire geographic range. At regional, local and individual ecosystem levels, individual plant species are typically distributed in more or less discrete populations or demes separated from one another by assemblages of other similarly distributed species. Each of these species, by virtue of particular life-history attributes – germination and establishment requirements, precocity, mean life span, sensitivity to competitive pressures and so forth – possesses particular spatial and temporal scales over which new populations develop, flourish and ultimately become extinct. When combined the products of these distinctive individual patterns are multispecies communities in which the positions of individual species within the community may wax and wane at rates quite separate from one another. As a consequence, communities typically vary dynamically over a wide range of spatial and temporal scales.

This inherent heterogeneity of plant communities is the source of much of the diversity we aim to protect, and also of many of the difficulties that arise in formulating strategies to achieve that end. Clearly, establishing precisely what it is that we are trying to conserve and for what period of time is *the* central issue in any conservation strategy. Answers to those questions have immediate implications for the process of implementation – that is, what areas and of what size, must be protected to achieve our ideals.

9.2 WHAT ARE WE TRYING TO CONSERVE AND FOR HOW LONG?

Conservation practices developed for the protection of agriculturally important germplasm resources (gene banks and the like: Chapter 4), or in some cases rare and endangered wild species (the developing use of botanic gardens: Chapter 7), are markedly different from those needed for the protection of whole communities. Although the former strategies generate heated debate over matters of ownership, they are widely accepted as essential safeguards for the future. In contrast, for many sectors of society, one of the most contentious aspects of ecosystem conservation is the perceived alienation of large tracts of land or vegetation. To counter these arguments of the 'opportunity cost' of the development foregone, it is essential that we have a clear understanding of the purpose of particular conservation efforts and their expected benefits.

For what purposes do we aim to conserve entire communities and ecosystems? Is this simply seen as a blunderbuss approach whereby large numbers of species and diversity can be protected cheaply? Or have we a more scientific rationale aiming to protect particular vegetational associations and the interactions between species of all types (plant, fungal, bacterial and animal) as long-term evolutionary entities? Is interest essentially focused on one or two rare species with others being left to fend for themselves? Or are we formulating strategies to protect areas for any of a host of cultural reasons, or the maintenance of ecosystem services like nutrient cycling or hydrological processes that have an immediate return to human society? A clear consequence of the diversity of these potential aims is that no one plan can cater for all community or ecosystem based conservation efforts.

A question that inevitably complements that of purpose, is the question of 'the time scale of concern' (Frankel 1974). When we talk of conserving ecosystems and communities, do we envisage a process lasting over decades, centuries or millennia? Being a human activity it is not surprising that most people will tend to think of conservation in terms of events that encompass their own, their children's, or their children's children's life spans. However, even at current life expectancies for Western countries this represents little more than 140 years – a time period that is relatively ephemeral in communities where dominant trees may live for many hundreds of years. No generation can predict the behaviour and attitudes of those that follow it. However, by striving for reserve designs that can reasonably be expected to function effectively for the foreseeable future, we can place a strong obligation on future generations for their continued maintenance.

'Wherever possible' is a vital caveat to this strategy. Reserves of all shapes and sizes, even the smallest urban patches of bushland, may contribute materially to the security of major reserves through their sociological significance and educational value (Frankel 1970b; Goldsmith 1991). Indeed, in practical terms such reserves will be the vast majority of all future cases. In the broader sweep of time, though, very small reserves are especially vulnerable to catastrophic change induced by climatic calamities (e.g. drought, fire), alternative land uses and invasive species. As such they will frequently be little more than static museums whose direct contribution to the conservation of species and communities is slowly eroded. As a consequence, a major effort must continue to be devoted to conservation strategies which protect areas that inherently stand a reasonable chance of continuing, long-term evolutionary adjustment and change.

9.3 HOW DO WE DETERMINE WHAT AREAS TO PROTECT?

In recent years much has been written about reserve design (May 1975; Frankel & Soulé 1981; Margules *et al.* 1982; Margules & Nicholls 1988; Spellerberg 1991). This has often generated considerable argument, the origins of which frequently lie in a lack of clarity or precision as to the aims of particular reserves and their perceived lifespan. Clearly, all decisions concerning the conservation of biodiversity are inevitably shaped by an amalgam of ecological, economic, social and political circumstances (Frankel 1974). This is particularly so at the ecosystem, community and landscape levels where biological reasons for the siting, size, design and management of reserves often confront quite contrary economic imperatives. Although resolution of such conflicts through compromise of both scientific and economic ideals is increasingly the case, an understanding of the biological rationale behind 'ideal' reserve design is essential to ensure the best possible outcome and its subsequent effective management.

9.3.1 The ideal reserve model

Ideally, most reserves should be large. To maximize the probability of long-term protection, reserves need to be contiguous areas that maintain the integrity of the physical environment and the basic geochemical processes of the region in question (e.g. watersheds, drainage systems). Biologically, reserves need to be large to mimimize the impact of external variables on the long-term maintenance of the interspecific links that are ultimately responsible for the diversity and structure of specific communities. Such links frequently involve animal components (e.g. grazing herbivores, fruit and seed predators) whose space requirements and relationships may be quite different from those of plants. The ramifications of many of the resultant food webs to include large carnivores like bears, lions and wolves – vital regulatory forces acting on many herbivore populations – will extend the size of the areas needed to provide protection by many-fold.

Predatory species play a central role in determining the need for large reserves. For a number of biological and social reasons, most large carnivores are highly mobile with home ranges that extend over many square kilometres. In these areas they naturally occur at low population densities. Simply to cater for these movements, the inevitable occurrence of sequences of unfavourable years, and especially the long-term dangers of loss of fitness or sterility associated with inbreeding in small populations, large reserves are essential.

Just how large, 'large' may have to be, is indicated by a number of computations that dot the conservation literature. Thus in the western United States, predators like the mountain lion naturally occur at a density of one mature individual every 2600 hectares (Hornocher 1969), while individual grizzly bears apparently require 7600 hectares (Mann & Plummer 1993). When these figures are multiplied by 500 to generate what is widely regarded as the minimum effective population size (Franklin 1980; Soulé 1980), these areas expand to 1.3×10^6 and 3.8×10^6 hectares, respectively. However, if one accepts Soulé's (1980) proposition that as a rule of thumb the effective size of a population (N_e: see section 5.4.1) is only approximately one-third of its actual size, then these numbers further increase to 3.9×10^6 and 1.14×10^7 hectares. Reserves of this size simply do not exist anywhere in the United States. Indeed, in the world as a whole, where large predators exist on all continents other than Antarctica and Australia, only a handful of reserves over 2×10^6 hectares exist (IUCN 1975; but see Box 9.1).

Such calculations clearly indicate the importance of large areas for the continuing conservation of many large predators. However, as pointed out earlier (Chapter 5), this approach of estimating reserve size through population viability analysis (PVA) is fraught with difficulty. The results obtained and hence the reserve area needed to maintain such populations, are heavily dependent on a range of computational assumptions.

Of course, the loss of major carnivores or herbivores does not mean that all other species in the community will immediately cease to exist. However,

BOX 9.1. MAINTENANCE OF BIODIVERSITY THROUGH PROTECTION OF LARGE HERBIVORES AND CARNIVORES

The network of interspecific relationships that characterize natural communities inevitably ensures that changes in the presence or absence of some organisms may have substantial short or long-term consequences elsewhere in the community. Using elegant experiments involving the selective removal of predatory starfish, hence increasing or decreasing herbivory, Paine (1966) demonstrated the consequences of such effects in simple, yet real marine communities.

In terrestrial ecosystems similar interspecific interactions exist. However, the space requirements of large mammalian grazers and predators are

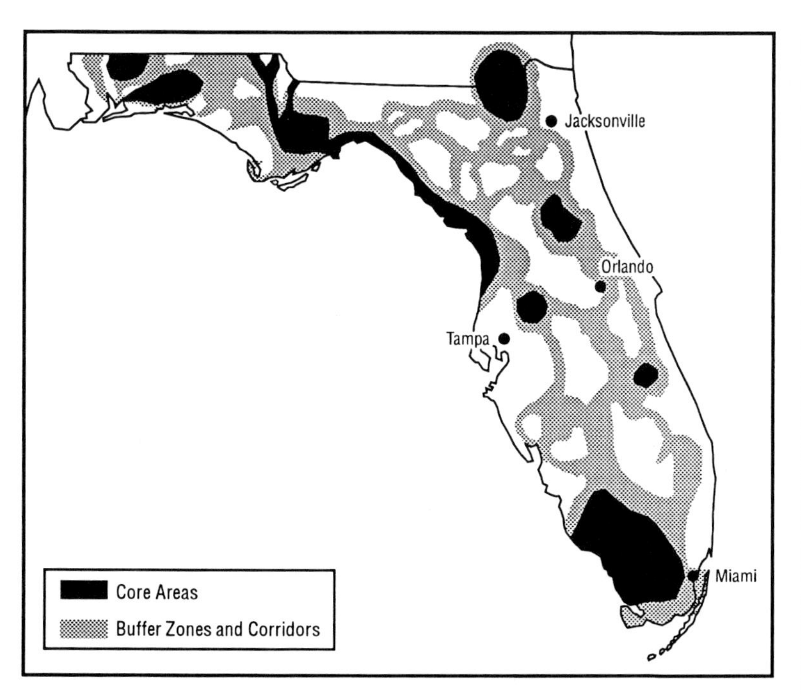

Figure 9.1. An extensive network of reserves and corridors is seen as the only way of keeping the Florida panther (figure reprinted with permission from C. C. Mann & M. L. Plummer (1993), Science **260**: *1868–71. Copyright 1993, American Association for the Advancement of Science).*

BOX 9.1. (*cont.*)

several orders of magnitude greater than those of sea urchins and molluscs on rocky shores. As a consequence, over the past 20 years biologists have come to recognize that relatively few reserves in the world are large enough to provide long-term protection for populations of large animals like elephants, lions, cheetahs, wolves and hunting dogs.

Various strategies have been proposed to circumvent this problem including the establishment of corridors between adjacent reserves (see section 9.3.2, p. 219). The most Utopian of these plans is the recently advanced Wildlands Project that proposes enormous changes to the pattern of land utilization in the contiguous United States in order to provide the conditions considered essential to protect North American biodiversity (Mann & Plummer 1993). The core of this proposal is the creation of large continuous reserves consisting of core areas, buffer zones and corridors catering specifically to the needs of large mammals, particularly carnivores. Combined, these areas would transform the face of America. For example, along the Oregon coast almost half of the land would be returned to wilderness or severely restricted in terms of human use. In Florida (see Figure 9.1), a web-like tracery of corridors crisscrossing the state is seen as the only means whereby the Florida panther can be saved from extinction.

because of the intricate relationships and links that stretch across the different trophic levels within a community, the loss of keystone species does spell the beginning of changes that are largely, if not totally, unforeseen. Examples of such effects are seen in the conversion of African grassland savannas to shrublands following the removal of elephants (Bond 1993); and in the sharp reductions in kelp beds as numbers of sea urchins soared following the near extermination of sea otters along the north Pacific coast of Canada (Duggins 1980). In time some new dynamic equilibrium will be achieved; however, this is likely to be quite different from the original community. Furthermore, in arriving at that new situation, many species may disappear; others may increase; while yet others may invade from elsewhere.

In general, attempts to accommodate the spatial requirements of the large mammals and birds of an area will usually more than adequately encompass the needs of all but the largest co-occurring plant species. However, claims like those of McNeely and his associates (1990) that 'many plants can survive for several centuries in a forest niche scarcely larger than the diameter of its leaf rosette' reflect a perception of the environmental requirements of plants that is completely wrong. Certainly, plants are markedly less vagile than animals; however, such views fail to recognize their often specialized germination, establishment, growth and reproductive requirements and the interdependence of the animal and plant components of most communities.

Large reserves carry with them a host of biological, physical and management advantages that are progressively eroded as reserves become smaller and more akin to islands in an inhospitable sea (see sections 10.3, 10.4). Bearing this in mind, how does the current distribution of reserve sizes compare with ambitious schemes like the Wildlands Project (Box 9.1)? Reserves of all sizes and degrees of protection exist, ranging from those of a hectare or so through to huge reserves covering millions of hectares. However, the size distribution is extremely uneven. A survey of 918 of the world's more significant nature reserves[1] (IUCN 1975) found that 76% of these reserves were less than 100,000 hectares in area while only 3.5% exceeded a million hectares (10,000 square kilometres).

More detailed surveys of the size of reserves in Australia and the United States confirm these patterns (Figure 9.2). In both cases, the majority of reserves are less than 1000 hectares in size. However, only the Australian data give an accurate picture of the relative distribution of different park sizes (the US data form but a small subset of all the parks occurring there: Schonewald-Cox 1983). In Australia, there are 3429 reserves conferring some form of government protection on 6.5% of the total land area. Seventy-four per cent of these reserves are less than 1000 hectares in size, while only 2.6% exceed 100,000 hectares. Conversely, though, these large parks account for a very substantial fraction of all the land protected. The nine largest reserves (all exceeding a million hectares) comprise 37.5% of the protected area.

The degree of protection provided by reserves varies widely, ranging from those which are largely free of human activity through to those in which controlled resource extraction may take place. Clearly, the more a reserve is subject to human activity the greater are the dangers of loss of biodiversity. However, neither those reserves that encompass both conservation and utilitarian elements, nor those that are very small (<100 hectares), should be dismissed out of hand. In many instances the former reserves may be all that can be achieved or may provide an effective buffer zone around some central, relatively undisturbed core (see section 10.3). Small reserves, on the other hand, have a very important educative role. By bringing a local focus to many conservation issues, such reserves heighten

[1] Note that these data are strongly biased towards the upper end of the reserve size spectrum.

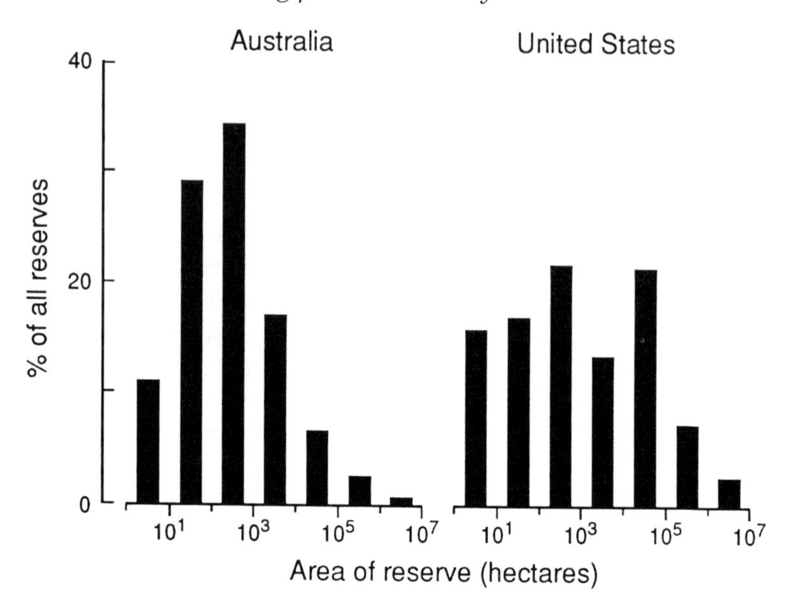

Figure 9.2. Frequency distribution of reserves in Australia and The United States according to size of the reserve. The Australian figure is based on 84% of all reserves (2878 of 3429 reserves: Hooy & Shaughnessy 1992); the US data on 320 parks (Schonewald-Cox 1983). [Data for US from Genetics and Conservation by Schonewald-Cox et al. *Copyright © 1983 by The Benjamin/Cummings Publishing Company USA. Reprinted by permission.]*

community awareness and hence assist the preservation of larger wilderness areas. Indeed, this is probably the most important long-term role for very small reserves. From a strict conservation point of view they will always be relatively vulnerable, needing more management per hectare than will larger reserves.

9.3.2 Reserve selection and design

Over most of the globe opportunities for the establishment of totally new community reserves that ascribe to the ideal model (continuous areas $\gg$ 100,000 hectares) have essentially disappeared and reserve selection has to occur in areas where habitats are already fragmented. What, if any are the principles that should guide selection under these less than ideal conditions?

Reserve selection and design can be broken down into distinct but complementary biological and physical components. The biological component assesses the relative conservation value of particular communities or suites of communities. The physical component, while superficially concerned with the size and shape of reserves, has major biological relevance for the future. It is this aspect of reserve design that most directly attempts to grapple with the longer-term integrity of communities.

Biological aspects of reserve design

For much of the past 25 years, considerations of the size, shape and other physical aspects of reserve design have been dominated by biogeographical ideas and theories. The most celebrated of these is the dynamic equilibrium theory of island biogeography (Preston 1962; MacArthur & Wilson 1963, 1967). This theory suggested that patterns of species occurrence and richness are the result of a dynamic equilibrium between rates of extinction and rates of immigration. For a particular island, the precise number of species at equilibrium depends mostly on island size ('area effect') and distance from the source of immigrants ('distance effect') (Frankel & Soulé 1981).

The obvious analogy between true islands and terrestrial nature reserves ('islands' of natural vegetation in an otherwise hostile landscape: Figure 9.3) was recognized at an early stage (May 1975). Since then consideration of area

Figure 9.3. Remnant patches of vegetation in an agricultural landscape. How closely do these mimic conditions on true islands? Photograph reproduced by kind permission of Colin Totterdale, Greening Australia Limited.

and distance effects has had profound consequences for the development of theory relevant to the design of nature reserves. Unfortunately, however, as several authors have noted (Wright & Hubbell 1983; Spellerberg 1991), aspects of the theory have often been accepted and applied uncritically to vegetational patches without regard for the non-trivial differences between these and real islands. This has lead to much argument (for example, the SLOSS debate: see below). Despite such problems, the MacArthur & Wilson model was a seminal step in the development of conservation biology. Its dual focus on current numbers and the future consequences of extinction and immigration rates, again stressed the different time scales on which conservation planning has to take place.

Short-term concerns and long-term protection

The desire to protect the world's biodiversity places two distinct requirements on reserves. First, reserves have the immediate function of protecting the maximum possible number of species and communities. Second, the same reserves are popularly expected to provide adequate conditions for the long-term survival of evolving communities. These two aims are not necessarily complementary and strategies devised for the immediate maximization of species numbers often appear to run counter to longer term aims. This is particularly reflected in aspects of the SLOSS (**S**ingle **L**arge **O**r **S**everal **S**mall) debate concerning the relative merits and disadvantages of reserve subdivision. Like many points of contention in conservation biology, the SLOSS debate has often been fuelled by confusion and misunderstanding about terms and ultimate intentions (Simberloff 1988).

Empirically, most studies have found a greater diversity of species in a combination of small reserves than in a single large one with the same total area (Soulé & Simberloff 1986; Quinn & Harrison 1988). These results are hardly surprising even if one assumes that such patches are set out in the same general area. Within single, large and apparently homogeneous areas, habitats vary subtly from point to point. Greater variation can therefore be expected in spatially more dispersed areas. In essence, by selectively dotting reserves around, samples of more physical environments with a wider range of habitats and communities can be encompassed. In addition, such dispersed reserves may provide some degree of protection from environmental catastrophes like fire and, for animals in particular, disease. Clearly, when applying the *immediate* objectives of safeguarding as many species as possible, large reserves are not always the most appropriate (Spellerberg 1991). However, as Simberloff (1988) points out, there may be situations where reserves are simply too small to sustain particular species or groups of species. As we have already noted, amongst animals, obvious examples are large grazing animals and predators; while among plants they may include large trees or species relying on conditions created by larger herbivores.

Giving immediate protection to threatened species and communities is an important part of a global response to the biodiversity crisis but this is only *part* of the minimum response needed. Without measures to ensure permanence, short-term species 'collecting' will be largely a futile gesture. Ideally we should aim for any reserve design that confers on the target community the biological propensity to propagate itself (and hence all its myriad interspecific parasitic, mutualistic and symbiotic interactions) in perpetuity. By focusing on long-term evolutionary reserves, aspects of the complex relationship between the biological, spatial and temporal needs of species within communities can best be illustrated. Many of these have general validity for all conservation efforts.

What, then, is the long-term fate of individual populations, entire species and hence, ultimately, of whole communities in given habitat patch sizes? Do physical features like reserve size, shape and degree of isolation have significant effects on important biological criteria like extinction and turnover rates, minimum population sizes and metapopulation structures? In other words, is there greater long-term evolutionary safety associated with reserves of particular sizes?

An obvious deduction from the species–area relationship that underpins the MacArthur–Wilson model is that a decrease in the size of a patch of vegetation will be accompanied by a loss of species and the eventual attainment of a new, lower equilibrium. The speed and extent to which this 'collapse' (Soulé *et al.* 1979) or 'relaxation' (Diamond 1972) in species numbers occurs is clearly of considerable importance in determining the long-term viability of communities. Attempts to measure these changes have largely focused on fluxes in the numbers of particular taxa (often birds) on series of land-bridge islands that became separated following the end of the last Ice Age (for example, Diamond 1972; Terborgh 1975).

More recent attempts to assess similar changes on artificially created islands only serve to illustrate the inherent differences between natural islands and recently created artificial islands or patches of vegetation ('terrestrial islands'). In the former situation the biota have developed and been shaped, often over millennia, by the prevailing physical environmental conditions. In contrast, artificial and terrestrial islands are often all that is left of previously more extensive stands and are now exposed to quite different environmental regimes (Box 9.2). Such fragments experience microclimatic effects markedly different from those encountered by large

BOX 9.2. VEGETATIONAL CHANGES RESULTING FROM FRAGMENTATION

An example of the consequences of the isolation of small tracts of vegetation from formerly extensive natural communities is seen in the changing patterns of tree diversity recorded on six small islands (all less than 1 ha in size) that were created by the filling of Lake Gatun in the Panama canal (Leigh *et al.* 1993). Forests on these islands are now more exposed than previously and have suffered more damage from the storms and dry season winds that sweep across the lake than have equivalent sized areas embedded in continuous forest on the adjoining mainland. Tree falls are common and the forests are generally drier as winds penetrate deep into the stand.

A combination of this increased exposure and isolation from the mainland has resulted in a decline in tree diversity as some species were lost and a subset of others have increased in prominence. Loss of species was attributed to changes in the degree of exposure while the increasing dominance of a restricted subset of other species was attributed to changes brought about by isolation and a concomitant loss of small seed-eating mammals.

tracts of undisturbed vegetation (Saunders *et al.* 1991). Air temperatures at the edge of remnants can be significantly higher than those found in the interior of the patch (Kapos 1989), and light can penetrate deep into the remnant edge, affecting the growth of existing species (Lovejoy *et al.* 1986) and the establishment of others. Unlike true islands, recently created terrestrial islands also face a substantial and continuing influx of exotic plants and animals (see section 10.4.2, p.250). The effect such organisms can have through changed competitive interactions is potentially enormous.

The physical consequences of interactions between terrestrial reserves and the surrounding environment may on their own be sufficiently detrimental to counsel against the establishment of very small reserves in all but special circumstances. However, there is a series of other equally compelling scientific reasons favouring larger rather than smaller reserves. These reasons revolve around the extent and likelihood of extinction events (see below).

Metapopulation[2] models

With the increasing realization of some of the inherent weaknesses in the equilibrium theory of island biogeography (Wilcox & Murphy 1985),

[2] A metapopulation is composed of a series of spatially discrete populations between which limited migration occurs.

metapopulation models have come to the fore as a means of understanding the consequences of terrestrial habitat fragmentation. These models focus on the dynamics of the demographic and genetic processes affecting individual species rather than simple changes in the number of species. One of the great strengths of this approach relative to the MacArthur–Wilson model is its explicit recognition of the effects of environmental heterogeneity both within and between patches (Merriam 1991).

The individuals of a metapopulation are distributed among a series of more or less discrete local populations, each of which has a finite life expectancy (see Chapter 5). In a dynamically stable community, as established populations disappear, new ones arise as colonists gain a foothold in favourable niches. Clearly, species whose metapopulations are suddenly and severely slashed in size run a real risk of failing to establish new populations as existing ones disappear. In these circumstances, the four sources of uncertainty (demographic, environmental, or genetic uncertainty, and natural catastrophes: see Chapter 5 for detailed discussion) with which they previously were able to cope may now, either through independent action, or more commonly by cumulative effect, be sufficient to tip particular populations over the abyss of extinction.

Measures of the rate of turnover (the combined effects of extinction and recolonization) of populations in habitat patches of varying size are essential to the development of a true appreciation of the vulnerability of individual species. When integrated across all species in a community this has direct implications for the vulnerability of the community as a whole. At the individual species level, turnover rates will determine whether we are essentially managing a single population with an unbroken genetic and temporal continuity, or whether a species is totally dependent on a metapopulation structure of ephemeral subpopulations (Soulé 1987b).

These two extremes imply very different amplitudes in the numbers of contrasting species and marked differences in their need for disturbance of the surrounding community. Species whose occupancy of a given site is relatively short rely on more frequent disturbance regimes than do those whose site occupancy is more protracted. Indeed, disturbance regimes that are optimal for one suite of species may be highly detrimental to others. Furthermore, different species will have metapopulations that occupy quite different spatial scales. Thus many small sedentary species have the inherent potential to thrive on restricted reserves while larger, more mobile or more diffusely dispersed organisms (for example, the larger vertebrates, or many rainforest trees) may require the protection afforded by substantially larger areas. As a result, a reserve that might be perfectly adequate for the long-term survival of one particular group of species, may be woefully inadequate for others. Identification of the latter group of organisms, their importance to the long-term functioning of ecosystems and requirements for their long-term management will ultimately determine the minimum size that reserves can be while still fulfilling the essential requirement of fostering the ecological and genetic coevolution of species assemblages in perpetuity.

Countering fragmentation – do corridors have any value?

Corridors of linking vegetation have been advanced as one option for countering the depressing reality that many existing reserves are too small to maintain viable populations of many species. In a general sense corridors may contribute to regional conservation systems by assisting in reducing wind and water erosion, by providing shelter and by increasing the aesthetic appearance of a landscape (Hobbs 1992; Figure 9.4). However, in the current context our interest is focused on the role corridors play in enhancing the long-term viability of populations of species and communities. Such benefits potentially could accrue by (i) reducing the effects of isolation on the genetic structure of populations and (ii) by assisting the recolonization of habitats in which pre-existing populations have gone extinct.

Figure 9.4. Ribbons of vegetation linking remnant patches of vegetation. Can these serve as effective corridors for the movement of animals and plants? Photograph reproduced by kind permission of Murray Fagg.

What evidence is there that such effects occur? Both potential benefits require movement – a feature for which there are currently no empirical data relevant to plant species. For animals, the situation is somewhat different although, even for these much more mobile organisms, the role of corridors is still equivocal (Saunders & Hobbs 1991; Simberloff *et al.* 1992). While examples exist of the importance of corridors in ensuring or promoting inter-patch movement (Henderson *et al.* 1985; Bennett 1990), their effectiveness is likely to vary greatly depending on the species involved. For some species corridors may be beneficial while for others they may act as 'demographic sinks', doing more harm than good (Soulé & Gilpin 1991).

Given the very slow rate of natural spread of all but weedy plant species, it is only through the movement of animals, fire or disease that corridors are likely to have any discernible effect on the structure and longevity of plant populations and communities in the foreseeable future. In this respect, one can envisage two extreme consequences of habitat fragmentation. On the one hand are the consequences that may be reflected in the demography of individual plant species that have beneficial associations with animals. Thus plant species that rely on mammals or birds for seed dispersal may be less severely affected where habitat fragments are linked by corridors than where they are completely isolated. Highly specialized associations provide extreme examples, perhaps the most obvious involving plants with complex floral morphologies and pollination mechanisms that restrict pollination success to particular species of insect or bird (Gilbert 1980). The local loss of these specialized pollinators may result in a long-term collapse of populations of the associated plant species unless the pollinator population is re-established. For at least some birds this may well be favoured by corridor links.

At the other extreme are the more diffuse effects that may occur as a result of the loss of grazing animals from reserve fragments that are, by themselves, too small to support viable populations of such species. In many communities the balance between grassland and shrub or tree cover is largely controlled by the activities of herbivores, with grazing suppressing the invasion of shrubs and trees (see section 8.3.1). Clearly, if such 'keystone' grazers are lost, substantial changes to the community will occur over the longer term. Conversely, if such grazers experience a local population explosion within isolated reserves (for example, as a result of prohibitions on hunting or a series of mild winters), then the consequences of the resultant overgrazing may be disastrous for the plant community. In both these cases, by linking isolated habitat fragments corridors may facilitate movement and hence ameliorate the detrimental effects of population extremes.

Although there are one or two well-documented cases of animals using corridors for exploratory movements (for example, cougars in the Santa Ana range of California: Beier 1993), to date evidence that corridors actually play any significant role in the movement of many species across the landscape is very restricted (Hobbs 1992). Given the competing demands on the conservation dollar and the likely high cost of acquisition of corridor land, this situation needs rectification. We need to know accurately whether corridors really have a general and important role to play or whether support for them is poorly placed. Such knowledge will only come from more careful, scientifically planned empirical and observational studies (Nicholls & Margules 1991). However, even the most detailed of observational studies will not prove useful in assessing the most extreme arguments favouring corridors. These see corridors as providing the routes along which whole biota may shift their geographic distribution in response to the projected effects of global warming (see section 10.2). Given that such movements could only occur over very many generations, the corridors needed for this purpose would have to be far more than narrow ribbons of vegetation joining larger reserves together (such as along highways and stream courses in an otherwise agricultural landscape). As Simberloff *et al.* (1992) have pointed out, corridors would have to function as both a travel route and a viable habitat. For movements that potentially could span hundreds of kilometres, the latter need generates impossible conservation requirements.

Again the evidence forces the conclusion that we must preserve areas as large as possible if we are to hand on to our children natural areas with truly viable long-term futures. Corridors provide no escape from this requirement!

Evaluating the conservation value

How should scientific decisions regarding the biological selection of areas for nature reserves be made? Even when we recognize that the primary objective of a reserve system is to encompass the biological diversity of a region, one of the basic problems that still has to be confronted is just what is meant by, and hence how do we measure, biodiversity. When comparing different landscapes, communities or ecosystems, do we regard all species as being of equal importance or do we give priority to particular taxa? If the latter, on what basis is such priority given and how is this weighting factored into any particular assessment? Furthermore, how do we integrate assessments based on individual species with the distribution of naturally occurring assemblages? Finally, do we assume that the selection of each new reserve is independent of pre-existing ones or do we aim to maximize the diversity of species and assemblages across a network of reserves in any general area?

Simple indices of biodiversity

A simple approach to the problem of measuring (and implicitly ranking) biodiversity has been the adoption of indices which summarize some aspect of the 'diversity value' of individual communities into a single number. These indices range in complexity of construction, from a simple census of species to those based on the sum, product or some other combination of two or more different values. Although the number and scope of such indices is constrained only by one's imagination, they can be grouped into broad categories reflecting their disciplinary origins. ***Ecological indices*** are the best known of these categories, encompassing direct measures of *species richness* – unweighted counts of the total number of species in an area – and *diversity indices* which weight species according to their relative abundance or rarity. Examples of the latter type include the Shannon–Weaver and Simpson indices (Pielou 1977).

Taxonomic indices are a second and more recently devised method of measuring diversity. The evolutionary histories of the species in question are developed as cladistic hierarchies which are then used to estimate the relationship between individual taxa. The more distinct a particular taxon is from others in the community, the greater weight is given to its conservation. Two different indices have been developed to promote this approach. *Taxic diversity indices* examine the topology or branching pattern of cladograms representing the relationships among the species and generate a weighted index based on the number of nodes or points of separation of individual taxa from the root of the cladogram (Vane-Wright *et al.* 1991). *Phylogenetic diversity indices*, on the other hand, reflect not only the topology of the cladogram but also the extent of separation since branches diverged (branch length) (Crozier 1992; Faith 1992). This approach favours the conservation of highly contrasted taxa.

While differing in detail, both taxic and phylogenetic diversity indices favour evolutionarily distinct species, a large proportion of which will be relictual taxa that represent unsuccessful evolutionary paths. Is it appropriate that such species should be accorded favoured treatment in biodiversity assessments? Indeed, an alternative view of these same indices would be to take a forward looking evolutionary viewpoint to construct ***evolutionary indices***, by arguing for bias in favour of those lineages that show evidence of recent diversification or speciation. Turning the extreme view of Vane-Wright *et al.* (1991) on its head, this would tend to favour actively evolving species like *Rubus fruticosus*, in which more than 200 microspecies have been recorded rather than evolutionary relicts like *Welwitschia*! Potentially such dynamic taxa may offer the best opportunities for future radiative evolution and hence should be given special status.

One feature that all currently popular indices fail to assess is the diversity of interactions that occur between the members of a dynamic community. An ***interaction index*** is needed which would fill this gap by measuring a vital part of the maintenance of community structure that cannot be assessed through simple inventories of its components. Indeed, the competitive, parasitic, mutualistic and symbiotic interactions that occur between species provide the biotic driving force for adaptation and change. An index measuring these interactions would have to reflect the number of such links, their specificity and their interdependence. A good example of the complexity that can be harboured and maintained by such links is afforded by the complex of associations found in galls caused by the rust pathogen *Uromycladium tepperianum* on various species of Australian *Acacia* (Figure 9.5). In addition to the plant–pathogen association itself, these galls are utilized by a range of at least 15 species of moths, weevils, beetles and mining insects. Apparently at least some of these moths and weevils breed only in these galls (New 1984).

Complex measures of biodiversity

A major flaw with indices of the type considered above is the inability of any single measure, whether it be ecologically, taxonomically or evolutionarily based, to reflect accurately such a multi-faceted attribute as the true diversity of particular communities. However, given the inevitability of the limited supply of resources for conservation, selection has to be made between competing areas. The problem lies not in the concept of ranking but in the unthinking application of simple diversity indices as ranking tools.

Ecological measures of species richness and abundance or rarity have certainly played, and continue to play, a significant role in reserve selection. Furthermore, they will always have more relevance at the community level than will measures of taxonomic or evolutionary diversity which tend to focus on small subsets of species. But even ecological indices are individually relatively weak. This has been illustrated by Pressey & Nicholls (1989), who determined the efficiency of a range of simple indices (various assessments of rarity and representation) to generate rankings of potential conservation sites. The effect of the assessment index was dramatic. For a set of wetland sites on the north coast of New South Wales, the diversity measure was very inefficient, with over 80% of sites having to be included to achieve five

Figure 9.5. Rust gall on Acacia dealbata *caused by the pathogen* Uromycladium tepperianum. *Galls are utilized by a wide range of insects several of which are restricted to this habitat.*

representations of all plant species. In contrast, selections based on a representation measure required the inclusion of just over 20% of sites.

Increasing awareness of the inefficiency of such simplified approaches (Margules 1986; Smith & Theberge 1986) has spurred the development of alternative procedures that attempt to reflect the multiple ways in which diversity can be expressed biologically in a set of communities. One of these approaches has been the application of multivariate methods like clustering and ordination to the analysis of species pattern. In an analysis of well-studied stream systems, Faith & Norris (1989) were able to develop a predictive ordination model that related community variation in macro-invertebrate communities to physical environmental variables. Such developments hold open the promise of using relatively easily obtained physical data to devise rapid, yet efficient, means of selecting conservation sites. This would streamline procedures by reducing the initial need for exhaustive biological surveys.

McKenzie *et al.* (1989) also used pattern analysis to identify a set of sites for potential inclusion in a reserve system. Their work is of particular interest as it was based on the relatively poorly known 220,000 km^2 Nullarbor region of southern Australia (by comparison the United Kingdom covers 241,000 km^2). The biological resources of the area were largely assessed in a single year-long survey, that utilized a partially stratified sampling procedure based on surficial geological types plus an initial reconnaissance of the area. The resultant data were used to generate contours of assemblage richness which reflected the proportion of the number of species in each assemblage. Subsequent limited field verification of the predicted patterns found correlations of better than 80% in 127 of 140 comparisons. The greatest departures from predicted patterns occur at the edge of the Nullarbor where influences from other regions impinge. This procedure would seem to have considerable promise for the assessing of large, poorly known regions although further testing on geologically and biologically more variable areas needs to be made.

An alternative strategy has developed that utilizes numerical algorithms to identify the minimum subset of sites needed, within a given larger group of sites, to ensure that particular ecological attributes under consideration (for example, the number of species or community types) are represented at least once somewhere within the reserve system (Kirkpatrick 1983; Margules & Nicholls 1987, 1988). In essence this is achieved by a stepwise iterative process that takes into account the attributes of sites already selected at the start of each new cycle of selection. These procedures are clearly superior to any system of simple indices (Pressey & Nicholls 1989) and hold considerable promise for rational evaluation. Furthermore, they may be constrained to reflect any particular requirements deemed necessary to the selection process (for example, existing reserves or sites of known special biological interest). Indeed, these methods have recently been further extended to take criteria for reserve design and land suitability into account (Bedward *et al.* 1992; Nicholls & Margules 1993; Box 9.3).

Finally, it must be restated that all points of view – scientific, economic, cultural, ethical – have a legitimate right to be heard in the selection of communities for conservation. However, *none* of these have by *right* an overriding say in the final decision. Ranking of the importance of the various arguments is necessary as is ranking of the communities to be protected. By using iterative processes like those developed by Margules and his

BOX 9.3. ITERATIVE RESERVE SELECTION PROCEDURES

The selection of reserves is fraught with difficulties arising from the clash of legitimate but conflicting land use strategies. Iterative selection procedures represent a major step towards the resolution of these problems as they possess three distinct advantages over previous methods.

BOX 9.3. (*cont.*)

- First, they are *explicit* in that selection rules have to be clearly stated before the operating algorithm can be developed.
- Second, they are *efficient* in that they select the minimum areas required to achieve the objectives embodied in the selection rules. This is clearly a practical necessity in a world of increasing alternative land-use demands.
- Third, they are *flexible* in that specific requirements can be built into the selection procedures. A recent advance in this respect has been the development of algorithms that reflect the widely accepted criterion that, all other things being equal, reserves should be as large as possible or at least concentrated together.

An example of the effect of incorporating a proximity requirement into the iterative selection algorithm is provided by the work of Nicholls and Margules (1993); see Figure 9.6. They simulated the selection of sufficient 9 km^2 reserves to include at least 10% of each of 31 forest communities from the south coast of New South Wales, Australia. Grouping of cells (B) increased the area needed over that in an ungrouped situation (A) by only 7% but achieved a far more manageable distribution of reserves.

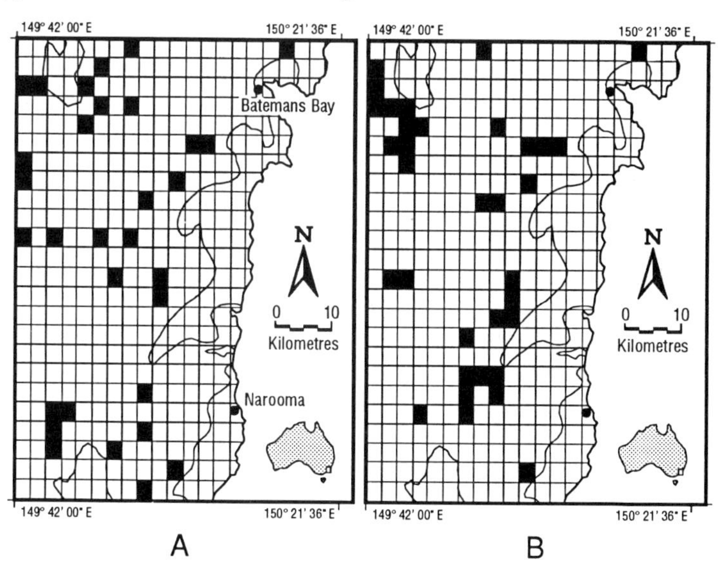

Figure 9.6. An upgraded reserve selection algorithm. Figure reprinted from Biological Conservation **64**, *A. O. Nicholls & C. R. Margules, pp. 165–9. Copyright 1993, with kind permission from Elsevier Science Ltd, The Boulevard, Langford Lane, Kidlington OX5 1GB, UK.*

colleagues it is becoming easier to maximize the achievement of competing goals and hence to preserve the widest range of biodiversity possible.

9.4 CONCLUSIONS – THE REALITY OF SAVING WHAT WE CAN

Consideration of the central issues of this chapter return us to one of the starting points of focus – the time scale of concern. It is essential that we keep in mind that the conservation of biodiversity must incorporate a time scale with two nodes of focus – the immediate and the long-term future. Obviously, many less than ideal decisions have to be made now if we are to counter the tidal wave of extinctions that threatens to engulf many biota. In doing this the gross disparity in wealth that occurs between different countries and regions of the world cannot be ignored. Conservation strategies that are appropriate for much of the developed world are often impossible luxuries in less developed areas. In the latter regions, fragmentation is almost inevitable and the ideals of the comfortable, Western conservationist must be tempered by a healthy dose of realism. The design of a biologically efficient system of reserves may be a relatively straightforward matter in regions where the distribution patterns of species are largely known. However, in much of the tropics this is not the case. With so many species threatened with extinction before they are even recognized by science, conservation efforts must single-mindedly focus on saving what is still available to be rescued. The urgency of the problem is such that choices between potential areas must be based on relatively simple and easily determined characters (for example, at the coarsest level, the number of different ecosystem or community types protected; at a finer level, the number of species of easily recognized structural or functional groups – trees, birds, mammals). After protection is gained, more exhaustive inventory surveys can then be usefully applied to generate the baseline data necessary for subsequent effective management. Indeed, the necessity of this pragmatic approach is rammed home by the sheer numbers of species in the world – estimates range between 3 and 40 million (Hodkinson & Casson 1991; May 1990, 1992), of which less than 1.4 million are formally known to science – our current inability to provide an accurate estimate of these, and the belief that the large majority occur in the tropical rainforests that are under intense threat.

Habitat fragmentation is also inevitable in economically richer parts of the world where even the possibility of setting aside new, large reserves is usually precluded by widespread land clearing and habitat destruction that occurred decades or centuries ago. Again, pragmatism must temper idealism. However, in this process continuing attention must be given to options that maximize the long-term viability of reserves.

Ultimately, however, unless we are able to ensure the protection of areas that are large enough to maintain the dynamic forces of species interaction, much of what may currently appear to have been saved may still be lost. For some communities, such areas may be quite small. In others, for example communities that have evolved in tandem with large grazing herbivores (for example, the veld of southern Africa), reserves will have to be large enough to allow for both the grazing pressures exerted by herds of these species and their natural migratory movements. The only possible alternative to such an approach is intensive, continuing monitoring and management. Such a strategy may be sufficient in a few situations but is unlikely ever to be a generally adequate alternative.

10

Managing plant community reserves

10.1 INTRODUCTION – IS MANAGEMENT NECESSARY?

In an ideal world, management would be an unnecessary aspect of most conservation strategies at the ecosystem level. Reserves would be large enough to absorb the effects of disturbances like storms and wildfires as part of a natural cycle of growth, decay and rejuvenation; while humans would be sufficiently in harmony with their surroundings to minimize their detrimental effects. In reality this is very rarely the case and effective long-term management is essential if the value of existing reserves, and the gains expected through the selection and establishment of new ones, are not to be eroded and eventually lost through the onslaught of a battery of degrading influences.

The importance of management is emphasized by the knowledge that the majority of reserves in at least the more developed parts of the world are less than 1000 hectares in size (see section 9.3.1). A sizeable fraction are considerably smaller. For example, even in a country as big and relatively unpopulated as Australia, 40% of all reserves are less than 100 hectares in size. Whether reserves in this size range have any long-term conservation value is debatable. Certainly, however, any continuing value they have is inextricably linked with their ability to maintain relatively intact ecosystems. Even where the sole function of reserve establishment is the protection of one particular species, long-term success is unlikely unless the community in which such a species naturally occurs is also maintained.

Few natural communities are 'compact' enough to fit *fully* and function within 100, 1000 or even 10,000 hectares. As a consequence, if the substantial fraction of the world's reserve system that falls within this size range is to have a long-term future, management is essential. Once the concept of management is accepted, questions immediately arise as to the focus of such management. In this chapter we explore the potential consequences of differences in the physical size and shape of individual reserves for the biological integrity of the communities they are intended to protect (see section 10.3). Unfortunately, however, threats to the conservation of biodiversity are not restricted to the alienation of land for alternative purposes. Changes abound within individual reserves and indicators are needed against which management practices can be assessed. This, then, leads to a consideration of the importance of individual species to the 'health' of a reserve's plant communities (see section 10.4.1). Fire, invasive species, grazing and disease are among the most important contributors to such changes and an understanding of their influence is essential in developing management protocols (see section 10.4.2).

First, though, we turn to a consideration of the most widespread and pervasive threat to the maintenance of existing reserves – the likelihood of world-wide changes in climate resulting from anthropogenic alterations in the underlying physical and chemical environment. Global climate change has many aspects- rising CO_2 levels, increasing temperatures, changing rainfall patterns, volumes and intensities, increasing UV penetration of the atmosphere. These changes have both direct and indirect effects on all plant species, and hence on the health and resilience of whole communities. While the causes of such changes are far beyond the control of individual reserve managers, the consequences are likely to be far more immediate and need consideration.

10.2 THE CONSEQUENCES OF GLOBAL CLIMATE CHANGE

Geology and climate are the major factors of the physical environment that inevitably determine the potential floristic composition of any given site. In the majority of cases, geological conditions remain virtually constant. In contrast, climate is far more labile and any sustained perturbation will alter successional dynamics that, because of the novel combination and intensity of constraints, may be quite different in both direction and intensity to those previously observed (Tilman 1993).

On a global scale, the potential effects of increasing temperatures may be at least partially predictable from extensive palaeopalynological studies that have documented the ebb and flow of the predominant vegetation across Europe and North America following the last glaciation. Thus the northern limit of all tree species appears to be governed by the 10 °C mean July isotherm. However, at these latitudes the boreal forests are almost exclusively coniferous as most deciduous trees also have an absolute minimum temperature requirement of $-40\,^{\circ}C$ (Woodward 1987). The boreal forests are somewhat unusual in that they are relatively unfragmented and their northern limits are largely unconstrained by alienation of surrounding lands. Potentially, then, increasing global temperatures will shift the $10\,^{\circ}C$ mean July isotherm north allowing an expansion of the forest into the tundra to compensate for concomitant losses along the southern flanks of its range. Elsewhere in the northern temperate zone, alienation of land for agricultural purposes precludes the possibility of such large-scale, natural floristic migrations. However, as the rate of climatic change is expected to be much greater in the future than in the past (Peters 1988), human intervention (for example, through the deliberate movement of propagules) may in any case be the only feasible means of response.

At a regional level, the end points of successional processes put in train by changing climatic patterns are likely to be far less predictable. At this level of scale communities tend to be smaller, more closely adapted to particular environmental combinations and, as a consequence, far more fragmented. This generates a complex patchwork in which there will be winners and losers among both the individual species and the communities of a region. Indeed, as Tilman has pointed out (1993), many existing species that are best suited to particular combinations of environmental parameters are currently likely to occur far away from where those conditions may come into existence in the future.

If this process of changing environmental conditions involves the loss of the existing environment, then without an effective means of crossing inhospitable terrain to a new favourable location, species and potentially whole communities are doomed to extinction. On the other hand, species and communities whose environment is simply extended would gain. Finally, new environments that open up well away from similar pre-existing environments will largely be colonized by relatively poorly adapted species from surrounding ecosystems.

These patterns are illustrated in Figure 10.1 where hypothetical climatic changes lead to rising temperatures and increasing aridity, cause an altitudinal migration in particular plant communities. For the subalpine forest on peak A this simply reduces the area of the forest. In contrast, however, for a similar community on peak B these changes are disastrous. There, the same climatic change is sufficient to move the subalpine–montane forest vegetation boundary above the altitude of the peak itself. In these circumstances, the entire subalpine forest community may be swept away, including any rare endemics.

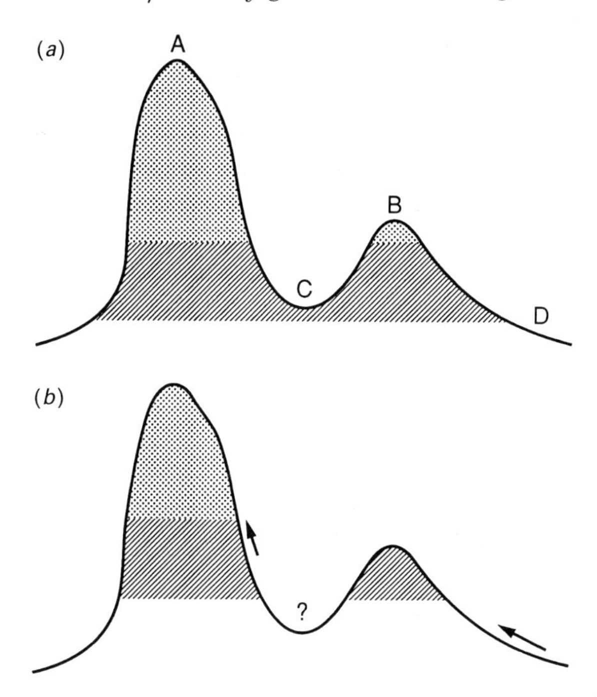

Figure 10.1. Changing patterns of vegetation distribution across a mountain range associated with a hypothetical increase in mean annual temperature. (a) *Pre-existing distribution;* (b) *changed distribution following temperature increases leading to loss of subalpine forests on peak B and the opening of a new environment in valley C that is physically separated from existing communities of a similar type D.*

An example of the consequences that such gross climatic changes may have is seen in the distribution of the mistletoe *Amyema dilatipes*, a local endemic that is confined to subalpine communities on Mt Wilhelm, Mt Kerigomna and Mt Michael in the Eastern Highlands District of Papua New Guinea (Barlow 1991; cf. Peak A, Figure 10.1). Putative hybrids between this species and lower elevation mistletoes (*A. pachypus* and *A. wichmannii*) are known from two slightly lower mountains in the region (cf. Peak B, Figure 10.1). The simplest explanation for the existence of these hybrids without their higher altitude parent is that they are biological memories of a time in the most recent cool phase 16,000–18,000 years ago when *A. dilatipes* had a wider distribution. If during that time, *A. dilatipes* had been restricted to the two lower peaks (a degree of narrow endemism not uncommon among tropical plants and animals; Wilson 1993), it would most probably be extinct by now.

The potential extent of the problem of extinction through climatic displacement has been demonstrated by Maddox and Morse (cited in Pitelka 1993). Making the simplifying assumption that the current distributions of 14,000 North American native plants are determined solely by climate, these workers have shown that for 7.6% of the species a simple 3 °C rise in average annual temperature results in a new predicted range that is totally discontinuous with the existing range. Eighty-three per cent of these species are already defined as rare by the US Nature Conservancy and most are particularly vulnerable to the consequences of global warming.

Changes in temperature are not the sole effect of global climate change. Rather, changes in temperature, moisture regimes and atmospheric chemistry (rising CO_2, etc.) will produce interactions with resource availability and alter disturbance regimes (e.g. increasing fire frequencies) to generate a complex array of ways in which global change may differentially affect individual plants and communities. Rochefort and Woodward (1992) have modelled the effect of such a complex set of interactions (involving changes in temperature, CO_2 concentration and precipitation). Their model suggested the possibility of an increase in diversity in a third of the floristic regions of the world and relatively little change elsewhere. However, within such broad changes individual species composition may change drastically.

The spectre of global change is an international problem that has forced the implementation of a range of policies at the highest political levels. In addition, however, is there any practical, regionally based management response to these potential changes? In essence, no! In the short term, global change seems certain to cause a decline in total species diversity (Hobbie *et al.* 1993). Individual species that are threatened may be saved through ex situ measures, but within particular ecosystems are doomed as the niche they currently occupy ceases to exist. At the community level, the picture is no less bleak. The possibility of moving whole communities and re-establishing them in new areas raises a host of ethical questions: for example, which communities are to be extinguished to make way for those that are moved? More prosaically, such wholesale rehabilitations have been tried too infrequently to ensure success. Currently, then, preparation for the potentially enormous changes that could ensue from global climate change must largely rest on developing a better understanding of the interactions that determine successional dynamics in any given community.

10.3 MAINTAINING THE PHYSICAL INTEGRITY OF RESERVES

Reserves do not occur in a vacuum. Increasingly they are isolated in landscapes dominated by distinctly different land uses – intensive cropping, grazing or forestry activities, urban development, extractive mining. This juxtaposition may lead to substantial changes in the hydrology of an area, to marked alterations in the local microclimate of the reserve itself, or increase the entry of invasive organisms of all sorts or the damage due to human interference.

For rainforest remnants in the Amazon (Lovejoy *et al.* 1984), or the tropical forests surrounding Barro Colorado Island in Panama (Leigh *et al.* 1993), increasing penetration of light and drying winds trigger changes that result in death of some tree species and the prolific growth of vines and other secondary rainforest species. In other areas illicit grazing, or the removal of timber for shelter and fuel may seriously degrade ecosystems. Yet elsewhere, highly developed agricultural lands frequently abut reserve boundaries leading to problems of invasive weeds, herbicide spray drift or eutrophication of watercourses. These situations exemplify the two major areas of any reserve – the *central core* and the *buffer zone*. The central core is the internal area of a park that is relatively protected from the outside world. In contrast, the buffer zone stands between the central core and the surrounding environment and is particularly exposed to many degrading influences. In many cases, the buffer zone lies within the reserve boundary and is hence a legally protected part of the reserve. Increasingly, however, attempts are being made to increase the effective size of reserves by acquiring buffer zones outside the reserve perimeter. In such areas limited harvesting of some species may be permitted.

The depth to which external influences penetrate into reserves varies with aspect and criterion used (Matlack 1993) but may range from 15 metres (Ranney *et al.* 1981) to 5 kilometres (Janzen 1986). Indeed, in some cases external anthropogenic changes of this type may threaten the immediate existence of entire communities. Thus Lake Toolibin, the last remaining perched freshwater wetland in Western Australia, is under immediate danger of destruction as a consequence of the increasing salinization of the entire catchment in which it is situated (Froend *et al.* 1987). Over the years, large-scale clearance of native vegetation has destroyed all other freshwater wetlands in this area.

Obviously, not all threats are so all-embracing nor are all areas of any given reserve equally at risk. Typically, border areas between reserves and surrounding land uses are under particular pressure. For any given reserve of regular shape and depth of buffer zone, the absolute area subject to edge effects increases with reserve size. However, the detrimental consequences of such effects are likely to be felt most strongly in small or oddly shaped reserves where a relatively higher proportion of the total area is potentially compromised (Box 10.1).

Similarly, not all reserves are equally vulnerable to edge effects. Reserves that incorporate whole communities, which have evolved over time as a defined patch, are likely to be more resilient than equivalent sized areas recently carved out of much larger, continuous stands of vegetation. Communities that are naturally delimited by sharp geological boundaries, for example, those on serpentine outcrops, or small patches of rainforest that develop on isolated volcanic extrusions in sedimentary rock in southern Queensland, fall into the former category. These communities have natural edges that have developed over time and have far more in common with real islands than do recently created habitat fragments where the raw edges of a mature community are exposed to dramatic environmental change.

Further complications may arise in situations where changes occurring

BOX 10.1. EDGE EFFECTS AND THE SIZE AND SHAPE OF RESERVES

Both the extent of spatial penetration of external effects (i.e. the width of the buffer zone) and the shape of the reserve itself have a marked effect on the total area of the buffer zone (Figure 10.2). For an idealized, circularly-shaped reserve, the 'real' area of a nominally 100 ha reserve with a 1 km wide buffer is only 68 ha. Such a reserve would have to exceed 5000 ha in area before 95% was within the central core. If the width of the buffer zone is increased to 2 km, the situation is far worse with only 42 ha of a 100 ha reserve falling within the central core. In a reserve of 5000 ha, 10% of the total area is still in the buffer zone.

The shape of the reserve is also of considerable significance. Circular and square-shaped parks differ little in the area lost to buffer zones. In contrast, a disproportionate fraction of rectangular reserves is 'lost' to buffer zones. With a 1 km buffer zone, only just over half of a 100 ha 2 × 1 rectangular reserve is within the central core. For a 4 × 1 rectangular reserve this figure falls to just 36%. Again as reserve sizes increase logarithmically the proportion of the reserve within the core region asymptotes to 100%. Even then, though, until reserve sizes exceed 10,000 and 30,000 ha, respectively, more than 5% of the area of these parks lies within the buffer zone.

BOX 10.1. (*cont.*)

These two features – size and shape – can be brought together in an area to perimeter length ratio (p/a) which provides a good measure of the extent to which the interior of most reserves is exposed to the surrounding environment (Diamond & May 1976; Schonewald-Cox & Bayless 1986).

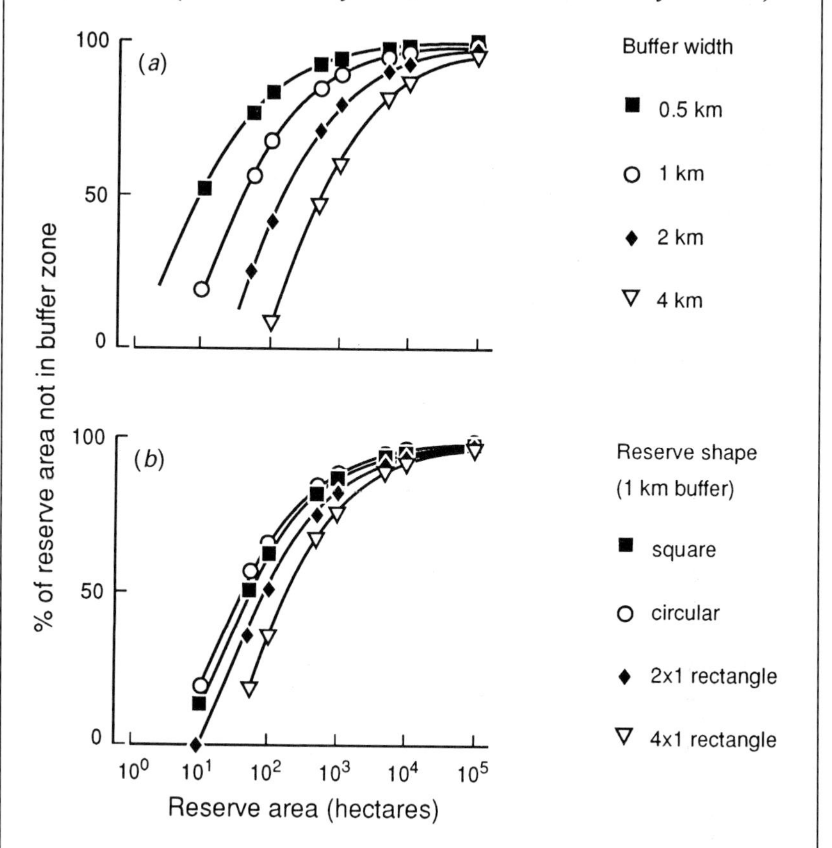

Figure 10.2. Effects of (a) different buffer widths and (b) reserve shapes on the percentage of a reserve within the central core.

within boundary zones, or in areas immediately outside reserves, differentially favour individual species. Such increases may be apparently benign as is the case in Amazonian rainforest patches where secondary vegetation developing along reserve edges favours an increase in the number and diversity of butterfly species (Lovejoy *et al.* 1984). However, in other situations changes have the potential to be extremely damaging. Thus along the boundary of the Olympic National Park, USA elk move out of the park into surrounding clear-cut forest patches to forage (Schonewald-Cox & Bayless 1986). If this leads to a long-term increase in the elk population the vegetation of the park itself could suffer.

The latter possibility clearly indicates the importance of management strategies that reduce the consequences of edge effects on the integrity of reserves. Other than the obvious ploy of increasing reserve size, this may be achieved through (i) extending the buffer zone beyond the formal boundaries of the reserve itself (thus making the reserve informally bigger); and (ii) reducing the degree of irregularity of the reserve outline. As seen by the elk example above, extending the buffer zone may have some problems. However, where reserves are surrounded by compatible land use practices, for example limited extractive forestry, these areas provide space for wide-ranging movement of wildlife and increase the total population size of most if not all of the inhabitants of the reserve. In addition, they are likely to reduce the influx of invasive organisms into the reserve itself.

10.4 MAINTAINING THE BIOLOGICAL INTEGRITY OF RESERVES

10.4.1 Biotic monitors of ecosystem health

How important are individual species to the health of communities? Are all species essentially equal, such that the loss of any one is of no more or less concern than the loss of any other species? Or are species unequal in their relative contribution to the long-term functioning of communities and their associated biogeochemical processes? If the latter view is correct, are some species in a community essentially redundant, being replaceable by other, co-occurring species? Or do they all make some significant contribution, however small? These sorts of questions have led to two extreme views of the role of species in ecosystem function. The first of these is Ehrlich and Ehrlich's (1981) 'rivet' hypothesis which largely sees each species as contributing uniquely to the integrity of communities in at least some small way. The alternative extreme, the 'redundant species' hypothesis (Lawton & Brown 1993) largely sees species number and identity as irrelevant, arguing that the essential ecological processes of communities can be maintained by a small subset of the species that are present. However, it should be noted that even the rivet hypothesis does concede the possibility that ecosystems 'tend to have redundant subsystems' (Ehrlich & Ehrlich 1981).

Ecological redundancy

A redundant species is one whose position, function or contribution to a community may be fulfilled by other co-occurring species. In its simplest form this concept may be illustrated by a process of removing species singly from a community and observing the consequences. Where the productivity of the community is affected only for a short time, then the removed species may be classed as redundant (Lawton & Brown 1993). However, the problem with using this principle as a guide in formulating management decisions is the impossibility of recognizing *a priori* which taxa are redundant.

Redundancy is not likely to be an immutable character of a species which is instantly recognizable. Rather, the contribution of individual species to communities may well vary depending upon environmental conditions, upon the temporal and spatial scale of assessment, and the identity of other community members. On a global scale it may be possible for the biogeochemical cycles of the planet to be maintained by fewer species than currently exist. Such a continuation of these basic cycles apparently occurred during and after the mass extinctions of the past (Lawton & Brown 1993). However, as these authors stressed there are major differences in relative rates of extinction between the past and present. Furthermore, we have no way of knowing what degree of disruption occurred to individual ecosystems during the mass extinctions.

At the scale of individual communities, two recent studies of the link between species richness and ecosystem function clearly suggest that reductions in biodiversity result in decreased stability and productivity. In one experiment Tilman and Downing (1994) compared both the ability of the vegetation on 207 grassland plots to resist the effects of a two-year drought and its resilience or rate of return of the community to pre-existing conditions once the drought had broken. Both these parameters showed an increasing but non-linear association with species richness (Figure 10.3) such that the incremental loss of species led to communities that were progressively less able to cope with the effects of drought. Naeem *et al.* (1994), on the other hand, used a quite different strategy – deliberately constructing multi-trophic level communities containing a total of 9, 15 or 31 species – to show that reduced levels of biodiversity may alter plant productivity, community respiration, decomposition and other biogeochemical-related processes of ecosystems.

In a conservation context the concept of redundancy is potentially dangerous. It seems to suggest that we need not worry about many species

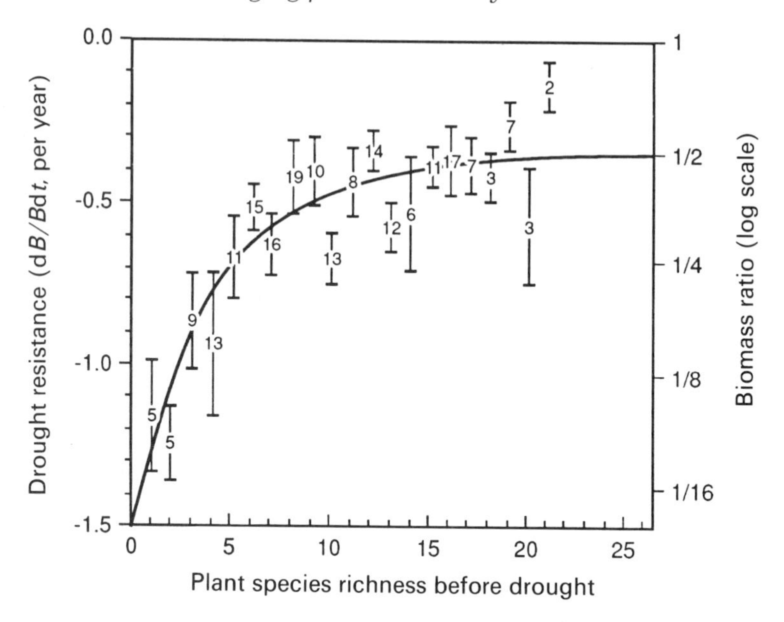

Figure 10.3. Relationship between grassland plots and plant species richness preceding a severe drought. Mean, standard error and number of plots with a given species richness are shown. Drought resistance was measured as 0.5([biomass$_{1988}$/biomass$_{1986}$]), where biomass$_{1988}$ was the total biomass of the plot at the height of the drought in 1988 and biomass$_{1986}$ was that for the period immediately preceding the drought. Biomass ratio (biomass$_{1988}$/biomass$_{1986}$; right-hand scale) shows the proportionate decrease in plant biomass. (Tilman & Downing 1994; reprinted with permission from Nature **367**: *363–5. Copyright 1994, Macmillan Magazines Limited.)*

extinctions because overall community processes may be little affected. However, in the overall struggle to maintain biodiversity there is no guarantee that the first wave of species lost from communities will necessarily be those that have co-occurring analogues.

Keystone species

The opposite concept to species redundancy is that of the 'keystone' species – a species whose presence is crucial to the maintenance of the structure and function of the community. Since its introduction to describe the role of predatory starfish in the intertidal zone (Paine 1966), the term keystone species has been extended to encompass a very wide range of species operating in a variety of ways at many different trophic levels (Bond 1993; Mills *et al.* 1993). Predators, prey, herbivores, pathogens, plants, pollinators and animals that alter the physical nature of the environment through engineering works have all been classed as keystone species.

The widespread use of the keystone epithet has been beneficial in focusing attention on the differing strengths of the interactions occurring between community members (Mills *et al.* 1993). Furthermore, the concept itself gives focus to what is almost certain to be true of most communities – that just one or a few species have a disproportionate and immediate effect on the community while the majority of others are relatively inert. However, because there is no obvious way by which keystone species can be recognized *a priori* (Bond 1993; Lawton & Brown 1993; Mills *et al.* 1993), their use in the development of conservation strategies will tend to be limited. Indeed, perhaps because of the cachet associated with the title, there seems to have been a tendency to elevate species to keystone status with little hard evidence of their pivotal effects on community structure and dynamics. Such species may be important, or even crucial to the survival of a few others, but whether their loss would trigger a cascade of effects on other members of an ecosystem (Bond 1993) is more doubtful.

Ecosystem management through species conservation

We advocate that communities should be the main focus of conservation efforts. This is the only way in which unique assemblages of particular species and the seemingly endless array of close symbiotic, mutualistic and parasitic associations can be maintained. In addition, such a 'catch-all' approach helps protect those elements of biodiversity that have still to be recognized and catalogued. This may not be of great concern if one takes a narrow view and concentrates on the conservation of specific groups or types of well-documented organisms: for example, large vertebrates. However, in many environments the level of knowledge and documentation is so incomplete that many vascular plants are likely to remain undiscovered let alone the myriad of insects and other invertebrates, fungi, bacteria and so forth that remain undescribed.

Despite a community approach, individual species still retain a position of considerable importance (see Chapter 6). Any management strategy must have standards against which to measure the performance of the community – that is, the extent to which it achieves or continues to achieve our preconceived notions of what represents a healthy community. In many situations, these standards or indicators will be individual species or groups of species, identified for one reason or another as being of importance. Obvious candidates in this regard are keystone species. However, because there is often no simple way to recognize the contribution particular species make to community structure, management guided by the health and well-being of keystone species may be restricted to particular, clear-cut situations involving large herbivores (e.g. elephants) or obviously important plants like figs (Terborgh 1986).

Other species that could be, and have been, used as environmental monitors are any rare or endangered taxa occurring in the area. This may be particularly appropriate for communities: (i) in the temperate zone where species diversity is generally lower than in tropical areas; (ii) that are themselves strictly limited in area (for example, many of the world's alpine communities); and/or (iii) that contain a high frequency of local endemics. Ironically, the incidence of particular invasive aliens that pose a potential or real threat to the integrity of a reserve may well provide the clearest indication of the extent and rate at which community degradation is occurring.

Depending on the ultimate conservation aims for particular reserves, such an approach of *ecosystem management through species conservation* can be supplemented by management practices deliberately aimed at the community level: for example, the judicious use of fire to ensure that certain percentages of the conserved area are at different stages of succession or are representative of different community types (but see the following section).

10.4.2 Abiotic and biotic threats

Natural disturbance is an important component in the dynamics of most plant communities. Storms that open gaps in a forest canopy, fires, or heavy rain that causes landslips or changes in watercourses, all 'reset' the ecological clock of a community providing opportunities for natural regenerative processes to begin. Ideally, conservation reserves should be big enough to allow natural disturbance regimes to operate and, hence, to support a mosaic of vegetation in different stages of community maturation. Conversely, however, excessive disturbance may alter communities by favouring a subset of their component species or by promoting the entry of alien species.

A wide range of physical (fire, flood, storm) and biological factors (grazing, disease) disturb communities. Here we consider the community-level consequences of fire, the most pervasive and potentially destructive of abiotic disturbances, and invasive species.

Fire

Fire is an integral part of the environment of very many plant and animal communities. It simultaneously occupies the role of both a significant abiotic threat and a feature vital to the maintenance of community diversity. Controlled and used intelligently, fire is an effective management tool; allowed to run free or used for purposes other than nature conservation, it poses a significant threat to the structure and composition of nature reserves.

The effects of fire depend on the frequency, intensity, season of burning and type of fire – that is, on the fire regime (Gill & Bradstock 1994). Many plants and animals that inhabit fire-prone environments have evolved a range of life-history attributes which enable them to thrive under particular fire regimes. However, such attributes reflect the long-term evolutionary effects of *particular* fire regimes and may provide little or no protection when conditions change and the frequency, intensity or timing of fires alters. Indeed, some species may rapidly become locally extinct if successive fires occur more frequently than the long-term pattern to which the species is adapted.

Particularly good examples are found among seeders (see Box 10.2) that

BOX 10.2. SEEDERS AND SPROUTERS – ALTERNATIVES STRATEGIES IN THE RESPONSE OF PLANTS TO FIRE

Two widespread, contrasting (and somewhat simplified) strategies adopted by plants in response to fire are exemplified by 'sprouters' and 'seeders' (Gill 1981). The former are species that are relatively tolerant of fire, regenerating from protected aerial shoots or from root systems. In contrast, the latter species are fire sensitive being readily killed by the destruction of exposed buds. These seeder species rely on seed held in the soil or on the plant for subsequent regeneration. Both these groups may be detrimentally affected by fire. However, seeder species are especially vulnerable to patterns of increasing fire frequency.

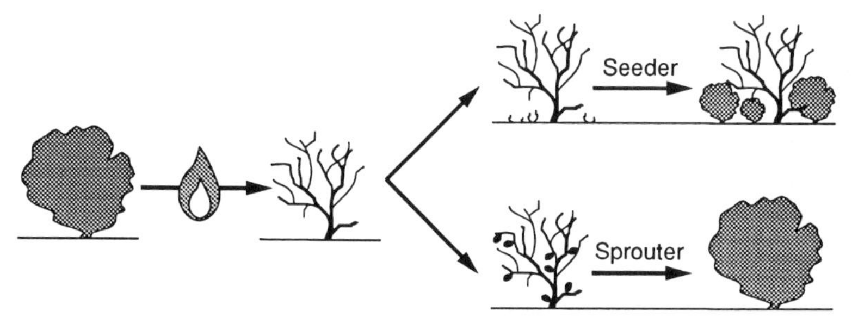

Figure 10.4. The response of seeders and sprouters to fire.

produce only short-lived seeds (Noble & Slatyer 1981). Successive fires occurring before plants reach reproductive maturity can eliminate them from the community. These effects are rarely restricted to just one or two species. Rather, changes in fire regimes particularly increasing frequency and intensity produce rapid changes in the composition of entire communities. This was demonstrated by Jackson (1968), who emphasized the importance of the time lapse between successive fires. Under 'normal' conditions a burnt community will regenerate itself. However, if the regenerating community is burnt again before it has had the opportunity to replenish seed reserves, then an entirely new suite of species will prosper. These fire-maintained communities are to be found in many parts of the world and include grasslands (for example, the prairies of North America: Kucera & Koelling 1964; the veld of southern Africa: Cowling *et al.* 1986), heathlands, and even some sclerophyll forest where fire-sensitive rainforest species are held in check by infrequent fires.

Fire frequency is not the only factor determining the diversity and composition of individual communities. Fire intensity is also of great importance, although again the consequences depend greatly on the type of community burnt. A good contrast is provided by the different responses of the coniferous forests of the boreal zone and many eucalypt-based sclerophyllous forests. In the former case, the dominant species are fire-sensitive but, paradoxically, rely on periodic fires sweeping through an area to provide the conditions needed for establishment of the next generation. In the latter case, fires of similar intensity frequently defoliate without killing the dominant trees. In fact, in such eucalypt communities intense fires tend to increase biotic and physical complexity (Catling 1994). Low-intensity fires, on the other hand, may reduce complexity by killing understorey shrubs without inducing seed germination. This is especially the case for species that rely on a hard-coated seed strategy for protection.

Similarly, a change in the season of burning can alter the composition and structure of plant communities. In southern Australia, spring fires tend to enhance shrub regeneration while autumn fires favour herbaceous species (Baird 1977; Christensen *et al.* 1981). In northern Australia fires are more common late in the dry season. However, those occurring after plant growth has resumed in the early part of the wet season may be responsible for marked changes in species composition. Thus grasses with shallow roots like *Themeda triandra* may suffer substantial mortality while others with deep rootstocks and protected apices (e.g. *Chrysopogon fallax*) are unaffected (Lazarides *et al.* 1965).

The effects of fire on plant communities do not stop at the immediate interaction. More subtle effects may eventuate from the indirect consequences of fire on the fauna (vertebrate and invertebrate) and microflora of a region. Intense fires may temporarily reduce the size of populations of many of these species leading to either increased or decreased soil microbial activity (Warcup 1981), changes in grazing patterns (Catling 1991) and changes in soil nutrient status (Humphreys & Craig 1981). The ultimate consequences of such effects on co-occurring plant communities is largely unknown. Further interactions may also occur between communities, invading alien species and fire. Many weeds manage to establish a foothold following a severe fire (see the following section). These may either lead to marked increases in fuel loads and hence an increased probability of more frequent, high-intensity fires (Christensen & Burrows 1986); or to a reduction in fires as the alien species hampers fire penetration (for example, *Lantana camara* on rainforest edges: Stocker & Mott 1981).

Fire is the only major environmental factor over which some human control can be exercised (Good 1981). As a consequence, it is a common management tool in regions of the globe where fires occur naturally. Unfortunately, though, no management plan, whether it incorporates a planned fire regime or not, can satisfy all conservation objectives and reserve users. Fuel reduction strategies achieved by the relatively frequent use of low-intensity burns may be necessary to protect life and property in areas adjacent to major urban developments, prime agricultural land, or in multiple-use parks and reserves. However, as detailed above, such prescribed burning rarely if ever mimics the natural fire regime of an area and will usually result in changes in the structure and diversity of both plant and animal communities.

Where reserves are remote from human habitation, fire regimes in keeping with those of the immediate evolutionary past may be possible. Even then the relatively limited size of most reserves restricts the available options. A mosaic of vegetation of differing age and structure is often advanced as the ideal way of maintaining the greatest possible plant and animal diversity. However, because of the unpredictable nature of fires this is often difficult to achieve even in the largest parks. Indeed, Good (1981) argued that reserve management should aim to maintain the majority of conservation areas in a mature condition as this retains the greatest range of management options, particularly in situations where accidental fires are commonplace. In small parks (the large majority; see Chapter 9) a mosaic approach is virtually impossible as the entire area may be burnt within a very short time.

The conflicting demands of conservation, recreation, maintenance of water catchment areas and limited extractive industries inevitably mean that all fire management plans will be compromises. What is essential, is to ensure the imposition of the minimum level of external interference necessary to achieve the desired aims. For individual parks this will mean a mixed strategy of controlled burning and wildfire control. The extent of the former may be influenced by the desire to protect specific rare plant or animal species or communities by deliberately excluding fire. Indeed, in many reserves the desire to protect particular rare species may be the major biological focus of conservation efforts. Gill & Bradstock (1994) have suggested that as a general rule-of-thumb for plants, reserves should be managed for the most sensitive species with the proviso that no reserve should be completely burnt at any one time. Alternatively, suites of fire-sensitive species can be used as indicators of the likely vulnerability of particular communities to fire (Gill & Nicholls 1989). However, an exclusive focus on a few rare, endangered or fire-sensitive species may ultimately place the whole reserve at risk if excessive fuel levels are allowed to develop. Compromises are inevitable and essential.

Invasive species

Invasive plants

Invasive plant species constitute a serious threat to the integrity and survival of many communities. Such species may be herbaceous or woody, annual or long-lived perennials. They may be exotic introductions from other parts of the globe, or native species that have been carried by humans into previously unexposed communities in the same general region.

In considering the movement of organisms into reserve areas and the threat they pose, it is important to make a distinction between presence and invasiveness. Overall, the extent of global movement of plants has, and continues to be, enormous. Exotic plants have been carried by humans into virtually all terrestrial environments with only the most extreme apparently suffering no introductions (for example, the Special Protected Areas of the maritime Antarctic: Usher & Edwards 1986). As a percentage of the total flora the extent of introductions varies widely from region to region, ranging from less than 10% (Java) to almost 50% of the total flora (New Zealand: Heywood 1989). Within regions some habitat types seem to be more susceptible to introduction than others, with island communities suffering the highest accretion of exotics while desert and savanna communities suffer the least (Usher 1988; Lonsdale 1992).

Exotics vary widely in their invasive capacity, aggressiveness and extent of interaction with other plants and processes within different ecosystems

(Heywood 1989). As a consequence, absolute numbers of species are a relatively poor indicator of the degree to which the vegetation of an area is likely to be affected by introductions. Indeed, the invasive species that threaten individual plant communities are only a small subset of all the exotic species that arrive in the area. Thus, of 109 alien species occurring in the Cape fynbos system (Wells *et al.* 1983), only 12 are responsible for significant community modification (Richardson *et al.* 1992; Moll & Trinder-Smith 1992). Even more dramatic are the several instances of communities that are being seriously degraded by the aggressive invasion of just one or two species. Examples of this type of situation include *Melaleuca quinquenervia* in the Florida Everglades (Ewel 1986), *Lonicera japonica* and *Hedera helix* on Theodore Roosevelt Island, USA (Thomas 1980), and *Thunbergia grandiflora* and *Cryptostegia grandiflora* in lowland forest in northern and central Queensland (Humphries *et al.* 1991).

What makes a reserve prone to invasion, or conversely, some plant species particularly efficient colonizers is very poorly understood. Indeed, despite decades of concerted effort, weed scientists are still unable to predict *a priori* which taxa will be highly invasive, which will merge unobtrusively into a community and which will struggle to survive. Criteria which seem highly relevant for one particular taxon (for example, annual habit with enormous seed production; *Echium plantagineum*) may be countered by other examples of perennials which rely exclusively on vegetative reproduction (e.g. *Oxalis pes-caprae*). Similarly, an understanding of the factors that make particular communities vulnerable to invasion is limited although disturbance of one kind or another seems critical. In an extensive analysis of the numbers of introduced species occurring in 41 southern African and 21 North American nature reserves, Macdonald *et al.* (1989) found that the only characteristic that was consistently correlated with the number of introduced plants was the number of visitors to the reserve. A similar correlation was found in a New Zealand survey of 234 small (<500 ha) reserves (Timmins & Williams 1991), where the principal characteristics influencing the number of problem weeds in reserves were proximity to urban areas, distance from roads and railway lines and recreational use. Reserve shape played a more minor role.

Simply by their presence, exotic species affect the diversity of communities in which they occur. In many cases, however, the consequences of invasion may be far reaching, affecting a wide range of ecosystem functions. These include the following.

1. Alterations in geomorphological processes. These may lead to long-term changes in stream flow, sand dune configuration or even the

persistence of tidal mud flats (Macdonald *et al.* 1989). A good example is seen in the changes that have occurred to sand dune systems in northern California and Oregon following invasion by the European grass *Ammophila arenaria* (Barbour & Johnson 1977).

2. Changes in nutrient cycles. The nitrogen status of low fertility soils may increase substantially as a result of invasion by plants which actively fix nitrogen in association with symbiotic organisms. Such a situation is occurring in the Hawaii Volcanoes National Park where the invasive actinorrhizal nitrogen-fixer *Myrica faya* is quadrupling nitrogen inputs into primary successional ecosystems (see Chapter 8; Vitousek 1990).
3. Changes in fire regimes. Invasive species, particularly grasses, often substantially increase the quantity and spatial continuity of fuel, leading to changes in the seasonal occurrence of fires (Macdonald *et al.* 1986) and increases in their number, intensity and spread (Humphries *et al.* 1991). This has been implicated in changing species composition in nature reserves in North and South America, Africa and Australia (Macdonald *et al.* 1988) and may lead to increases in fire intensity around rainforest edges and hence increasing fragmentation. Burdon & Chilvers (1994) have suggested that the invasion of eucalypt forest by *Pinus radiata* may have long-term fire implications (Figure 10.5).
4. Reductions in seedling recruitment. The direct effect of invasive plants on the recruitment of indigenous species is well illustrated by species that form dense monospecific stands that suppress existing plants and prevent recruitment. The smothering effects of vines, honeysuckles and ivies have been recognized in both tropical and temperate ecosystems (Humphries *et al.* 1991); while the dense shading effects of species like *Rhododendron ponticum* (Usher 1988) and *Mimosa pigra* (Braithwaite *et al.* 1989) have been documented in Ireland and northern Australia, respectively.

Invasive animals

To this point we have only considered the consequences of invasive vascular plants. However, as previously documented (see section 8.3.1, p.197), many studies attest to the ways in which grazing animals contribute to the vegetational structure and diversity of the communities of which they are naturally part. In invasive situations these effects are often accentuated as a consequence of the invasion of communities devoid of large mammalian grazers (for example, many island communities invaded by goats: Coblentz 1978; Hamann 1979); or of communities previously exposed to quite

Figure 10.5. Invasive spread of Pinus radiata *into eucalypt dry sclereophyll forest in southeast Australia (reprinted with permission from J. J. Burdon & G. A. Chilvers,* Oecologia **97***: 419–23. 1994 © Springer-Verlag).*

different grazing techniques (pulling and tearing versus cutting). Furthermore, invasions that present management problems are invariably coupled with excessive population sizes. The resultant intense grazing may lead to a combination of biological and physical effects, including: (i) a decline or loss of palatable species, (ii) rapid increases in unpalatable ones; (iii) increased soil disturbance; (iv) changes in soil structure; (v) increased erosion; and (vi) damage to watercourses. Examples of these processes are associated with the introduction and spread of rabbits, goats, water buffalo, deer and pigs in many parts of the world. Where uncontrolled these processes may completely destroy the existing community. Moreover, intense grazing may exacerbate the rate of invasion of exotic plants or lead to conditions that alter existing fire regimes. Some of the subtleties of the interactions that occur between grazing and plant community composition were considered in Chapter 8 and are not revisited here.

Invasive pathogens

The threat posed by introduced pathogens is also of considerable importance. Initially pathogens may be less obvious than invasive vascular plants or grazing animals but, because their effects are virtually irreversible, their growth rates are enormous and the potential for their eradication is negligible, invasions of such organisms are potentially far more dangerous. As detailed in Chapter 8, the effects of *Phytophthora cinnamomi* on heath and forest communities in southwest Western Australia provide the best example of the damage that such introductions can wreak. However, many other examples exist of exotic pathogens invading, or indigenous pathogens increasing greatly in activity, and subsequently having a marked effect on individual species or communities (for example, *Cryphonectria parasitica* devastating American chestnut populations; ash yellows, beech bark disease: see Burdon 1993 for review). These effects may go well beyond the direct effect of the pathogen on its host species. Thus the decline of the American chestnut as a result of the depredations of *C. parasitica* has led to the disappearance of at least five microlepidopteran species (Pyle *et al.* 1981) and the suggestion that the decline in chestnuts has indirectly increased the rate of increase of *Ceratocystis fagacearum* (oak wilt disease: Quimby 1982).

Equally worrying is the real prospect that further such invasions, each with major implications, may yet occur. A particular example of great current concern is that posed by the rust pathogen *Puccinia psidii*, a native of South America where it naturally occurs on guava. Epidemics on eucalypts growing in plantations in Brazil clearly indicate the potential for this pathogen to spread widely and attack many species of the dominant tree genus in Australia. The full extent of the threat this pathogen poses to species in the family Myrtaceae has yet to be determined but it may have enormous effects on both Australian and South African ecosystems should it be introduced.

Control

The control of invasive species frequently poses a quandary for management. To maintain the diversity and long-term functioning of natural communities, conservation efforts must recognize the need for, and maintain, natural disturbance processes. Paradoxically, however, disturbance is very often the catalyst that precipitates and accelerates the invasion of reserves by exotic species. Inevitably then, managers have to establish control priorities by ranking invasive species according to the threat they pose to community structure and function, to endemic species, to general landscape aesthetics or to any other management aims. By taking a preventative stance managers can minimize the effects of invasive species. Exotic species need to be monitored so that invasive species can be identified rapidly and control measures initiated before the invasive process progresses too far. Manual or chemical control may provide temporary respite but in the face of large weed populations must be maintained for very long periods of time. In fact, biological control may be the only long-term solution to the control of many invasive species.

Biological control has already become an integral part of the armoury of weed control in many agricultural situations. In contrast, its role in the control of weeds in more natural situations has received relatively little attention until lately. However, one of the most successful instances of biological control – that of the use of *Cactoblastis cactorum* and other insects to control huge cactus infestations in Australia – also served to control these invasive species in many natural communities (Mann 1970). More recently, the rust pathogen *Uromycladium tepperianum* (see Figure 9.5) has been released in an attempt to control invasive Australian species of *Acacia* in South Africa (Morris 1987); and Waterhouse (1994) has argued that there are good to excellent prospects for reducing the weediness of *Mimosa pigra* in northern Australia through the release of a group of insects and pathogens. However, biological control should not be viewed as a 'magic bullet' cure. In general, it is a slow and expensive process that carries no guarantee of success (Crawley 1989).

In essence, rational management of invasive species can be summarized under the general concepts of: (i) prevention; (ii) early intervention; (iii) treatment of causes of invasion; and (iv) rehabilitation of 'liberated' areas. All four of these activities may have to be carried out at the level of the individual reserve. In addition, however, consideration of both the prevention of invasions and treatment of their causes must be pursued at broader community and governmental levels to ensure that potentially invasive species are not introduced into a region in the first place (Humphries *et al.* 1991).

10.5 CONCLUSIONS – CAN ALL COMMUNITIES BE SAVED?

'There are no hopeless cases, only people without hope and expensive cases' (Soulé 1987c). While Soulé presented this argument in the context of individual species conservation, is it also a fair reflection of the reality of ecosystem conservation, or is there truly a stage beyond which there is no salvation?

One of the most notable features of very many plant communities is their great resilience and ability to resume successional processes even after release from long-term disturbances like grazing that may have been intense for decades. Despite this, however, in some situations perturbations may have been so gross and the environmental consequences so pervasive, that there is virtually no possibility of returning to the original community. Examples of such possibilities may arise as a result of major edaphic changes caused by the effects of air pollution. Thus, marked changes in soil acidification produced by air-borne pollutants have led to magnesium, calcium, potassium and iron deficiencies of sufficient magnitude to cause widespread forest death in central Europe (Schulze & Gerstberger 1993). Less commonly they may result from persistent biotic effects, like the devastating effects wrought by *Phytophthora cinnamomi* in western Australia (Wills 1993). It is also possible that the incremental local or regional loss of too many of the original component species of a community could result in a new assemblage unable to return to its original type.

Even if species assemblages can be reconstructed through transplantation from other sites, from botanic gardens or germplasm banks, can we realistically hope to reproduce the vanished community, or at best will we produce no more than an ersatz framework? Although we may manage to re-establish the vascular plant component of a community, the likelihood of re-establishing the myriad of other species and interactions of long-term vital significance is remote. Certainly, we have little experience of rehabilitation of whole communities. The experience gleaned from the reintroduction of individual species clearly indicates that this is 'an expensive, high risk option that commits personnel to long-term monitoring and management' (Maunder 1992). Where whole community restoration has been attempted this has been in communities of limited complexity. In the case of dry tropical forest (Janzen 1986) and rainforest (Nepstad *et al.* 1990) restoration has relied on recolonization from remaining fragments of the pre-existing community. Without such help, restoration of most communities to anything like their former glory will be impossible.

An idea that has gained some currency with respect to the ultimate rehabilitation of highly fragmented communities has been the corridor concept (see Chapter 9). In theory this is an attractive approach, however, in practice corridors are often extremely long and narrow and their value for all but the most mobile animals must be questioned (see Figure 9.4). Indeed, from the perspective of the conservation of plant biodiversity their value is totally unproven. They may assist in promoting bird-mediated seed dispersal or pollinator movement from patch to patch; equally, however, they may assist disease and fire movement between fragments, or be the source of exotic plant invasions.

One particularly difficult question to answer is whether there is a point in the decline of a community at which we should abandon conservation efforts. There is no single answer to this question. Ideally, we might aspire to save all. However, if our aim is *solely* to conserve biodiversity, such a catch-all may not be necessary or justifiable. A rational, scientifically based decision would have to rest on an assessment of the importance of the community, its current health and its likely resistance to further change. It would require answers to questions like: does the community harbour species that are unique or is it an amalgam of species that can be found in a range of other communities? To what extent has deterioration already led to loss of both the conspicuous components of the community – the vascular plants and vertebrates – and its less obvious members – the pollinators, the herbivorous insects, the pathogens and the symbionts? What is the long-term prognosis for the community if we assume no further intervention other than cessation of the immediate threat? In all these respects small remnants of what was formerly a widespread community may well be less able to cope than communities that have always been very restricted in distribution (see section 10.3).

Ultimately, decisions concerning conservation are rarely, if ever, made on scientific grounds alone. Which communities to save will depend on priorities determined by any one, or a combination of scientific, aesthetic, economic or social grounds. In these circumstances, communities that have strong protagonists are likely to receive more attention. Unfortunately, many others are probably doomed through benign neglect. The challenge for the conservation biologist is to replace such neglect with informed scientific decision making.

Conclusions

The last sentence of the first, introductory, chapter of this book reads as follows:

> While it is at the species level that human needs, interests and expectations are directly identified and safeguarded, diversity itself, the essence of life, can only be preserved in the natural communities that we and our descendants safeguard and, we hope, cherish.

This sentence indicates the directions and levels of biological conservation. They are:

1. *Plants of use, interest or concern to humans.* These include domesticates, their wild relatives; wild species used as raw materials in industry, medicine, etc.; species of scientific, aesthetic, social interest; threatened and endangered species. Conservation is mainly *ex situ*, landraces in some regions, and some wild species, *in situ*.
2. *Communities in their natural habitats.* All communities contain both known and as yet unknown components constituting a reserve of diversity. On a global scale, nature reserves should be representative of at least the major ecosystems. Individually they should conform to recognized requirements of size and management. Special emphasis may be given to species that are ecologically significant (e.g. keystone species – mainly animals), and rare and endangered species.

The two categories complement each other. The first is directed towards survival and advancement of the dominant species, the second towards the survival of life as a whole. The first is subject to management and selection by the one species, the second to natural selection alone. The first is subject to short-term impacts from social, economic or scientific developments, the second to slower and potentially profound impacts resulting from global change. The first is orientated towards one species – a transient objective on a geological time scale; the second has a background of 100 million years and a prospect of millions more.

Both, the species serving humanity and the communities safeguarding life and its diversity, are of immense value. The highest priority the human species can confer must go to their preservation, based on exploration and research on how best to manage, develop, use and preserve this incomparable heritage that has become mankind's responsibility.

References

Abbott, D. C., Brown, A. H. D. & Burdon, J. J. (1992). Genes for scald resistance from wild barley (*Hordeum vulgare* ssp. *spontaneum*) and their linkage to isozyme markers. *Euphytica* **61**: 225–31.

Abdalla, F. H. & Roberts, E. H. (1968). Effects of temperature, moisture and oxygen on the induction of chromosome damage in seeds of barley, broad beans, and peas during storage. *Annals of Botany* **32**: 119–36.

Abdalla, F. H. & Roberts, E. H. (1969). The effects of temperature and moisture on the induction of genetic changes in seeds of barley, broad beans, and peas during storage. *Annals of Botany* **32**: 153–67.

Abebe, D. & Hagos, E. (1991). Plants as a primary source of drugs in the traditional health practices of Ethiopia. In *Plant genetic resources of Ethiopia*, ed. J. M. M. Engels, J. G. Hawkes & Melaku Worede, pp. 101–13. Cambridge University Press, Cambridge.

Adams, R. P. (1993). The conservation and utilization of genes from endangered and extinct plants: DNA bank-net. In *Proceedings of the Twentieth Stadler Symposium: Gene conservation and exploitation*, ed. J. P. Gustafson, R. Appels & P. Raven, pp. 35–52. Plenum Press, New York.

Adams, R. P., Miller, J. S., Golenberg, E. M. & Adams, J. E. (1994). *Conservation of plant genes II. Monographs in systematic botany from the Missouri Botanical Garden*, Vol. 48.

Ågren, J. & Schemske, D. W. (1993). Outcrossing rate and inbreeding depression in two annual monoecious herbs, *Begonia hirsuta* and *B. semiovata*. *Evolution* **47**: 125–35.

Akerele, O., Heywood, V. & Synge, H. (1991). *The conservation of medicinal plants*. Cambridge University Press, Cambridge.

Akihama, T. & Nakajima, K. (ed.) (1978). *Long term preservation of favourable germ plasm in arboreal crops*. Fruit Tree Research Station, M. A. F. Fujimoto, Japan.

Akihama, T., Omura, M. & Kozaki, I. (1978). Further investigation of freeze-drying for deciduous fruit tree pollen. In *Long term preservation of favourable germ plasm in arboreal crops*, ed. T. Akihama & K. Nakajima, pp. 1–7. Fruit Tree Research Station, M. A. F. Fujimoto, Japan.

Alexander, M. P. & Ganeshan, S. (1989). Preserving viability and fertility of tomato and eggplant pollen in liquid nitrogen. *Indian Journal of Plant Genetic Resources* **2**: 140–44.

Allard, R. W. (1970). Population structure and sampling methods. In *Genetic resources in plants – their exploration and conservation*, ed. O. H. Frankel & E. Bennett. *IBP Handbook No. 11*, pp. 97–107. Blackwell Scientific Publications, Oxford.

Allard, R. W. (1975). The mating system and microevolution. *Genetics* **79** [Suppl.]: 115–26.

Allard, R. W. (1988). Genetic changes associated with the evolution of adaptedness in cultivated plants and their wild progenitors. *Journal of Heredity* **79**: 225–38.

Allard, R. W. (1990). The genetics of host–pathogen coevolution: implications for genetic resources conservation. *Journal of Heredity* **81**: 1–6.

Allard, R. W. & Adams, J. (1969). Population studies in predominantly self-pollinating species. XIII. Intergenotypic competition and population structure in barley and wheat. *American Naturalist* **103**: 621–45.

Altieri, M. A. & Merrick, L. C. (1987). *In situ* conservation of crop genetic resources through maintenance of traditional farming systems. *Economic Botany* **41**: 86–96.

Altman, D. W., Stelly, D. M. & Kohel, R. J. (1987). Introgression of the glanded-plant and glandless-seed trait from *Gossypium sturtianum* Willis into cultivated upland cotton using ovule culture. *Crop Science* **27**: 880–4.

Ameha, M. (1991). Significance of Ethiopian coffee genetic resources to coffee improvement. In *Plant genetic resources of Ethiopia*, ed. J. M. M. Engels, J. G. Hawkes & Melaku Worede, pp. 354–9. Cambridge University Press, Cambridge.

Anikster, Y. & Noy-Meir, I. (1991). The wild-wheat field laboratory at Ammiad. *Israel Journal of Botany* **40**: 351–62.

Appels, R. & Baum, B. (1991). Evolution of the *Nor* and *5SDna* loci in the Triticeae. In *Molecular systematics of plants*, ed. P. S. Soltis, D. E. Soltis & J. J. Doyle, pp. 92–116. Sinauer Associates, Sunderland, Mass.

Ashri, A. (1971). Evaluation of the world collection of safflower, *Carthamus tinctorius* L. I. Reaction to several diseases and associations with morphological characters in Israel. *Crop Science* **11**: 253–7.

Ashri, A. (1973). *Divergence and evolution in the safflower genus* Carthamus L. Final research report, P. L. 480. US Department of Agriculture, Washington, DC.

Ashri, A. (1975). Evaluation of the germplasm collection of safflower, *Carthamus tinctorius* L. V. Distribution and regional divergence for morphological characters. *Euphytica* **24**: 651–9.

Ashri, A. (1989). Major gene mutations and domestication of plants. In *Plant domestication by induced mutation*, pp. 3–9. International Atomic Energy Agency, Vienna.

Ashton, P. S. (1988). Conservation of biological diversity in botanical gardens. In *Biodiversity*, ed. E. O. Wilson, pp. 269–78. National Academy Press, Washington, DC.

Austin, R. B., Bingham, J., Blackwell, R. D., Evans, L. T., Ford, M. A., Morgan, C. L. & Taylor, M. (1980). Genetic improvements in winter wheat yields since 1900 and associated physiological changes. *Journal of Agricultural Science, Cambridge* **94**: 675–89.

Bachmann, K. (1994). Tansley Review No. 63: Molecular markers in plant ecology. *New Phytologist* **126**: 403–18.

Baird, A. M. (1977). Regeneration after fire in King's Park, Perth, Western Australia. *Journal, Royal Society of Western Australia* **60**: 1–22.

Balick, M. J. (1990). Ethnobotany and the identification of therapeutic agents from the rainforest. *CIBA Foundation Symposia* **154**: 22–39.

Barbour, M. G. & Johnson, A. F. (1977). Beach and dune. In *Terrestrial vegetation of California*, ed. M. G. Barbour & J. Major, pp. 223–61. John Wiley, New York.

Barlow, B. A. (1991). Conspectus of the genus *Amyema* Tieghem (Loranthaceae). *Blumea* **36**: 293–381.

Barnes, P. W., Flint, S. D. & Caldwell, M. M. (1987). Photosynthesis damage and protective pigments in plants from a latitudinal arctic/alpine gradient exposed to supplemental UV-B radiation in the field. *Arctic and Alpine Research* **19**: 21–7.

Barrett, S. C. H. & Kohn, J. R. (1991). Genetic and evolutionary consequences of small population size in plants: implications for conservation. In *Genetics and conservation of rare plants*, ed. D. A. Falk & K. E. Holsinger, pp. 3–30. Oxford University Press, New York.

Bedward, M., Pressey, R. L. & Keith, D. A. (1992). A new approach for selecting fully representative reserve networks: addressing efficiency, reserve design and land suitability with an iterative analysis. *Biological Conservation* **62**: 115–25.

Begon, M., Harper, J. L. & Townsend, C. R. (1986). *Ecology: Individuals, populations and communities*. Blackwell Scientific Publications, Oxford.

Beier, P. (1993). Determining minimum habitat areas and habitat corridors for cougars. *Conservation Biology* **7**: 94–108.

Bennett, A. F. (1990). Habitat corridors and the conservation of small mammals in a fragmented forest environment. *Landscape Ecology* **4**: 109–22.

Bennett, E. (1965). Plant introduction and genetic conservation: genecological aspects of an urgent world problem. *Scottish Plant Breeding Station Records*, pp. 27–113.

Bennett, E. (ed.) (1968). *Record of the FAO/IBP Technical Conference on the Exploration, Utilization and Conservation of Plant Genetic Resources*, 1967. FAO, Rome.

Bennington, C. C., McGraw, J. B. & Vavrek, M. C. (1991). Ecological and genetic variation in seed banks. II. Phenotypic and genetic differences between young and old subpopulations of *Luzula parviflora*. *Journal of Ecology* **79**: 627–43.

Benz, B. F. (1988). *In situ* conservation of the genus *Zea* in the Sierra de Manantlan biosphere reserve. In *Recent advances in the conservation and utilization of genetic resources: Proceedings of the global maize germplasm workshop*, pp. 59–69. CIMMYT, Mexico.

Blixt, S. (1982). The pea model for documentation of genetic resources. In *Documentation of genetic resources: a model*, ed. S. Blixt & J. T. Williams, pp. 3–24. IBPGR, Rome.

Bond, W. J. (1993). Keystone species. In *Biodiversity and ecosystem function*, ed. E. D. Schulze & H. A. Mooney, pp. 237–53. Springer-Verlag, Berlin.

Bothmer, R. von, Jacobsen, N., Baden, C., Jørgensen, R. B. & Linde-Laursen, I. (1991). *An ecogeographical study of the genus Hordeum*. IBPGR, Rome.

Bowers, M. A. & Sacchi, C. F. (1991). Fungal mediation of a plant–herbivore interaction in an early successional plant community. *Ecology* **72**: 1032–7.

Boyce, M. S. (1992). Population viability analysis. *Annual Review of Ecology and Systematics* **23**: 481–506.

Bradshaw, A. D. (1984). Ecological significance of genetic variation between populations. In *Perspectives on plant population ecology*, ed. R. Dirzo & J. Sarukhán, pp. 213–28. Sinauer Associates, Sunderland, Mass.

Bradshaw, A. D. & McNeilly, T. (1991). Evolutionary response to global climatic change. *Annals of Botany* **67**: 5–14.

Braithwaite, R. W., Lonsdale, W. M. & Estbergs, J. A. (1989). Alien vegetation and native biota in tropical Australia: the impact of *Mimosa pigra*. *Biological Conservation* **48**: 189–210.

Breese, E. L. (1989). *Regeneration and multiplication of germplasm resources in seed genebanks: the scientific background*. IBPGR, Rome.

Breiman, A., Bogher, M., Sternberg, H. & Graur, D. (1991). Variability and uniformity of mitochondrial DNA in populations of putative diploid ancestors of common wheat. *Theoretical and Applied Genetics* **82**: 201–8.

Brown, A. H. D. (1978). Isozymes, plant population genetic structure and genetic conservation. *Theoretical and Applied Genetics* **52**: 145–57.

Brown, A. H. D. (1989a). The case for core collections. In *The use of plant genetic resources*, ed. A. H. D. Brown, O. H. Frankel, D. R. Marshall & J. T. Williams, pp. 136–56. Cambridge University Press, Cambridge.

Brown, A. H. D. (1989b). Core collections: a practical approach to genetic resources management. *Genome* **31**: 818–24.

Brown, A. H. D. (1992). Human impact on plant gene pools and sampling for their conservation. *Oikos* **63**: 109–18.

Brown, A. H. D. & Briggs, J. D. (1991). Sampling strategies for genetic variation in *ex situ* collections of endangered plant species. In *Genetics and conservation of rare plants*, ed. D. A. Falk & K. E. Holsinger, pp. 99–119. Oxford University Press, New York.

Brown, A. H. D., Burdon, J. J. & Grace, J. P. (1990). Genetic structure of *Glycine canescens*, a perennial relative of soybean. *Theoretical and Applied Genetics* **79**: 729–36.

Brown, A. H. D., Grant, J. E., Burdon, J. J., Grace, J. P. & Pullen, R. (1984). Collection and utilization of wild perennial *Glycine*. In *Proceedings of the world soybean research conference III*, ed. R. Shibbles, pp. 345–52. Westview Press, Boulder, Colorado.

Brown, A. H. D. & Moran, G. F. (1981). Isozymes and the genetic resources of forest trees. In *Isozymes of North American forest trees and forest insects*, ed. M. T. Conkle, pp. 1–10. Pacific south West Foest and Range Experiment Station Technical Report No. 48. US Department of Agriculture, Washington, DC.

Brown, A. H. D. & Munday, J. (1982). Population genetic structure and optimal sampling of land races of barley from Iran. *Genetica* **58**: 85–96.

Brown, A. H. D. & Schoen, D. J. (1992). Plant population genetic structure and biological conservation. In *Conservation of biodiversity for sustainable development*, ed. O. T. Sandlund, K. Hindar & A. H. D. Brown, pp. 88–104. Scandinavian University Press, Oslo.

Brown, J. S. (1990). Pathogenic variation among isolates of *Rhynchosporium secalis* from barley grass growing in south east Australia. *Euphytica* **50**: 81–9.

Brown, V. K. (1990). Insect herbivory and its effect on plant succession. In *Pests, pathogens and plant communities*, ed. J. J. Burdon & S. R. Leather, pp. 275–88. Blackwell Scientific Publications, Oxford.

Brown, V. K. & Gange, A. C. (1992). Secondary plant succession: how is it modified by insect herbivory? *Vegetatio* **101**: 3–13.

Brown, W. L. (1983). Genetic diversity and genetic vulnerability – an appraisal. *Economic Botany* **37**: 4–12.

Brücher, H. (1968). Südamerika als Herkunftsraum von Nutzpflanzen. In *Biogeography and Ecology in South America*, Vol. 1, ed. E. J. Fittkau *et al.*, pp. 251–301. W. Junk, The Hague.

Brush, S. B. (1989). Rethinking crop genetic resources conservation. *Conservation Biology* **3**: 19–29.

Brush, S. B. (1991). A farmer-based approach to conserving crop germplasm. *Economic Botany* **45**: 153–65.

Burdon, J. J. (1980). Variation in disease-resistance within a population of *Trifolium repens*. *Journal of Ecology* **68**: 737–44.

Burdon, J. J. (1987). *Diseases and plant population biology*. Cambridge University Press, Cambridge.

Burdon, J. J. (1993). The role of parasites in plant populations and communities. In *Biodiversity and ecosystem function*, ed. E.-D. Schulze & H. A. Mooney, pp. 165–79. Springer-Verlag, Berlin.

Burdon, J. J. (1994). The distribution and origin of genes for race specific resistance to *Melampsora lini* in *Linum marginale*. *Evolution* (in press).

Burdon, J. J. & Chilvers, G. A. (1994). Demographic changes and the development of competition in a native Australian eucalypt forest invaded by exotic pines. *Oecologia* **97**: 419–23.

Burdon, J. J. & Jarosz, A. M. (1989). Disease in mixed cultivars, composites, and natural plant populations: some epidemiological and evolutionary consequences. In *Plant population genetics, breeding, and genetic resources*, ed. A. H. D. Brown, M. T. Clegg, A. L. Kahler & B. S. Weir, pp. 215–28. Sinauer Associates, Sunderland, Mass.

Burdon, J. J. & Thompson, J. N. (1995). Changed patterns of resistance in a population of *Linum marginale* attacked by the rust pathogen *Melampsora lini*. *Journal of Ecology* **82** (in press).

Byerlee, D. & Moya, P. (1993). *Impacts of international wheat breeding research in the developing world, 1966–1990*. CIMMYT, Mexico.

Byrne, M. & Moran, G. F. (1994). Population divergence in the chloroplast genome of *Eucalyptus nitens*. *Heredity*, **73**: 18–28.

Cameron, D. F. (1983). To breed or not to breed. In *Genetic resources of forage plants*, ed. J. G. McIvor & R. A. Bray, pp. 237–50. CSIRO, Melbourne.

Carlson, K. D., Knapp, S. A., Thompson, A. E., Brown, J. H. & Joliff, G. D. (1992). Nature's abundant variety: new oilseed crops on the horizon. In *1992 Yearbook of agriculture, new crops, new uses, new markets*, pp. 124–33. US Department of Agriculture, Washington, DC.

Carson, H. L. (1983). The genetics of the founder effect. In *Genetics and conservation. A reference for managing wild animal and plant populations*, ed. C. M. Schonewald-Cox, S. M. Chambers, B. MacBryde & W. L. Thomas, pp. 189–200. The Benjamin/Cummings Publishing Company, Menlo Park, California.

Catling, P. C. (1991). Ecological effects of prescribed burning practices on the mammals of south-eastern Australia. In *Conservation of Australia's forest fauna*, ed. D. Lunney, pp. 353–64. Royal Zoological Society of NSW, Mosman.

Catling, P. C. (1994). Bushfires and prescribed burning: protecting native fauna. *Search* **25**: 37–40.

Ceccarelli, S. (1994). Specific adaptation and breeding for marginal environments. *Euphytica* (in press).

Ceccarelli, S. & Grando, S. (1991). Selection environment and environmental sensitivity in barley. *Euphytica* **57**: 157–67.

Ceccarelli, S., Valkoun, J., Erskine, W., Weigand, S., Miller, R. & Van Leur, J. A. G. (1992). Plant genetic resources and plant improvement as tools to develop sustainable agriculture. *Experimental Agriculture* **28**: 89–98.

Chalmers, K. J., Sprent, J. I., Simons, A. J., Waugh, R. & Powell, W. (1992). Patterns of genetic diversity in a tropical tree legume (*Gliricidia*) revealed by RAPD markers. *Heredity* **69**: 465–72.

Chang, T. T. (1976a). The rice cultures. *Philosophical Transactions of the Royal Society of London, Series B*, **275**: 143–57.

Chang, T. T. (1976b). Rice. In *Evolution of crop plants*, ed. N. W. Simmonds, pp. 98–104. Longman, London.

Chang, T. T. (1976c). The origin, evolution, cultivation, dissemination and diversification of Asian and African rices. *Euphytica* **25**: 425–41.

Chang, T. T. (1984). Conservation of rice genetic resources: luxury or necessity? *Science* **224**: 251–6.

Chang, T. T. (1989a). The case for large collections. In *The use of plant genetic resources*, ed. A. H. D. Brown, O. H. Frankel, D. R. Marshall & J. T. Williams, pp. 123–35. Cambridge University Press, Cambridge.

Chang, T. T. (1989b). The management of rice genetic resources. *Genome* **31**: 825–833.

Chapman, C. G. D. (1989). Collection strategies for the wild relatives of field crops. In *The use of plant genetic resources*, ed. A. H. D. Brown, O. H. Frankel, D. R. Marshall & J. T. Williams, pp. 263–79. Cambridge University Press, Cambridge.

Charlesworth, D. & Charlesworth, B. (1987). Inbreeding depression and its evolutionary consequences. *Annual Review of Ecology and Systematics* **18**: 237–68.

Chiariello, N., Hickman, J. C. & Mooney, H. A. (1982). Endomycorrhizal role for interspecific transfer of phosphorus in a community of annual plants. *Science* **217**: 941–3.

Chickwendu, V. E. & Okezie, C. E. A. (1989). Factors responsible for the ennoblement of African yams: inferences from experiments in yam domestication. In *Foraging and farming: the evolution of plant exploitation*, ed. D. R. Harris & G. C. Hillman, pp. 344–57. Unwin Hyman, London.

Chin, H. F. & Pritchard, H. W. (1988). *Recalcitrant seeds – a status report, including a bibliography 1979–87*. IBPGR, Rome.

Christensen, P. E. & Burrows, N. D. (1986). Fire: an old tool with a new use. In *Ecology of biological invasions: an Australian perspective*, ed. R. H. Groves & J. J. Burdon, pp. 97–105. Australian Academy of Science, Canberra.

Christensen, P., Recher, H. & Hoare, J. (1981). Responses of open forests (dry sclerophyll forests) to fire regimes. In *Fire and the Australian biota*, ed. A. M. Gill, R. H. Groves & I. R. Noble, pp. 367–93. Australian Academy of Science, Canberra.

CIMMYT (1991). *Annual Report, 1991*. CIMMYT, Mexico.

Clark, R. L. (1989). Seed maintenance and storage. In *The National Plant Germplasm System of the United States*, ed. J. Janick, *Plant Breeding Reviews* **7**: 95–110. Timber Press, Portland, Oregon.

Clausen, J., Keck, D. D. & Hiesey, W. M. (1948). *Experimental studies on the nature of species. III. Environmental responses of climatic races of* Achillea. Carnegie Institute Washington, Publication 581.

Clegg, M. T. (1989). Molecular diversity in plant populations. In *Plant population genetics, breeding, and genetic resources*, ed. A. H. D. Brown, M. T. Clegg, A. L. Kahler & B. S. Weir, pp. 98–115. Sinauer Associates, Sunderland, Mass.

Clegg, M. T. & Allard, R. W. (1972). Patterns of genetic differentiation in the slender wild oat species *Avena barbata*. *Proceedings of the National Academy of Sciences USA* **69**: 1820–4.

Clegg, M. T. & Durbin, M. L. (1990). Molecular approaches to the study of plant biosystematics. *Australian Systematic Botany* **3**: 1–8.

Coblentz, B. E. (1978). The effects of feral goats (*Capra hireus*) on island ecosystems. *Biological Conservation* **13**: 279–86.

Collins, D. J., Culvenor, C. C. J., Lamberton, J. A., Loder, J. W. & Price, J. R. (1990). *Plants for medicines: a chemical and pharmacological survey of plants in the Australian region*. CSIRO, Melbourne.

Connell, J. H. & Slatyer, R. O. (1977). Mechanisms of succession in natural communities and their role in community stability and organization. *American Naturalist* **111**: 1119–44.

Cook, S. A., Copsey, A. D., & Dickman, A. W. (1989). Response of *Abies* to fire and *Phellinus*. In *The evolutionary ecology of plants*, ed. J. H. Bock & Y. B. Linhart, pp. 363–92. Westview Press, Boulder, Colorado.

Cowling, R. M., Pierce, S. M. & Moll, E. J. (1986). Conservation and utilization of south coast Renosterveld, an endangered South African vegetation type. *Biological Conservation* **37**: 363–77.

Cox, P. A., Elmqvist, T., Pierson, E. D. & Rainey, W. E. (1991a). Flying foxes as pollinators and seed dispersers in Pacific island ecosystems. *US Department of the Interior, Fish & Wildlife Service. Biological Report* **90**: 18–23.

Cox, P. A., Elmqvist, T., Pierson, E. D. & Rainey, W. E. (1991b). Flying foxes as strong interactors in south Pacific island ecosystems: a conservation hypothesis. *Conservation Biology* **5**: 448–54.

Crawford, T. J. (1984). What is a population? In *Evolutionary ecology*, ed. B. Shorrocks, pp. 135–74. British Ecological Society Symposium No. 23. Blackwell Scientific Publications, Oxford.

Crawley, M. J. (1983). *Herbivory: the dynamics of animal–plant interactions*. Blackwell Scientific Publications, Oxford.

Crawley, M. J. (1989). Insect herbivores and plant population dynamics. *Annual Review of Entomology* **34**: 531–64.

Crawley, M. J. (1989). The success and failure of weed biocontrol using insects. *Biocontrol News and Information* **10**: 213–24.

Cromarty, A. S., Ellis, R. H. & Roberts, E. H. (1982). *The design of seed storage facilities for genetic conservation*. IBPGR, Rome.

Cropper, S. C. (1993). *Management of endangered plants*. CSIRO, Melbourne.

Crow, J. F. & Denniston, C. (1988). Inbreeding and variance effective population numbers. *Evolution* **42**: 482–95.

Crow, J. F. & Kimura, M. (1970). *An introduction to population genetics theory*. Harper & Row, New York.

Crozier, R. H. (1992). Genetic diversity and the agony of choice. *Biological Conservation* **61**: 11–15.

Cullis, C. A. (1985) Sequence variation and stress. In *Genetic flux in plants*, ed. B. Hohn & E. S. Dennis, pp. 157–68. Springer-Verlag, Vienna.

Cunningham, A. B. (1990). People and medicines: the exploitation and conservation of traditional Zulu medicinal plants. *Mitteil. Inst. Allegemeine Bot.* **23b**: 979–90.

Dale, J. L. (1988). The status of disease indexing and the international distribution of banana germplasm. In *Conservation and movement of vegetatively propagated germplasm: in vitro culture and disease aspects*. IBPGR, Rome.

Dallas, J. F. (1988). Detection of DNA 'fingerprints' of cultivated rice by hybridization with a human minisatellite DNA probe. *Proceedings of the National Academy of Sciences USA* **85**: 6831–5.

Darlington, C. D. (1969). The silent millennia in the origin of agriculture. In *The domestication and exploitation of plants*, ed. P. J. Ucko & G. W. Dimbleby, pp. 67–72. Duckworth, London.

DeMauro, M. M. (1993). Relationship of breeding system to rarity in the Lakeside Daisy (*Hymenoxys acaulis* var. *glabra*). *Conservation Biology* **7**: 542–50.

Dennis, E. S. & Peacock, W. J. (1984). Knob heterochromatin homology in maize and its relatives. *Journal of Molecular Evolution* **20**: 341–50.

Diamond, J. M. (1972). Biogeographic kinetics: estimation of relaxation times for avifaunas of southwest Pacific islands. *Proceedings of the National Academy of Sciences USA* **69**: 3199–203.

Diamond, J. M. & May, R. M. (1976). Island biogeography and the design of natural reserves. In *Theoretical ecology: principles and applications*, ed. R. M. May, pp. 163–86. Saunders, Philadelphia.

Dickie, J. B., Linington, S. & Williams, J. T. (ed.) (1984). *Seed management techniques for genebanks*. IBPGR, Rome.

Dickman, A. (1992). Plant pathogens and long-term ecosystem changes. In *The fungal community: its organization and role in the ecosystem*, ed. G. C. Carroll & D. T. Wicklow, 2nd ed., pp. 499–520. Marcel Dekker, New York.

Dinoor, A. (1975). Evaluation of sources of disease resistance. In *Crop genetic resources for today and tomorrow*, ed. O. H. Frankel & J. G. Hawkes. *International Biological Programme 2*, pp. 201–10. Cambridge University Press, Cambridge.

Dinoor, A. (1976). Germplasm and the phytopathologist. *Plant Genetic Resources* **32**: 36–8.

Dinoor, A. & Eshed, N. (1984). The role and importance of pathogens in natural plant communties. *Annual Review of Phytopathology* **22**: 443–66.

Dobzhansky, Th. (1970). *Genetics of the evolutionary process*. Columbia University Press, New York.

Doebley, J. (1989). Isozymic evidence and the evolution of crop plants. In *Isozymes in plant biology*, ed. D. E. Soltis & P. S. Soltis, pp. 165–91. Dioscorides Press, Portland, Oregon.

Doebley, J. (1990). Molecular evidence and the evolution of maize. *Economic Botany* **44** (3 supplement): 6–27.

Doebley, J. F., Goodman, M. M. & Stuber, C. W. (1984). Isozymatic variation in *Zea* (Gramineae). *Systematic Botany* **9**: 203–18.

Doebley, J. F., Goodman, M. M. & Stuber, C. W. (1985). Isozyme variation in the races of maize in Mexico. *American Journal of Botany* **72**: 629–39.

Dole, J. & Ritland, K. (1993). Inbreeding depression in two *Mimulus* taxa measured by multigenerational changes in the inbreeding coefficient. *Evolution* **47**: 361–73.

Dong, J. & Wagner, D. B. (1993). Taxonomic and population differentiation of mitochondrial diversity in *Pinus banksiana* and *Pinus contorta*. *Theoretical and Applied Genetics* **86**: 573–8.

Dorofeev, V. F. (1975). Evaluation of material for frost and drought resistance in wheat breeding. In *Crop genetic resources for today and tomorrow*, ed. O. H. Frankel & J. G. Hawkes. *International Biological Programme 2*, pp. 211–22. Cambridge University Press, Cambridge.

Doyle, J. J., Doyle, J. L. & Brown, A. H. D. (1990). Chlorophlast DNA polymorphism and phylogeny in the B genome of *Glycine* subgenus *Glycine* (Leguminosae). *American Journal of Botany* **77**: 772–82.

Drury, W. H. (1974). Rare species. *Biological Conservation* **6**: 162–9.

Duggins, D. O. (1980). Kelp beds and sea otters: an experimental approach. *Ecology* **61**: 447–53.

Duvick, D. N. (1984). Genetic diversity in major farm crops on the farm and in reserve. *Economic Botany* **38**: 161–78.

Dvorak, J. (1989). Evolution of multigene families: the ribosomal RNA loci of wheat and related species. In *Plant population genetics, breeding, and genetic resources*, ed. A. H. D. Brown, M. T. Clegg, A. L. Kahler & B. S. Weir, pp. 83–97. Sinauer Associates, Sunderland, Mass.

Ehrlich, P. R. & Ehrlich, A. H. (1981). *Extinction: the causes and consequences of the disappearance of species*. Random House, New York.

Eldridge, K. G. (1990). Conservation of forest genetic resources with particular reference to *Eucalyptus* species. *Commonwealth Forestry Review* **69**: 45–53.

Eldridge, K., Davidson, J., Harwood, C. & van Wyk, G. (1993). *Eucalypt domestication and breeding*. Clarendon Press, Oxford.

Ellis, R. H., Hong, T. D. & Roberts, E. H. (1985). *Handbook of seed technology for genebanks*. (2 vols). IBPGR, Rome.

Ellstrand, N. C. & Elam, D. R. (1993). Population genetic consequences of small population size: implications for plant conservation. *Annual Review of Ecology and Systematics* **24**: 217–42.

Epperson, B. K. (1989). Spatial patterns of genetic variation within plant populations. In *Plant population genetics, breeding, and genetic resources*, ed. A. H. D. Brown, M. T. Clegg, A. L. Kahler & B. S. Weir, pp. 229–53. Sinauer Associates, Sunderland, Mass.

Evans, L. T. (1993). *Crop evolution, adaptation and yield*. Cambridge University Press, Cambridge.

Ewel, J. J. (1986). Invasibility: lessons from south Florida. In *The ecology of biological invasions of North America and Hawaii*, ed. H. A. Mooney & J. A. Drake, pp. 214–30. Springer-Verlag, New York.

Ewens, W. J. (1972). The sampling theory of selectively neutral alleles. *Theoretical Population*

Biology **3**: 87–112.

Ewens, W. J. (1990). The minimum viable population size as a genetic and a demographic concept. In *Convergent issues in genetics and demography*, ed. J. Adams, D. A. Lam, A. I. Hermalin & P. E. Smouse, pp. 307–16. Oxford University Press, New York.

Ewens, W. J., Brockwell, P. J., Gani, J. M. & Resnick, S. I. (1987). Minimum viable population size in the presence of catastrophes. In *Viable populations for conservation*, ed. M. E. Soulé, pp. 59–68. Cambridge University Press, Cambridge.

Faith, D. P. (1992). Conservation evaluation and phylogenetic diversity. *Biological Conservation* **61**: 1–10.

Faith, D. P. & Norris, R. H. (1989). Correlation of enrironmental variables with patterns of distribution and abundance of common and rare freshwater macroinvertebrates. *Biological Conservation* **50**: 77–98.

Falk, D. A. (1990). Integrated strategies for conserving plant genetic diversity. *Annals of the Missouri Botanical Garden* **77**: 38–47.

Falk, D. A. (1992). From conservation biology to conservation practices: strategies for protecting plant diversity. In *Conservation Biology: the theory and practice of nature conservation, preservation and management*, ed. P. L. Fiedler & S. K. Jain, pp. 397–431. Chapman & Hall, New York.

Falk, D. A. & Holsinger, K. E. (ed.) (1991). *Genetics and conservation of rare plants*. Oxford University Press, New York.

FAO [Food and Agriculture Organization] (1973). *Report of the Fourth Session of the FAO Panel of Experts on animal genetic resources*. FAO, Rome.

FAO (1975). *Report of the Sixth Session of the FAO Panel of Experts on plant exploration and introduction*. FAO, Rome.

FAO (1989). *Plant genetic resources: Their conservation in situ for human use*. FAO, Rome.

FAO (1991). Strategies for the establishment of a network of *in situ* conservation areas. *Forest Genetic Resources Information* **19**: 3–8. FAO, Rome.

Farnsworth, N. R. (1988). Screening plants for new medicines. In *Biodiversity*, ed. E. O. Wilson, pp. 83–97. National Academy Press, Washington, DC.

Fenster, C. B. & Ritland, K. (1992). Chloroplast DNA and isozyme diversity in two *Mimulus* species (Scrophulariaceae) with contrasting mating systems. *American Journal of Botany* **79**: 1440–7.

Fiedler, P. L. & Jain, S. K. (1992). *Conservation biology: the theory and practice of nature conservation, preservation and management*. Chapman & Hall, New York.

Flannery, K. V. (1969). Origins and ecological effects of early domestications in Iran and the Near East. In *The domestication and exploitation of plants*, ed. P. J. Ucko & G. W. Dimbleby, pp. 73–100. Duckworth, London.

Ford, E. B. (1971). *Ecological Genetics*, 3rd edn. Chapman and Hall, London.

Ford-Lloyd, B. V. & Jackson, M. (1986). *Plant genetic resources: an introduction to their conservation and use*. Edward Arnold, London.

Frankel, O. H. (1970a). Genetic conservation in perspective. In *Genetic resources in plants – their exploration and conservation*, ed. O. H. Frankel & E. Bennett. *IBP Handbook No. 11*, pp. 469–89. Blackwell Scientific Publications, Oxford.

Frankel, O. H. (1970b). Variation – the essence of life. *Proceedings of the Linnean Society of New South Wales* **95**: 158–69.

Frankel, O. H. (1974). Genetic conservation: our evolutionary responsibility. *Genetics* **78**: 53–65.

Frankel, O.H. (1975). Genetic resources centres – a co-operative global network. In *Crop genetic resources for today and tomorrow*, ed. O. H. Frankel & J. G. Hawkes, pp. 473–81. Cambridge University Press, Cambridge.

Frankel, O. H. (1978). Germplasm 'preservation'. *Plant Genetic Resources Newsletter* **34**: 18–19.

Frankel, O. H. (1983). The place of management in conservation. In *Genetics and conservation: a reference for managing wild animal and plant populations*, ed. C. M. Schonewald-Cox, S. M. Chambers, B. MacBryde & W. L. Thomas, pp. 1–14. The Benjamin/Cummings Publishing Company, Menlo Park, Calif.

Frankel, O. H. (1985). Genetic resources: the founding years. *Diversity* **7**: 26–9.

Frankel, O. H. (1986a). Genetic resources: the founding years. II. The movement's constituent assembly. *Diversity* **8**: 30–2.

Frankel, O. H. (1986b). Genetic resources: the founding years. III. The long road to the International Board. *Diversity* **9**: 30–3.

Frankel, O. H. (1987). Genetic resources: the founding years. IV. After twenty years. *Diversity* **11**: 25–7.

Frankel, O. H. & Bennett, E. (1970a). Genetic resources – introduction. In *Genetic resources in plants – their exploration and conservation*, ed. O. H. Frankel & E. Bennett. *IBP Handbook No. 11*, pp. 7–17. Blackwell Scientific Publications, Oxford.

Frankel, O. H. & Bennett, E. (ed.) (1970b). *Genetic resources in plants – their exploration and conservation. IBP Handbook No. 11*. Blackwell Scientific Publications, Oxford.

Frankel, O. H. & Brown, A. H. D. (1984). Current plant genetic resources – a critical appraisal. *Proceedings XV International Congress of Genetics*, pp. 3–11. Oxford and IBH Publishing Co, New Delhi.

Frankel, O. H. & Hawkes, J. G. (ed.) (1975). *Crop genetic resources for today and tomorrow. International Biological Programme 2*. Cambridge University Press, Cambridge.

Frankel, O. H. & Soulé, M. E. (1981). *Conservation and evolution*. Cambridge University Press, Cambridge.

Franklin, I. A. (1980). Evolutionary change in small populations. In *Conservation biology: an evolutionary–ecological perspective*, ed. M. E. Soulé & B. A. Wilcox, pp. 135–50. Sinauer Associates, Sunderland, Mass.

Frese, L. (1992). Progress report of the International Database for *Beta* (IDBB). In *International Beta genetic resources network*. IBPGR, Rome.

Froend, R. H., Heddle, E. M., Bell, D. T. & McComb, A. J. (1987). Effects of salinity and waterlogging on the vegetation of Lake Toolibin, Western Australia. *Australian Journal of Ecology* **12**: 281–98.

Gange, A. C., Brown, V. K. & Farmer, L. M. (1990). A test of mycorrhizal benefit in an early successional plant community. *New Phytologist* **115**: 85–91.

Garcia, A. (1992). Conserving the species-rich meadows of Europe. *Agriculture, Ecosystems and Environment* **40**: 219–32.

Gaut, B. S. & Clegg, M. T. (1993). Molecular evolution of the *Adh1* locus in the genus *Zea*. *Proceedings of the National Academy of Sciences USA* **90**: 5095–9.

Gentry, A. H. (1993). Tropical forest biodiversity and the potential for new medicinal plants. *ACS Symposium Series* **534**: 13–24.

Gepts, P. & Clegg, M. T. (1989). Genetic diversity in pearl millet [*Pennisetum glaucum* (L.) R. Br.] at the DNA sequence level. *Journal of Heredity* **80**: 203–8.

Geric, I., Zlokolica, M. & Geric, M. (1989). Races and populations of maize in Yugoslavia. Isozyme variation and genetic diversity. *Systematic and ecogeographic studies on crop genepools 3*. IBPGR, Rome.

Gilbert, L. E. (1980). Food web organization and the conservation of neotropical diversity. In *Conservation biology: an evolutionary–ecological perspective*, ed. M. E. Soulé & B. A. Wilcox, pp. 11–34. Sinauer Associates, Sunderland, Mass.

Giles, B. E. (1984). A comparison between quantitative and biochemical variation in the wild barley *Hordeum murinum*. *Evolution* **38**: 34–41.

Gill, A. M. (1981). Adaptive responses of Australian vascular plant species to fires. In *Fire and the Australian biota*, ed. A. M. Gill, R. H. Groves & I. R. Noble, pp. 243–72. Australian Academy of Science, Canberra.

Gill, A. M. & Bradstock, R. A. (1994). Extinction of biota by fire. In *Conserving biodiversity: threats and solutions*. New South Wales National Parks & Wildlife Service.

Gill, A. M. & Nicholls, A. O. (1989). Monitoring fire-prone flora in reserves for nature conservation. In *Fire management on native conservation lands*, ed. N. Burrows, L. McCaw & G. Friend, pp. 137–51. W. A. Department Conservation & Land Management, Occasional Paper No. 1/89.

Gilpin, M. E, (1991). The genetic effective size of a metapopulation. *Biological Journal of the Linnean Society* **42**: 165–75.

Given, D. R. (1981). Threatened plants of New Zealand: documentation in a series of islands. In *The biological aspects of rare plant conservation*, ed. H. Synge, pp. 67–79. John Wiley, Chichester.

Gliessman, S. R., Garcia, E. R. & Amador A. M. (1981). The biological basis for the application of traditional agricultural technology in the management of tropical agro-ecosystems. *Agro-ecosystems* **7**: 173–85.

Goldsmith, F. B. (1991). The selection of protected areas. In *The scientific management of temperate communities for conservation*, ed. I. F. Spellerberg, F. B. Goldsmith & M. G. Morris, pp. 273–91. Blackwell Scientific Publications, Oxford.

Good, R. B. (1981). The role of fire in conservation reserves. In *Fire and the Australian biota*, ed. A. M. Gill, R. H. Groves & I. R. Noble, pp. 529–49. Australian Academy of Science, Canberra.

Goodman, M. M. (1988). US maize germplasm: origins, limitations and alternatives. In *Recent advances in the conservation and utilization of genetic resources*. Proceedings of the Global Maize Germplasm Workshop, Mexico.

Grant, V. (1981). *Plant speciation*, 2nd edn. Columbia University Press, New York.

Green, A. G. (1984). The occurrence of ricinoleic acid in *Linum* seed oils. *Journal of the American Oil Chemists Society* **61**: 939–40.

Green, A. G. (1986a). A mutant genotype of flax (*Linum usitatissimum* L.) containing very low levels of linolenic acid in its seed oil. *Canadian Journal of Plant Science* **66**: 499–503.

Green, A. G. (1986b). Genetic control of polyunsaturated fatty acid biosythesis in flax (*Linum usitatissimum*) seed oil. *Theoretical and Applied Genetics* **72**: 654–61.

Grime, J. P., Mackey, J. M. L., Hillier, S. H. & Read, D. J. (1987). Floristic diversity in a model system using experimental microcosms. *Nature* **328**: 420–2.

Grubb, P. J. (1977). The maintenance of species-richness in plant communities: the importance of the regeneration niche. *Biological Review* **52**: 107–45.

Guzman, E. D. de (1975). Conservation of vanishing timber species in the Philippines. In *South East Asian Plant Genetic Resources*, ed. J. T. Williams, C. H. Lamoureux & N. Wullijarni-Soetjipto, pp. 198–204. LIPI, Bogor.

Haines, A.L. (1974). *Yellowstone National Park*. National Parks Service, US Department of the Interior, Washington, DC.

Haldane, J. B. S. (1949). Disease and evolution. *La Ricerca Scientifica* **19**: 1–11.

Hamann, O. (1979). Regeneration of vegetation on Santa Fe and Pinta Islands, Galapagos, after the eradication of goats. *Biological Conservation* **15**: 215.

Hamann, O. (1991). The Joint IUCN-WWF Plants Conservation Programme and its interest in medicinal plants. In *The conservation of medicinal plants*, ed. O. Akerele, V. Heywood & H. Synge, pp. 13–22. Cambridge University Press, Cambridge.

Hamilton, W. D., Axelrod, R. & Tanese, R. (1990). Sexual reproduction as an adaptation to resist parasites. *Proceedings of the National Academy of Sciences USA* **87**: 3566–73.

Hamon, S. & van Sloten, D. H. (1989). Characterization and evaluation of okra. In *The use of plant genetic resources*, ed. A. H. D. Brown, O. H. Frankel, D. R. Marshall & J. T. Williams, pp. 173–96. Cambridge University Press, Cambridge.

Hamrick, J. L. & Godt, M. J. W. (1989). Allozyme diversity in plant species. In *Plant population genetics, breeding, and genetic resources*, ed. A. H. D. Brown, M. T. Clegg, A. L. Kahler & B. S. Weir, pp. 43–63. Sinauer Associates, Sunderland, Mass.

Hanson, J. (1985). *Procedures for handling seeds in genebanks*. IBPGR, Rome.

Harlan, J. R. (1951). Anatomy of gene centers. *American Naturalist* **85**: 97–103.

Harlan, J. R. (1956). Distribution and utilization of natural variability in cultivated plants. *Brookhaven Symposia in Biology* **9**: 191–206.

Harlan, J. R. (1971). Agricultural origins: centers and noncenters. *Science* **174**: 468–74.

Harlan, J. R. (1975a). *Crops and man*. American Society of Agronomy, Madison, Wisconsin.

Harlan, J. R. (1975b). Our vanishing genetic resources. *Science* **188**: 618–21.

Harlan, J. R. (1976). Genetic resources in wild relatives of crops. *Crop Science* **16**: 329–33.

Harlan, J. R. (1983). The scope for collection and improvement of forage plants. In *Genetic resources of forage plants*, ed. J. G. McIvor & R. A. Bray, pp. 3–14. CSIRO, Melbourne.

Harlan, J. R. (1989). The tropical African cereals. In *Foraging and farming: the evolution of plant exploitation*, ed. D. R. Harris, & G. C. Hillman, pp. 335–43. Unwin Hyman, London.

Harlan, J. R. & de Wet, J. M. J. (1971). Toward a rational classification of cultivated plants. *Taxon* **20**: 509–17.

Harlan, J. R., de Wet, J. M. J. & Price, E. G. (1973). Comparative evolution of cereals. *Evolution* **27**: 311–25.

Harlan, J. R. & Zohary, D. (1966). Distribution of wild wheats and barley. *Science* **153**: 1074–80.

Harley, J. L. (1970). The importance of micro-organisms to colonising plants. *Transactions of the Botanical Society of Edinburgh* **41**: 65–70.

Harper, J. L. (1971). Grazing, fertilizers and pesticides in the management of grasslands. In *The scientific management of animal and plant communities for conservation*, ed. E. Duffey & A. S. Watt, pp. 15–31. Blackwell Scientific Publications, Oxford.

Harper, J. L. (1977). *Population biology of plants*. Academic Press, London.

Harrington, J. F. (1970). Seed and pollen storage for conservation of plant gene resources. In *Genetic resources in plants – their exploration and conservation*, ed. O. H. Frankel & E. Bennett. *IBP Handbook No. 11*, pp. 501–21. Blackwell Scientific Publications, Oxford.

Harris, D. R. (1969). Agricultural systems, ecosystems and the orgins of agriculture. In *The domestication and expoitation of plants and animals*, ed. P. J. Ucko & G. W. Dimbleby, pp. 3–15. Duckworth, London.

Harris, D. R. (1990). Vavilov's concept of centres of origin of cultivated plants: its genesis and its influence on the study of agricultural origins. *Biological Journal of the Linnean Society* **39**: 7–16.

Harris, D. R. & Hillman, G. C. (1989). Introduction. In *Foraging and farming: the evolution of plant exploitation*, ed. D. R. Harris, & G. C. Hillman, pp. 1–8. Unwin Hyman, London.

Hartl, D. L. & Clark, A. G. (1989). *Principles of population genetics*, 2nd edn. Sinauer Associates, Sunderland, Mass.

Hawkes, J. G. (1969). The ecological background of plant domestication. In *The domestication and exploitation of plants*, ed. P. J. Ucko & G. W. Dimbleby, pp. 17–29. Duckworth, London.

Hawkes, J. G. (1970). Potatoes. In *Genetic resources in plants – their exploration and conservation*, ed. O. H. Frankel & E. Bennett, *IBP Handbook No. 11*, pp. 311–19. Blackwell Scientific Publications, Oxford.

Hawkes, J. G. (1990). *The potato: evolution, biodiversity and genetic resources*. Belhaven Press, London.

Hawkes, J. G. (1991). International workshop on dynamic *in-situ* conservation of wild relatives of major cultivated plants: summary of final discussion and recommendations. *Israel Journal of Botany* **40**: 529–36.

Hawkes, J. G. (1992). Gene banking strategies for botanic gardens. In *Ex situ conservation in botanical gardens*, ed. O Hamann, *Opera Botanica* **113**: 15–17.

Hawksworth, D. L. (1992). Biodiversity in microorganisms and its role in ecosystem function. In *Biodiversity and global change*, ed. O. T. Solbrig, H. M. van Emden & P. G. W. J. van Oordt, pp. 83–93. IUBS Press, Paris.

Hedrick, P. W. & Millar, P. S. (1992). Conservation genetics: techniques and fundamentals. *Ecological Applications* **2**: 30–46.

Hedrick, P. W. (1983). *Genetics of populations*. Science Books International, Boston.

Henderson, M., Merriam, G. & Wegner, J. (1985). Patchy environments and species survival: chipmunks in an agricultural mosaic. *Biological Conservation* **31**: 95–105.

Hermsen, J. G. Th. (1989). Current use of potato collections. In *The use of plant genetic resources*, ed. A. H. D. Brown, O. H. Frankel, D. R. Marshall & J. T. Williams, pp. 68–87. Cambridge University Press, Cambridge.

Heywood, J. S. (1986). The effect of plant size variation on genetic drift in populations of annuals. *American Naturalist* **127**: 851–61.

Heywood, V. H. (1989). Patterns, extents and modes of invasions by terrestrial plants. In *Biological invasions: a global perspective*, ed. J. A. Drake, H. A. Mooney, F. di Castri, R. H. Groves, F. J. Kruger, M. Regmánek & M. Williamson, pp. 31–60. John Wiley, Chichester.

Heywood, V. (H.) (1991). Botanic gardens and the conservation of medicinal plants. In *The conservation of medicinal plants*, ed. O. Akerele, V. Heywood & H. Synge, pp. 213–28. Cambridge University Press, Cambridge.

Heywood, V. H. (1992). Conservation of germplasm of wild plant species. In *Conservation of biodiversity for sustainable development*, ed. O. T. Sandlund, K. Hindar & A. H. D. Brown, pp. 189–203. Scandinavian University Press, Oslo.

Hill, W. G. & Robertson, A. (1968). Linkage disequilibrium in finite populations. *Theoretical and Applied Genetics* **38**: 226–31.

Hillman, G. C. & Davies, M. S. (1990). 6. Domestication rates in wild-type wheats and barley under primitive cultivation. *Biological Journal of the Linnean Society* **39**: 39–78.

Hobbie, S. E., Jensen, D. B. & Chapin, F. S., III. (1993). Resource supply and disturbance as controls over present and future plant diversity. In *Biodiversity and ecosystem function*, ed. E.-D. Schulze & H. A. Mooney, pp. 385–408. Springer-Verlag, Berlin.

Hobbs, R. J. (1992). The role of corridors in conservation: solution or bandwagon. *Trends in Ecology and Evolution* **7**: 389–92.

Hodkinson, I. D. & Casson, D. (1991). A lesser prediliction for bugs: Hemiptera (Insecta) diversity in tropical rain forests. *Biological Journal of the Linnean Society* **43**: 101–9.

Holcomb, J., Tolbert, T. M. & Jain, S. K. (1977). A diversity analysis of genetic resources in rice. *Euphytica* **26**: 441–50.

Holden, J. H. W. (1984). The second ten years. In *Crop genetic resources: conservation and evaluation*, ed. J. H. W. Holden & J. T. Williams, pp. 277–85. George Allen and Unwin, London.

Holm, L. G., Plucknett, D. L., Pancho, J. V. & Herberger, J. P. (1977). *The world's worst weeds: distribution and biology*. University Press of Hawaii, Honolulu.

Holsinger, K. E. & Gottlieb, L. D. (1991). Conservation of rare and endangered plants: principles and prospects. In *Genetics and conservation of rare plants*, ed. D. A. Falk & K. E. Holsinger, pp. 195–208. Oxford University Press, New York.

Hooy, T. & Shaughnessy, G. (ed.) (1992). *Terrestrial and marine protected areas in Australia (1991)*. Australian National Parks and Wildlife Service, Canberra.

Hope-Simpson, J. (1940). Studies of the vegetation of the English chalk. VI. Late stages in succession leading to chalk grassland. *Journal of Ecology* **28**: 386–402.

Horn, H. S. (1974). The ecology of secondary succession. *Annual Review of Ecology and Systematics* **5**: 25–37.

Hornocher, M. G. (1969). Winter territoriality in mountain lions. *Journal of Wildlife Management* **33**: 457–64.

Horovitz, A. & Feldman, M. (1991). Evaluation of the wild-wheat study at Ammiad. *Israel Journal of Botany* **40**: 501–8.

Howard, B. H. (1975). Possible long-term cold storage of woody plant material. In *Crop genetic resources for today and tomorrow*, ed. O. H. Frankel and J. G. Hawkes. *International Biological Programme 2*, pp. 359–67. Cambridge University Press, Cambridge.

Hubbell, S. P. & Foster, R. B. (1986). Commonness and rarity in a neotropical forest: implications for tropical tree conservation. In *Conservation biology: the science of scarcity and diversity*, ed. M. E. Soulé, pp. 205–31. Sinauer Associates, Sunderland, Mass.

Huenneke, L. F. (1991). Ecological implications of genetic variation in plant populations. In *Genetics and conservation of rare plants*, ed. D. A. Falk & K. E. Holsinger, pp. 31–44. Oxford University Press, New York.

Huff, D. R., Peakall, R. & Smouse, P. E. (1993). RAPD variation within and among natural populations of outcrossing buffalograss [*Buchloë dactyloides* (Nutt.) Englelm.]. *Theoretical and Applied Genetics* **86**: 927–34.

Huffaker, C. B. & Kennett, C. E. (1959). A ten year study of vegetational changes associated with biological control of Klamath weed. *Journal of Range Management* **12**: 69–82.

Humphreys, F. R. & Craig, F. G. (1981). Effects of fire on soil chemical, structural and hydrological properties. In *Fire and the Australian biota*, ed. A. M. Gill, R. H. Groves & I. R. Noble, pp. 177–200. Australian Academy of Science, Canberra.

Humphries, S. E., Groves, R. H. & Mitchell, D. S. (1991). Plant invasions of Australian ecosystems. In *Plant invasions: the incidence of environmental weeds in Australia*, pp. 1–127. Kowari 2. Australian National Parks and Wildlife Service, Canberra.

Hurka, H. (1994). Conservation genetics and the role of botanical gardens. In *Conservation genetics*, ed. V. Loeschcke, J. Tomiuk & S. K. Jain, pp. 371–80. Birkhäuser Verlag, Basel.

IARI [Indian Agricultural Research Institute] (1972). *Collection and study of cultivated and wild rices from North East India. USPL 480 Project, Final Report*. IARI, New Delhi.

IBP [International Biological Programme] (1966). Plant gene pools. *IBP News* **5**: 48–51.

IBPGR [International Board for Plant Genetic Resources] (1980). *Fruits*. IBPGR. Rome.

IBPGR (1981*a*). *Vegetables*. IBPGR. Rome.

IBPGR (1981*b*). *Root and tuber crops*. IBPGR. Rome.

IBPGR (1985). *Cost-effective long-term seed stores*. IBPGR, Rome.

IBPGR (1989). *Annual Report 1988*. IBPGR, Rome.

IBPGR (1991a). *Descriptors for Beta*. IBPGR/CGN, Rome.

IBPGR (1991b). *Descriptors for annual Medicago*. IBPGR, Rome.

Iltis, H. H., Doebley, J. F., Guzman, M. R. & Pazy, B. (1979). *Zea diploperennis* (Gramineae): a new teosinte from Mexico. *Science* **203**: 186–8.

Ingram, C. B. & Williams, J. T. (1984). *In situ* conservation of wild relatives of crops. In *Crop genetic resources: conservation and evaluation*, ed. J. H. W. Holden & J. T. Williams, pp. 163–79. Allen & Unwin, London.

IPGRI (1993). *Diversity for development. The strategy of the International Plant Genetic Resources Institute*. IBPGR, Rome.

IRRI (1975). *Annual Report for 1974*. International Rice Research Institute, Los Baños, Philippines.

IUCN/WWF (1988). *The joint IUCN–WWF plants conservation programme*. IUCN/WWF Gland,

Switzerland.

IUCN (1975). *United Nations list of national parks and equivalent reserves.* International Union for Conservation Of Nature and Natural Resources, Morges, Switzerland.

IUCN/BGCS (1987). The *International Transfer Format for botanic garden records.* Hunt Institute for Botanical Documentation, Pittsburgh.

Jackson, P. S. W. (ed.) (1989). *Resolutions of the Second International Botanic Gardens Conservation Congress.* IUCN Botanic Gardens Congress. Kew Green, Richmond, Surrey.

Jackson, P. S. W. (1991). Developing a world network of botanic gardens. In *Proceedings of the conference – Protective custody? – Ex situ plant conservation in Australia.* Australian National Botanic Gardens, Canberra.

Jackson, W. D. (1968). Fire, air, water and earth – an elemental ecology of Tasmania. *Proceedings, Ecological Society of Australia* **3**: 9–16.

Jain, S. K. (1975). Genetic reserves. In *Crop genetic resources for today and tomorrow,* ed. O. H. Frankel & J. G. Hawkes, *International Biological Programme 2,* pp. 379–96. Cambridge University Press, Cambridge.

Jain, S. K. (1975). Population structure and the effects of breeding system. In *Crop genetic resources for today and tomorrow,* ed. O. H. Frankel & J. G. Hawkes, pp. 15–36. Cambridge University Press, Cambridge.

Jain, S. [K.] (1989). Dichotomy of major genes and polygenes. In *Plant domestication by induced mutation,* pp. 11–27. International Atomic Energy Agency, Vienna.

Jain, S. K. (1994). Genetics and demography of rare plants and patchily distributed colonizing species. In *Conservation genetics,* ed. V. Loeschcke, J. Tomiuk & S. K. Jain, Birkhauser Verlag, Basel.

Jain, S. K. & Suneson, C. A. (1966). Increased recombination and selection in barley populations carrying a male sterility factor. I. Quantitative variability. *Genetics* **54**: 1215–24.

Jain, S. K., Qualset, C. O., Bhatt, G. M. & Wu, K. K. (1975). Geographical patterns of phenotypic diversity in a world collection of durum wheats. *Crop Science* **15**: 700–4.

Jana, S. & Pietrzak, L. N. (1988). Comparative assessment of genetic diversity in wild and primitive cultivated barley in a centre of diversity. *Genetics* **119**: 981–90.

Janick, J. (1990). Afterword – Symposium Summary. In *Advances in new crops,* ed. J. Janick & J. E. Simon, pp. 539–40. Timber Press, Portland, Oregon.

Janick, J. & Simon, J. E. (ed.) (1990). *Advances in new crops.* Timber Press, Portland, Oregon.

Janzen, D. H. (1986). The eternal external threat. In *Conservation biology: the science of scarcity and diversity,* ed. M. E. Soulé, pp. 286–303. Sinauer Associates, Sunderland, Mass.

Janzen, D. H. (1992). A south–north perspective on science in the management, use, and economic development of biodiversity. In *Conservation of biodiversity for sustainable development,* ed. O. T. Sandlund, K. Hindar & A. H. D. Brown, pp. 27–52. Scandinavian University Press, Oslo.

Jarosz, A. M. & Burdon, J. J. (1991). Host–pathogen interactions in natural populations of *Linum marginale* and *Melampsora lini*: II. Local and regional variation in patterns of resistance and racial structure. *Evolution* **45**: 1618–27.

Jayakar, S. D. (1970). A mathematical model for interaction of gene frequencies in a parasite and its host. *Theoretical Population Biology* **1**: 140–64

Johannsen, W. (1909). *Elements der exakten Erblichkeitslehre.* Gustav Fischer, Jena.

Johnston, M. O. (1992). Effects of cross and self-fertilization on progeny fitness in *Lobelia cardinalis* and *L. siphilitica. Evolution* **46**: 688–702.

Jones, M. G. (1933). Grassland management and its influence on the sward. *Empire Journal of Experimental Agriculture* **1**: 43–57.

Kapos, V. (1989). Effects of isolation on the water status of forest patches in the Brazilian

Amazon. *Journal of Tropical Ecology* **5**: 173–85.
Karnoven, P. & Savolainen, O. (1993). Variation and inheritance of ribosomal DNA in *Pinus sylvestris* L. (Scots pine). *Heredity* **71**: 614–22.
Karron, J. D. (1991). Patterns of genetic variation and breeding systems in rare plant species. In *Genetics and conservation of rare plants*, ed. D. A. Falk & K. E. Holsinger, pp. 87–98. Oxford University Press, New York.
Kartha, K. K. (1986). Production and indexing of disease-free plants. In *Plant tissue culture and its agricultural applications*, ed. L. A. Withers & P. G. Alderson, pp. 219–38. Butterworths, London.
Kato-Y, T. A. (1984). Chromosome morphology and the origin of maize and its races. In M. K. Hecht, B. Wallace & G. T. Prance (ed.), *Evolutionary Biology* **17**: 219–53. Plenum Press, New York.
Keim, P., Diers, B. W., Olson, T. C. & Shoemaker, R. C. (1990). RFLP mapping in soybean: association between marker loci and variation in quantitative traits. *Genetics* **126**: 735–42.
Khush, A. S. & Brar, D. S. (1991). Genetics of resistance to insects in crop plants. *Advances in Agronomy* **45**: 223–74.
Kimura, M. & Crow, J. F. (1964). The number of alleles that can be maintained in a finite population. *Genetics* **49**: 725–38.
King, L. M. & Schaal, B. A. (1990). Genotypic variation within asexual lineages of *Taraxacum officinale*. *Proceedings of the National Academy of Sciences USA* **87**: 998–1002.
Kirkpatrick, J. B. (1983). An iterative method for establishing priorities for the selection of nature reserves: an example from Tasmania. *Biological Conservation* **25**: 127–34.
Kloppenburg, J. R., Jr & Kleinman, D. L. (1987). The plant germplasm controversy: analysing empirically the distribution of the world's plant genetic resources. *BioScience* **37**: 190–8.
Knowles, P. F. (1969). Centers of plant diversity and conservation of germplasm: safflower. *Economic Botany* **23**: 324–9.
Konopka, J. & Hanson, J. (ed.) (1985). *Information handling systems for genebank management*. IBPGR. Rome.
Krupa, S. V. & Kickert, R. N. (1989). The greenhouse effect: impacts of ultraviolet-B (UV-B) radiation, carbon dioxide ($C0_2$) and ozone (O_3) on vegetation. *Environmental Pollution* **61**: 263–93.
Kucera, C. L. & Koelling, M. (1964). The influence of fire on composition of central Missouri prairie. *American Midland Naturalist* **72**: 142–7.
Kulman, H. M. (1971). Tree defoliation by insects. *Annual Review of Entomology* **16**: 289–324.
Ladizinsky, G. (1985). Founder effect in crop-plant evolution. *Economic Botany* **39**: 191–9.
Ladizinsky, G. (1989). Origin and domestication of the Southwest Asian grain legumes. In *Foraging and farming: the evolution of plant exploitation*, ed. D. R. Harris, & G. C. Hillman, pp. 374–89. Unwin Hyman, London.
Lagudah, E. S. (1986). Influence of the D genome on some aspects of wheat quality. PhD thesis, University of Melbourne.
Lagudah, E. S. & Halloran G. M. (1989). Phylogenetic relationships of *Triticum taushii* the D genome donor to heaxaploid wheat. 3. Variation in, and the genetics of, seed esterases (Est-S). *Theoretical and Applied Genetics* **77**: 851–6.
Lagudah, E. S., Appels, R., McNeil, D. & Schachtman, D. P. (1993). Exploiting the diploid 'D' genome chromatin for wheat improvement. In *Gene conservation and exploitation*, ed. J. P. Gustafson *et al.*, pp. 87–107. Plenum Press, New York.
Lande, R. & Barrowclough, G. F. (1987). Effective population size, genetic variation, and their use in population management. In *Viable populations for conservation*, ed. M. E. Soulé, pp. 87–123. Cambridge University Press, Cambridge.
Latter, B. D. H. (1973). The island model of population differentiation: a general solution.

Genetics **73**: 147–57.

Lawton, J. H. & Brown, V. K. (1993). Redundancy in ecosystems. In *Biodiversity and ecosystem function*, ed. E.-D. Schulze & H. A. Mooney, pp. 255–69. Springer-Verlag, Berlin.

Lazarides, M., Norman, M. J. T. & Perry, R. A. (1965). *Wet season development pattern of some native grasses in Katherine, N. T.* CSIRO Australia, Division of Land Use Research, Regional Survey, Technical Paper No. 26.

Ledig, F. T. (1988). The conservation of diversity in forest trees. *BioScience* **38**: 471–9.

Ledig, F. T. (1992). Human impacts on genetic diversity in forest ecosystems. *Oikos* **63**: 87–108.

Leigh E. G. Jr, Wright, S. J., Herre, E. A. & Putz, F. E. (1993). The decline of tree diversity on newly isolated tropical islands: a test of a null hypothesis and some implications. *Evolutionary Ecology* **7**: 76–102.

Leigh, J. H. & Briggs, J. D. (ed.) (1992). *Threatened Australian plants: overview and case studies.* Australian National Parks and Wildlife Service, Canberra.

Leigh, J. H., Wimbush, D. J., Wood, D. H., Holgate, M. D., Slee, A. V., Stanger, M. G. & Forrester, R. I. (1987). Effects of rabbit grazing and fire on a subalpine environment. I. Herbaceous and shrubby vegetation. *Australian Journal of Botany* **35**: 433–64.

Lenné, J. M. & Wood, D. (1991). Plant diseases and the use of wild germplasm. *Annual Review of Phytopathology* **29**: 35–63.

Leonard, K. J. (1977). Selection pressures and plant pathogens. *Annals of the New York Academy of Sciences* **287**: 207–22.

Leppik, E. E. (1970). Gene centres of plants as sources of disease resistance. *Annual Review of Phytopathology* **8**: 232–44.

Lester, R. N. (1989). Evolution under domestication involving disturbance of genic balance. *Euphytica* **44**: 125–32.

Levings, C. S. (1990). The Texas cytoplasm of maize: cytoplasmic male sterility and disease susceptiblility. *Science* **250**: 942–7.

Liu, Z. & Furnier, G. R. (1993). Comparison of allozyme, RFLP, and RAPD markers for revealing genetic variation within and between trembling aspen and bigtooth aspen. *Theoretical and Applied Genetics* **87**: 97–105.

Lloyd, D. G. & Bawa, K. S. (1984). Modification of the gender of seed plants in varying conditions. *Evolutionary Biology* **17**: 255–338.

Loechshke, V., Tomiuk, J. & Jain, S. K. (1994). *Conservation genetics*. Birkhäuser Verlag, Basel.

Londo, G. (1974). Successive mapping of dune slack vegetation. *Vegetatio* **29**: 51–61.

Lonsdale, W. M. (1992). The impact of weeds in national parks. In *Proceedings, 1st International Weed Control Congress*, ed. J. H. Combellack, K. J. Levick, J. Parsons & R. G. Richardson, pp. 145–9. Weed Science Society of Victoria, Melbourne.

Love, S. L., Rhodes, B. B. & Moyer, J. W. (1987). *Meristem-tip culture and virus indexing of sweet potatoes*. IBPGR, Rome.

Lovejoy, T. E., Rankin, J. M., Bierregaard, R. O. Jr, Brown, K. S. Jr, Emmons, L. H. & Van der Voort, M. E. (1984). Ecosystem decay of Amazon forest remnants. In *Extinctions*, ed. M. H. Nitecki, pp. 295–325. Chicago University Press, Chicago.

Lovejoy, T. E., Bierregaard, R. O. Jr, Rylands. A. B., Malcolm, A. B., Quintela, C. E., Harper, L. H., Brown, K. S. Jr, Powell, A. H., Powell, G. V. N., Schubart, H. O. R. & Hays, M. B. (1986). Edge and other effects of isolation on Amazon forest fragments. In *Conservation biology: the science of scarcity and diversity*, ed. M. Soulé, pp. 257–85. Sinauer Associates, Sunderland, Mass.

Lucas, G. U. (1976). Conservation: recent developments in international co-operation and legislation. In *Conservation of threatened plants*, ed. J. B. Simmons, R. I. Beyer, P. E. Brandham, G. U. Lucas & T. H. Parry, pp. 271–7. Plenum Press, New York.

MacArthur, R. H. & Wilson, E. O. (1963). An equilibrium theory of insular zoogeography. *Evolution* **17**: 373–87.

MacArthur, R. H. & Wilson, E. O. (1967). *The theory of island biogeography*. Princeton University Press, Princeton, NJ.

McBrien, H., Harmsen, R. & Crowder, A. (1983). A case of insect grazing affecting plant succession. *Ecology* **64**: 1035–9.

Macdonald, I. A. W., Graber, D. M., De Benedetti, S., Groves, R. H. & Fuentes, E. R. (1988). Introduced species in nature reserves in Mediterranean-type climatic regions of the world. *Biological Conservation* **44**: 37–66.

Macdonald, I. A. W., Loope, L. L., Usher, M. B. & Hamann, O. (1989). Wildlife conservation and the invasion of nature reserves by introduced species: a global perspective. In *Biological invasions: a global perspective*, ed. J. A. Drake, H. A. Mooney, F. di Castri, R. H. Groves, F. J. Kruger, M. Regmánek & M. Williamson, pp. 215–55. John Wiley, Chichester.

Macdonald, I. A. W., Powrie, F. J. & Siegfried, W. R. (1986). The differential invasion of southern Africa's biomes and ecosystems by alien plants and animals. In *The ecology and management of biological invasions in southern Africa*, ed. I. A. W. Macdonald, F. J. Kruger & A. A. Ferrar, pp. 209–25. Oxford University Press, Cape Town.

McGranahan, G. H., Hansen, J. & Shaw, D. V. (1988). Inter- and intraspecific variation in California black walnuts. *Journal American Society of Horticultural Science* **113**: 760–5.

McIvor, J. G. & Bray, R. A. (1983). *Genetic resources of forage plants*. CSIRO, Melbourne.

McKenzie, N. L., Belbin, L., Margules, C. R. & Keighery, G. J. (1989). Selecting representative reserve systems in remote areas: a case study in the Nullarbor region, Australia. *Biological Conservation* **50**: 239–61.

Macnair, M. R. (1983). The genetic control of copper tolerance in the yellow monkey flower, *Mimulus guttatus*. *Heredity* **50**: 283–93.

McNeely, J. A., Miller, K. R., Reid, W. V., Mittermeier, R. A. & Werner, T. B. (1990). *Conserving the world's biological diversity*. IUCN, Gland, Switzerland.

McNeely, J. A. & Thorsell, J. A. (1991). Enhancing the role of protected areas in conserving medicinal plants. In *The conservation of medicinal plants*, ed. O. Akerele, V. Heywood & H. Synge, pp. 199–212. Cambridge University Press, Cambridge.

Mangelsdorf, P. C. (1966). Genetic potentials for increasing yields of food crops and animals. *Proceedings of the National Academy of Sciences USA* **56**: 370–5.

Mann, C. C. & Plummer, M. L. (1993). The high cost of biodiversity. *Science* **260**: 1868–71.

Mann, J. (1970). *Cacti naturalized in Australia and their control*. Department of Lands, Queensland.

Margules, C. R. (1986). Conservation evaluation in practice. In *Wildlife conservation evaluation*, ed. M. B. Usher, pp. 297–314. Chapman and Hall, London.

Margules, C. R., Higgs, A. J. & Rafe, R. W. (1982). Modern biogeographic theory: are there any lessons for nature reserve design? *Biological Conservation* **24**: 115–28.

Margules, C. R. & Nicholls, A. O. (1987). Assessing the conservation value of remnant habitat 'islands': mallee patches on the western Eyre peninsula. In *Nature conservation: the role of remnants of native vegetation*, ed. D. A. Saunders, G. W. Arnold, A. A. Burbidge & A. J. M. Hopkins, pp. 89–102. Surrey Beatty & Sons, Chipping Norton, New South Wales.

Margules, C. R. & Nicholls, A. O. (1988). Selecting networks of reserves to maximise biological diversity. *Biological Conservation* **43**: 63–76.

Marshall, D. R. (1977). The advantages and hazards of genetic homogeneity. In *The genetic basis of epidemics in agriculture*, ed. P. R. Day, *Annals of the New York Academy of Sciences* **287**: 1–20.

Marshall, D. R. (1989). Crop genetic resources: Current and emerging issues. In *Plant population genetics, breeding and genetic resources*, ed. A. H. D. Brown, M. T. Clegg, A. L. Kahler & B. S. Weir, pp. 367–88. Sinauer Associates, Sunderland, Mass.

Marshall, D. R. & Brown, A. H. D. (1975). Optimum sampling strategies in genetic conservation. In *Crop genetic resources for today and tomorrow*, ed. O. H. Frankel & J. G. Hawkes, *International Biological Programme 2*, pp. 53–80. Cambridge University Press, Cambridge.

Martin, F. W. (1975). The storage of germplasm of tropical roots and tubers in the vegetative form. In *Crop genetic resources for today and tomorrow*, ed. O. H. Frankel & J. G. Hawkes, *International Biological Programme 2*, pp. 369–77. Cambridge University Press, Cambridge.

Matlack, G. R. (1993). Microenvironment variation within and among forest edge sites in the eastern United States. *Biological Conservation* **66**: 185–94.

Maunder, M. (1992). Plant re-introduction: an overview. *Biodiversity and conservation* **1**: 51–61.

Maunder, M. (1994). Botanic gardens: future challenges and responsibilities. *Biodiversity and Conservation* **3**: 97–103.

May, R. M. (1975). Island biogeography and the design of wildlife reserves. *Nature* **254**: 177–8.

May, R. M. (1990). How many species? *Philosophical Transactions of the Royal Society, London B* **330**: 293–304.

May, R. M. (1992). Past efforts and future prospects towards understanding how many species there are. In *Biodiversity and global change*, ed. O. T. Solbrig, H. M. van Emden & P. G. W. J. van Oordt, pp. 71–81. IUBS Press, Paris.

Mayr, E. (1963). *Animal species and evolution*. Harvard University Press, Cambridge, Mass.

Menges, E. S. (1990). Population viability analysis for an endangered plant. *Conservation Biology* **4**: 52–62.

Menges, E. S. (1991). The application of minimum viable population theory to plants. In *Genetics and conservation of rare plants*, ed. D. A. Falk & K. E. Holsinger, pp. 45–61. Oxford University Press, New York.

Merriam, G. (1991). Corridors and connectivity: animal populations in heterogeneous environments. In *Nature conservation: the role of corridors*, ed. D. Saunders & R. Hobbs, pp. 133–42. Surrey Beatty & Sons, Chipping Norton, New South Wales.

Mikola, J. (1982). Bud-set phenology as an indicator of climatic adaption of Scots pine in Finland. *Silva Fennica* **16**: 178–84.

Miles, J. (1987). Vegetation succession: past and present perceptions. In *Colonization, succession and stability*, ed. A. J. Gray, M. J. Crawley & P. J. Edwards, pp. 1–29. Blackwell Scientific Publishers, Oxford.

Millar, C. I. & Libby, W. J. (1991). Strategies for conserving clinal, ecotypic, and disjunct population diversity in widespread species. In *Genetics and conservation of rare plants*, ed. D. A. Falk & K. E. Holsinger, pp. 149–70. Oxford University Press, New York.

Millar, C. I. & Westfall, R. D. (1992). Allozyme markers in forest genetic conservation. In *Population genetics of forest trees*, ed. W. T. Adams, S. H. Strauss, D. L. Copes & A. R. Griffin, pp. 347–71. Kluwer Academic Publishers, Dordrecht, The Netherlands.

Miller, J. C. & Tanksley, S. D. (1990). RFLP analysis of phylogenetic relationships and genetic vaiation in the genus *Lycopersicon*. *Theoretical and Applied Genetics* **80**: 437–48.

Mills, L. S., Soulé, M. E. & Doak, D. F. (1993). The keystone-species concept in ecology and conservation. *BioScience* **43**: 219–24.

Moll, E. J. & Trinder-Smith, T. (1992). Invasion and control of alien woody plants in the Cape Peninsula Mountains, South Africa – 30 years on. *Biological Conservation* **60**: 135–43.

Moran, G. F. & Bell, J. C. (1987). The origin and genetic diversity of *Pinus radiata* in Australia. *Theoretical and Applied Genetics* **73**: 616–22.

Morris, M. J. (1987). Biology of the Acacia gall rust, *Uromycladium tepparianum*. *Plant Pathology* **36**: 100–6.

Moss, C. E. (1900). Changes in the Halifax flora during the last century and a quarter. *Naturalist, Hull,* 165–72.

Moss, J. P., Rao, V. R. & Gibbons, R. W. (1989). Evaluating the germplasm of groundnut (*Arachis hypogaea*) and wild *Arachis* species at ICRISAT. In *The use of plant genetic resources*, ed. A. H. D. Brown, O. H. Frankel, D. R. Marshall & J. T. Williams, pp. 212–34. Cambridge University Press, Cambridge.

Mosseler, A., Egger, K. N. & Hughes, G. A. (1992). Low levels of genetic diversity in red pine confirmed by random amplified polymorphic DNA markers. *Canadian Journal of Forest Research* **22**: 1332–7.

Munck, L. (1972). Improvement of nutritional value in barley. *Hereditas* **72**: 1–128.

Murphy, H. C., Wahl, I., Dinoor, A., Miller, J. D., Morey, D. D., Luke, H. H., Sechler, D. & Reyes, L. (1967). Resistance to crown rust and soil borne mosaic virus in *Avena sterilis*. *Plant Disease Reporter* **51**: 120–4.

Naeem, S., Thompson, L. J., Lawler, S. P., Lawton, J. H. & Woodfin, R. M. (1994). Declining biodiversity can alter the performance of ecosystems. *Nature* **368**: 734–7.

National Academy of Sciences (1975). *Underexploited tropical plants with promising economic value*. National Academy of Sciences, Washington, DC.

National Research Council (1972). *Genetic vulnerability of major crops*. National Academy of Sciences, Washington, DC.

National Research Council (1989). *Lost crops of the Incas: little-known plants of the Andes with promise for worldwide cultivation*. National Academy Press, Washington, DC.

National Research Council (1991). *Managing global genetic resources: forest trees*. National Academy Press, Washington, DC.

National Research Council (1993). The conservation of genetic stock collections. In *Managing global genetic resources*, pp. 219–38. National Academy Press, Washington, DC.

Nei, M. (1973). Analysis of gene diversity in subdivided populations. *Proceedings of the National Academy of Sciences USA* **70**: 3321–3.

Nepstad, D., Uhl, C. & Serrao, E. A. S. (1990). Surmounting barriers to forest regeneration in abandoned, highly degraded pastures. In *Alternatives to deforestation: steps towards sustainable utilization of Amazon forests*, ed. A. B. Anderson. Columbia University Press, New York.

Nevo, E. (1983). The evolutionary significance of genetic diversity: ecological, demographic and life history correlates. In *Evolutionary dynamics of genetic diversity*, ed. G. S. Mani, pp. 13–165. Springer-Verlag, Berlin.

New, T. R. (1984). *A biology of Acacias*. Oxford University Press, Melbourne.

Nicholls, A. O. & Margules, C. R. (1991). The design of studies to demonstrate the biological importance of corridors. In *Nature conservation 2: the role of corridors*, ed. D. A. Saunders & R. J. Hobbs, pp. 49–61. Surrey Beatty & Sons, Chipping Norton, New South Wales.

Nicholls, A. O., & Margules, C. R. (1993). An upgraded reserve selection algorithm. *Biological Conservation* **64**: 165–9.

Nichols, P. G. H. & Francis, C. M. (1993). Sardinian subterranean clovers – their importance to Australia. In *Proceedings of the 10th Australian Plant Breeding Conference* **2**: 19–20.

Noble, I. R. & Slatyer, R. O. (1981). Concepts and models of succession in vascular plant communities subject to recurrent fire. In *Fire and the Australian biota*, ed. A. M. Gill, R. H. Groves & I. R. Noble, pp. 311–35. Australian Academy of Science, Canberra.

Nunney, L. & Campbell, K. A. (1993). Assessing minimum viable population size: demography meets population genetics. *Trends in Ecology and Evolution* **8**: 234–9.

Oka, H. I. (1974). Experimental studies on the origin of cultivated rice. *Genetics* **78**: 475–86.

Oka, H. I. (1988). *Origin of cultivated rice.* Japan Scientific Societies Press, Tokyo. Elsevier, Amsterdam.

Oldfield, M. L. (1989). *The value of conserving genetic resources.* Sinauer Associates, Sunderland, Mass.

Oldfield, M. L. & Alcorn, J. B. (1987). Conservation of traditional agroecosytems. *BioScience* **37**: 199–208.

O'Malley, D. M. & Bawa, K. S. (1987). Mating system of a tropical rain forest tree species. *American Journal of Botany* **74**: 1143–9.

Oremus, P. A. I. & Otten, H. (1981). Factors affecting growth and nodulation of *Hippophae rhamnoides* L. ssp. *rhamnoides* in soils from two successional stages of dune formation. *Plant and Soil* **63**: 317–31.

Osterman, J. C. & Dennis, E. S. (1989). Molecular analysis of the ADH1-C^{m} allele of maize. *Plant Molecular Biology* **13**: 203–12.

Paine, R. T. (1966). Food web complexity and species diversity. *American Naturalist* **100**: 65–75.

Palmberg-Lerche, C. (1992). Criteria on choice of species for conservation: woody plants. In *Conservation biology: a training manual for biological diversity and genetic resources*, ed. P. Kapoor-Vijay & J. White, pp. 51–60. Commonwealth Secretariat, London.

Palmberg-Lerche, C. (1993). International programmes for the conservation of forest genetic resources. In *Proceedings of the International Symposium on Genetic Conservation and Production of Tropical Forest Seed.* [Chiang Mai, Thailand. June 14–16, 1993.] ASEAN/ CANADA Forest Tree Seed Centre, Muak Lek, Thailand.

Palmer, J. D. (1988). Intraspecific variation and multicircularity in *Brassica* mitochondrial DNAs. *Genetics* **118**: 341–51.

Panis, B., Dhed's, D. & Swennen, R. (1992). Freeze-preservation of embryogenic *Musa* suspension cultures. In *Conservation of plant genes. DNA banking and* in vitro *biotechnology*, ed. R. P. Adams & J. E. Adams, pp. 183–95. Academic Press, San Diego.

Paroda, R. S., Kapoor, P., Arora, R. K. & Mal, B. (ed.) (1988). *Life support plant species. Diversity and conservation.* National Bureau of Plant Genetic Resources, New Delhi.

Parsons, W. T. & Cuthbertson, E. G. (1992). *Noxious weeds of Australia.* Inkata Press, Melbourne.

Paterson, A. H., Lander, E. S., Hewitt, J. D., Peterson, S., Lincoln, S. E. & Tanksley, S. D. (1988). Resolution of quantitative traits into Mendelian factors, using a complete linkage map of restriction fragment length polymorphisms. *Nature* **335**: 721–6.

Pavlik, B. M, Nickrent, D. L. & Howald, A. M. (1993). The recovery of an endangered plant. I. Creating a new population of *Amsinckia grandiflora. Conservation Biology* **7**: 510–26.

Peacock, W. J. (1989). Molecular biology and genetic resources. In *The use of plant genetic resources*, ed. A. H. D. Brown, O. H. Frankel, D. R. Marshall & J. T. Williams, pp. 363–76. Cambridge University Press, Cambridge.

Peeters, J. P. & Martinelli, J. A. (1989). Hierarchical cluster analysis as a tool to manage variation in germplasm collections. *Theoretical and Applied Genetics* **78**: 42–8.

Person, C. (1966). Genetic polymorphism in a parasitic system. *Nature* **212**: 266–7

Peters II, R. L. (1988). The effect of global climatic change on natural communities. In *Biodiversity*, ed. E. O. Wilson, pp. 450–61. National Academy Press, Washington, DC.

Pickersgill, B. (1989). Cytological and genetic evidence on the domestication and diffusion of crops within the Americas. In *Foraging and farming: the evolution of plant exploitation*, ed. D. R. Harris, & G. C. Hillman, pp. 426–439. Unwin Hyman, London.

Pielou, E. C. (1977). *Mathematical ecology*. John Wiley, New York.

Pitelka, L. F. (1993). Biodiversity and policy decisions. In *Biodiversity and ecosystem function*, ed. E.-D. Schulze & H. A. Mooney, pp. 481–93. Springer-Verlag, Berlin.

Prescott-Allen, R. & Prescott-Allen, C. (1988). *Genes from the wild: using wild genetic resources for food and raw materials*, 2nd edn. Earthscan Publications, London.

Pressey, R. L. & Nicholls, A. O. (1989). Efficiency in conservation evaluation: scoring versus iterative approaches. *Biological Conservation* **50**: 199–218.

Preston, F. W. (1962). The canonical distribution of commonness and rarity: Part 1. *Ecology* **43**: 185–215.

Price, J. R., Lamberton, J. A. & Culvenor, C. C. J. (1993). The Australian Phytochemical Survey: historical aspects of the CSIRO search for new drugs in Australian Plants. *Historical Records of Australian Science* **9**: 335–56.

Price, M. V. & Wasser, N. M. (1979). Pollen dispersal and optimal outcrossing in *Delphinium nelsoni*. *Nature* **277**: 294–7.

Primack, R. B. (1993). *Essentials of conservation biology*. Sinauer Associates, Sunderland, Mass.

Principe, P. P. (1991). Valuing the biodiversity of medicinal plants. In *The conservation of medicinal plants*, ed. O. Akerele, V. Heywood & H. Synge, pp. 79–124. Cambridge University Press, Cambridge.

Prober, S. & Brown, A. H. D. (1994). Conservation of the grassy white box woodlands. I. Population genetics and fragmentation of *Eucalyptus albens* Benth. *Conservation Biology* (in press).

Pyle, R., Bentzien, M. & Opler, P. (1981). Insect conservation. *Annual Review of Entomology* **26**: 233–58.

Qualset, C. O. (1975). Sampling germplasm in a center of diversity: an example of disease resistance in Ethiopian barley. In *Crop genetic resources for today and tomorrow*, ed. O. H. Frankel & J. G. Hawkes, *International Biological Programme 2*, pp. 81–96. Cambridge University Press, Cambridge.

Quimby, P. C. (1982). Impact of diseases on plant populations. In *Biological control of weeds with plant pathogens*, ed. R. Charudattan & H. L. Walker, pp. 47–60. John Wiley, New York.

Quinn, J. F. & Harrison, S. P. (1988). Effects of habitat fragmentation and isolation on species richness: evidence from biogeographic patterns. *Oecologia* **75**: 132–40.

Rabinowitz, D. (1981). Seven forms of rarity. In *The biological aspects of rare plant conservation*, ed. H. Synge, pp. 205–17. John Wiley, New York.

Rabinowitz, D., Cairns, S. & Dillon, T. (1986). Seven forms of rarity and their frequency in the flora of the British Isles. In *Conservation biology: the science of scarcity and diversity*, ed. M. E. Soulé, pp. 182–204. Sinauer Associates, Sunderland, Mass.

Randhawa, N. S. (1988). Concept of life support species for emergency and extreme environmental conditions. In *Life support plant species. Diversity and conservation*, ed. R. S. Paroda, P. Kapoor, R. K. Arora & B. Mal, pp. 1–8. National Bureau of Plant Genetic Resources, New Delhi.

Ranney, J. W., Bruner, M. C. & Levenson, J. B. (1981). The importance of edge in the structure and dynamics of forest islands. In *Forest island dynamics in man-dominated landscapes*, ed. R. L. Burgess & D. M. Sharpe, pp. 67–95. Springer-Verlag, New York.

Raven, P. H. (1976). Ethics and attitudes. In *Conservation of threatened plants*, ed. J. B. Simmons *et al.*, pp. 155–79. Plenum Press, New York.

Raven, P. H. (1987). The scope of the plant conservation problem world-wide. In *Botanic gardens and the world conservation strategy*, ed. D. Bramwell, O. Hamann, V. Heywood & H. Synge, pp. 19–29. Academic Press, London.

Raybould, A. F. & Gray, A. J. (1993). Genetically modified crops and hybridization with wild relatives: a UK perspective. *Journal of Applied Ecology* **30**: 199–219.

Read, D. J. (1991). Mycorrhizas in ecosystems – nature's response to the 'law of the minimum'. In *Frontiers in mycology*, ed. D. L. Hawksworth, pp. 101–30. CAB International, London.

Readshaw, J. L. & Mazanec, Z. (1969). Use of growth rings to determine past phasmatid defoliation of Alpine Ash forests. *Australian Forestry* **33**: 29–36.

Richardson, D. M., Macdonald, I. A. W., Holmes, P. M. & Cowling, R. M. (1992). Plant and animal invasions. In *The ecology of fynbos: nutrient, fire and diversity*, ed. R. M. Cowling, pp. 271–308. Oxford University Press, Cape Town.

Richardson, S. D. (1970). Gene pools in forestry. In *Genetic resources in plants – their exploration and conservation*, ed. O. H. Frankel & E. Bennett, *IBP Handbook No. 11*, pp. 353–65. Blackwell Scientific Publications, Oxford.

Rick, C. M. (1973). Potential genetic resources in tomato species: clues from observations in native habitats. In *Genes, enzymes, and populations*, ed. A. M. Srb, pp. 255–69. Plenum Press, New York.

Rick, C. M. (1982). The potential of exotic germplasm for tomato improvement. In *Plant improvement and somatic cell genetics*, ed. I. K. Vasil, W. R. Scowcroft & K. J. Frey, pp. 1–28. Academic Press, New York.

Rick, C. M., Fobes, J. F. & Holle, M. (1977). Genetic variation in *Lycopersicon pimpinellifolium*: evidence of evolutionary change in mating systems. *Plant Systematics and Evolution* **127**: 139–70.

Rieseberg, L. H. (1991). Hybridization in rare plants: insights from case studies in *Cercocarpus* and *Helianthus*. In *Genetics and conservation of rare plants*, ed. D. A. Falk & K. E. Holsinger, pp. 171–81. Oxford University Press, New York.

Riley, R., Chapman, V. & Johnson, R. (1968). Introduction of yellow rust resistance of *Aegilops comosa* into wheat by genetically induced homoeologous recombination. *Nature* **217**: 383–4.

Ritland, K. (1989). Gene identity and the genetic demography of plant populations. In *Plant population genetics, breeding, and genetic resources*, ed. A. H. D. Brown, M. T. Clegg, A. L. Kahler & B. S. Weir, pp. 181–99. Sinauer Associates, Sunderland, Mass.

Roberts, E. H. (1975). Problems of long-term storage of seed and pollen for genetic resources conservation. In *Crop genetic resources for today and tomorrow*, ed. O. H. Frankel & J. G. Hawkes, *International Biological Programme 2*, pp. 269–95. Cambridge University Press, Cambridge.

Roberts, E. H. (1984). Monitoring seed viability in genebanks. In *Seed management techniques for genebanks*, ed. J. B. Dickie, S. Linington & J. T. Williams, pp. 268–77. International Board for Plant Genetic Resources, Rome.

Roberts, E. H. (1989). Seed storage for genetic conservation. *Plants Today* **2**: 12–17.

Roberts, E. H., Abdalla, F. H. & Owen, R. J. (1967). Nuclear damage and the ageing of seeds, with a model for seed survival curves. *Symposium of the Society for Experimental Biology* **21**: 65–100.

Roberts, E. H., King, M. W. & Ellis, R. H. (1984). Recalcitrant seeds: their recognition and storage. In *Crop genetic resources: conservation and evaluation*, ed. J. H. W. Holden & J. T. Williams, pp. 38–52. Allen and Unwin, London.

Roche, L. & Dourojeanni, M. J. (1992). Criteria for selection of conservation areas for forest genetic resources. In *Conservation biology: a training manual for biological diversity and genetic resources*, ed. P. Kapoor-Vijay & J. White, pp. 61–70. Commonwealth Secretariat, London.

Rochefort, L. & Woodward, F. I. (1992). Effects of climate change and a doubling of CO_2 on vegetational diversity. *Journal of Experimental Botany* **43**: 1169–80.

Room, P. M., Harley, K. L. S., Forno, I. W. & Sands, D. P. A. (1981). Successful biological control of the floating weed *Salvinia*. *Nature* **294**: 78–80.

Roos, E. E. (1989). Long term seed storage. In *The National Plant Germplasm System of the United States*, ed. J. Janick, *Plant Breeding Reviews* **7**: 129–58. Timber Press, Portland, Oregon.

Sakai, S., Doi, Y. & Nakayama, A. (1978). Changes in regenerative ability and carbohydrate reserve of tea root during a long-term storage for maintenance of useful germplasm. In *Long-term preservation of favourable germ plasm in arboreal crops*, ed. T. Akihama & K. Nakajima, pp. 71–9. Fruit Tree Research Station, M. A. F. Fujimoto, Japan.

Sakai, A. & Noshiro, M. (1975). Some factors contributing to the survival of crop seeds cooled to the temperature of liquid nitrogen. In *Crop genetic resources for today and tomorrow*, ed. O. H. Frankel & J. G. Hawkes, *International Biological Programme 2*, pp. 317–26. Cambridge University Press, Cambridge.

Salisbury, E. J. (1924). Changes in the Hertfordshire flora: a consideration of the influence of man. *Transactions of the Hertfordshire Natural History Society Field Club* **18**: 51–68.

Sampson, J. F., Hopper, S. D. & Coates, D. J. (1990). *Eucalyptus rhodantha*. Western Australian Management Program No. 3: Department of Conservation and Land Management, Perth.

Sarukhán, J. Martínez-Ramos, M. & Piñero, D. (1984) The analysis of demographic variability at the individual level and its populational consequences. In *Perspective in plant population ecology*, ed. R. Dirzo & J. Sarukhán, pp. 83–106. Sinauer Associates, Sunderland, Mass.

Saunders, D. A. & Hobbs, R. J. (1991). The role of corridors in conservation: what do we know and where do we go? In *Nature conservation 2: the role of corridors*, ed. D. A. Saunders & R. J. Hobbs, pp. 421–7. Surrey Beatty & Sons, Chipping Norton, New South Wales.

Saunders, D. A., Hobbs, R. J. & Margules, C. R. (1991). Biological consequences of ecosystem fragmentation: a review. *Conservation Biology* **5**: 18–32.

Sayed, H. I. (1985). Diversity of salt tolerance in a germplasm collection of wheat (*Triticum* spp.). *Theoretical and Applied Genetics* **69**: 651–7.

Schaal, B. A. (1980). Measurement of gene flow in *Lupinus texensis*. *Nature* **284**: 450–1.

Schaal, B. A., O'Kane, S. L. & Rogstad, S. H. (1991). DNA variation in plant populations. *Trends in Ecology and Evolution* **6**: 329–32.

Schneeberger, R. G. & Cullis, C. A. (1992). Intraspecific 5S-ribosomal DNA gene variation in flax, *Linum usitatissimum* (Linaceae). *Plant Systematics and Evolution* **183**: 265–80.

Schoen, D. J. & Brown, A. H. D. (1991). Intraspecific variation in population gene diversity and effective population size correlates with the mating system in plants. *Proceedings of the National Academy of Sciences USA* **88**: 4494–7.

Schoen, D. J. & Stewart, S. C. (1987). Variation in male fertilities and pairwise mating probabilities in *Picea glauca*. *Genetics* **116**: 141–52.

Schonewald-Cox, C. M (1983). Conclusions: Guidelines to management: a beginning attempt. In *Genetics and conservation: a reference for managing wild animal and plant populations*, ed. C. M. Schonewald-Cox, S. M. Chambers, B. MacBryde & L. Thomas, pp. 414–45. Benjamin/Cummings Publishing Company, Menlo Park, Calif.

Schonewald-Cox, C. M. & Bayless, J. W. (1986). The boundary model: a geographical analysis of design and conservation of nature reserves. *Biological Conservation* **38**: 305–22.

Schonewald-Cox, C. M., Chambers, S. M., MacBryde, B. & Thomas, W. L. (ed.) (1983). *Genetics and conservation. A reference for managing wild animal and plant populations*. The Benjamin/Cummings Publishing Company, Menlo Park, Calif.

Schultes, R. E. & Raffauf, R. F. (1990). *The healing forest: medicinal and toxic plants of the Northwest Amazonia*. Dioscorides Press, Portland, Oregon.

Schulze, E.-D. & Gerstberger, P. (1993). Functional aspects of landscape diversity: a Bavarian example. In *Biodiversity and ecosystem function*, ed. E.-D. Schulze & H. A. Mooney,

pp. 453–66. Springer-Verlag, Berlin.
Scowcroft, W. R. (1984). *Genetic variability in tissue culture: impact on germplasm conservation and utilization*. IBPGR, Rome.
Sears, E. R. (1956). The transfer of leaf-rust resistance from *Aegilops umbellulata* to wheat. *Brookhaven Symposia in Biology* **9**: 1–21.
Sécond, G. (1982). Origin of the genic diversity of cultivated rice (*Oryza* spp.): study of the polymorphism scored at 40 isozyme loci. *Japanese Journal of Genetics* **57**: 25–67.
Seetharaman, R., Sharma, S. D. & Shastry, S. V. S. (1972). Germplasm conservation and use in India. In *Rice breeding*, pp. 187–200. International Rice Research Institute, Los Baños, Philippines.
Seidler, R. J. & Levin, M. (1994). Potential ecological and nontarget effects of transgenic plant gene products on agriculture, silviculture, and natural ecosystems: a general introduction. *Molecular Ecology* **3**: 1–3.
Seiler, G. J. (1992). Utilization of wild sunflower species for the improvement of cultivated sunflower. *Field Crops Research* **30**: 195–230.
Shaffer, M. L. (1981). Minimum population sizes for species conservation. *Bioscience* **31**: 131–4.
Shaffer, M. L. (1987). Minimum viable populations: coping with uncertainty. In *Viable populations for conservation*, ed. M. E. Soulé, pp. 59–68. Cambridge University Press, Cambridge.
Shankar, V. (1988). Life support species in the Indian Thar desert. In *Life support plant species. Diversity and conservation*, ed. R. S. Paroda, P. Kapoor, R. K. Arora & B. Mal, pp. 37–46. National Bureau of Plant Genetic Resources, New Delhi.
Sheail, J. (1982). Wild plants and the perception of land-use change in Britain: an historical perspective. *Biological Conservation* **24**: 129–46.
Shearer, B. L. & Tippett, J. T. (1989). *Jarrah dieback: the dynamics and management of* Phytophthora cinnamomi *in the jarrah (*Eucalyptus marginata*) forest of south-western Australia*. (Research Bulletin No. 3) CALM, Como, Western Australia.
Simberloff, D. (1988). The contribution of population and community biology to conservation science. *Annual Review of Ecology and Systematics* **19**: 473–511.
Simberloff, D., Farr, J. A., Cox, J. & Mehlman, D. W. (1992). Movement corridors: conservation bargains or poor investments. *Conservation Biology* **6**: 493–504.
Simmonds, N. W. (1962). Variability in crop plants, its use and conservation. *Biological Reviews* **37**: 442–65.
Simmonds, N. W. (1966). Studies of the tetraploid potatoes. III. Progress in the experimental re-creation of the Tuberosum Group. *Journal of the Linnean Society (Botany)* **59**: 279–88.
Simmonds, N. W. (1976). Sugarcanes *Saccharum* (Gramineae – Andropogoneae). In *Evolution of crop plants*, ed. N. W. Simmonds, pp. 104–8. Longman, London.
Simmons, J. B., Beyer, R. I., Brandham, P. E., Lucas, G. Ll. & Parry, V. T. H. (ed.) (1976). *Conservation of threatened plants*. Plenum Press, New York.
Simpson, M. J. A. & Withers, L. A. (1986). *Characterization of plant genetic resources using isozyme electrophoresis: a guide to the literature*. IBPGR, Rome.
Singh, Bhag (1974). Current maize cultivars of north eastern Himalayan region. *SABRAO Journal* **6**: 229–35.
Singh, Bhag (1977). Evaluation and use of Indian primitive cultivars in the improvement of maize. In *3rd International Congress of the Society for the Advancement of Breeding Research in Asia and Oceania* (SABRAO) pp.1(a) 9–12. SABRAO, Canberra.
Singh, R. & Axtell, J. D. (1973). High lysine mutant gene (*hl*) that improves protein quality and biological value of grain sorghum. *Crop Science* **13**: 535–9.
Slatkin, M. (1987). Gene flow and the geographic structure of natural populations. *Science* **236**: 787–92.

Slatkin, M. & Barton, N. H. (1989). A comparison of three indirect methods for estimating average levels of gene flow. *Evolution* **43**: 1349–68.

Sloten, D. H. van (1990). IBPGR and the challenges of the 1990s: a personal point of view. *Diversity* **6**(2): 36–9.

Smith, P. G. R. & Theberge, J. B. (1986). A review of criteria for evaluating natural areas. *Environmental Management* **10**: 715–34.

Smith, R. W. (1988). The place of life support species in hostile or risk-prone environments: an overview. In *Life support plant species. Diversity and conservation*, ed. R. S. Paroda, P. Kapoor, R. K. Arora & B. Mal, pp. 14–19. National Bureau of Plant Genetic Resources, New Delhi.

Soltis, D. E., Soltis, P. S. & Milligan, B. G. (1992). Intraspecific chloroplast DNA variation: systematic and phylogenetic implications. In *Molecular systematics of plants*, ed. P. S. Soltis, D. E. Soltis & J. J. Doyle, pp. 117–50. Sinauer Associates, Sunderland, Mass.

Sossou, J., Karunaratne, S. & Kovoor, A. (1987). Collecting palm: *in vitro* explanting in the field. *Plant Genetic Resources Newsletter* **69**: 7–18.

Soulé, M. E. (1980). Thresholds for survival: maintaining fitness and evolutionary potential. In *Conservation biology: an evolutionary–ecological perspective*, ed. M. E. Soulé & B. A. Wilcox, pp 119–33. Sinauer Associates, Sunderland, Mass.

Soulé, M. E. (ed.) (1987a). *Viable populations for conservation*. Cambridge University Press, Cambridge.

Soulé, M. E. (1987b). Introduction. In *Viable populations for conservation*, ed. M. E. Soulé, pp. 1–10. Cambridge University Press, Cambridge.

Soulé, M. E. (1987c). Where do we go from here? In *Viable populations for conservation*, ed. M. E. Soulé, pp. 175–83. Cambridge University Press, Cambridge.

Soulé, M. E. & Gilpin, M. E. (1991). The theory of wildlife corridor capability. In *Nature conservation 2: the role of corridors*, ed. D. A. Saunders & R. J. Hobbs, pp. 3–8. Surrey Beatty & Sons, Chipping Norton, New South Wales.

Soulé, M. E. & Simberloff, D. S. (1986). What do genetics and ecology tell us about the design of nature reserves? *Biological Conservation* **35**: 19–40.

Soulé, M. E., Wilcox, B. A. (ed.). (1980). *Conservation biology: an evolutionary–ecological perspective*. Sinauer Associates, Sunderland, Mass.

Soulé, M. E., Wilcox, B. A. & Holtby, C. (1979). Benign neglect: a model of faunal collapse in the game reserves of East Africa. *Biological Conservation* **15**: 259–72.

Sousa, W. P. (1984). The role of disturbance in natural communities. *Annual Review of Ecology and Systematics* **15**: 353–91.

Southwood, T. R. E., Brown, V. K. & Reader, P. M. (1979). The relationships of plant and insect diversities in succession. *Biological Journal of the Linnean Society of London* **12**: 327–48.

Spellerberg, I. F. (1991). Biogeographical basis for conservation. In *The scientific management of temperate communities for conservation*, ed. I. F. Spellerberg, F. B. Goldsmith & M. G. Morris, pp. 293–322. Blackwell Scientific Publications, Oxford.

Srivastava, J. P. & Damania, A. B. (1989). Use of collections in cereal improvement in semi-arid areas. In *The use of plant genetic resources*, ed. A. H. D. Brown, O. H. Frankel, D. R. Marshall & J. T. Williams, pp. 88–104. Cambridge University Press, Cambridge.

Stalker, H. T. (1980). Utilization of wild species for crop improvement. *Advances in Agronomy* **33**: 111–47.

Stephenson, S. L. (1986). Changes in a former chestnut-dominated forest after a half century of succession. *American Midland Naturalist* **116**: 173–9.

Stocker, G. C. & Mott, J. J. (1981). Fire in the tropical forests and woodlands of northern Australia. In *Fire and the Australian biota*, ed. A. M. Gill, R. H. Groves & I. R. Noble, pp. 427–39. Australian Academy of Science, Canberra.

Stolen, O. (1965). Investigations on the tolerance of barley varieties to high hydrogen-ion concentration in soil. *Royal Veterinary and Agricultural College Yearbook 1965*, Copenhagen.

Strauss, S. H., Hong, Y.-P. & Hipkins, V. D. (1993). High levels of population differentiation for mitochondrial DNA haplotypes in *Pinus radiata, muricata*, and *attenuata*. *Theoretical and Applied Genetics* **86**: 605–11.

Stushnoff, C. & Fear, C. (1985). *The potential use of* in vitro *storage for temperate fruit germplasm*. IBPGR, Rome.

Subrahmanyam, P., Rao, V. R., McDonald, D., Moss, J. P. & Gibbons, R. W. (1989). Origin of resistances to rust and late leaf spot in peanut (*Arachis hypogaea*, Fabaceae). *Economic Botany* **43**: 444–55.

Sytsma, K. J. & Schaal, B. A. (1990). Ribosomal DNA variation within and among individuals of *Lisianthius* (Gentianaceae) populations. *Plant Systematics and Evolution* **170**: 97–106.

Tansley, A. G. & Adamson, R. S. (1925). Studies of the vegetation of the English chalk. III. The chalk grasslands of the Hampshire–Sussex border. *Journal of Ecology* **13**: 177–223.

Teramura, A. H., Ziska, L. H. & Sztein, A. E. (1991). Changes in growth and photosynthetic capacity of rice with increased UV-B radiation. *Physiologia Plantarum* **83**: 373–80.

Terborgh, J. (1975). Island biogeography and conservation: strategy and limitations. *Science* **193**: 1029–30.

Terborgh, J. (1986). Keystone plant resources in the tropical forest. In *Conservation biology: the science of scarcity and diversity*, ed. M. E. Soulé, pp. 330–44. Sinauer Associates, Sunderland, Mass.

Thibodeau, F. R. & Falk, D. A. (1987). Building a national *ex-situ* conservation network – the U. S. Center for Plant Conservation. In *Botanic gardens and the world conservation strategy*, ed. D. Bramwell, O. Hamann, V. Heywood & H. Synge, pp. 285–94. Academic Press, London.

Thomas, L. K. (1980). The impact of three exotic plant species on a Potomac Island. *National Park Service Scientific Monograph Series* **13**: 1–179.

Thompson, J. N. & Burdon, J. J. (1992). Gene-for-gene coevolution between plants and parasites. *Nature* **360**: 121–5.

Thompson, P. A. (1975). The collection, maintenance and environmental importance of the genetic resources of wild plants. *Environmental Conservation* **2**: 223–8.

Thompson, P. A. (1976). Factors involved in the selection of plant resources for conservation as seed in gene banks. *Biological Conservation* **10**: 159–67.

Thompson, P. A. & Brown, G. E. (1972). The seed unit at the Royal Botanic Gardens, Kew. *Kew Bulletin* **26**: 445–56.

Tilman, D. (1993). Community diversity and succession: the roles of competition, dispersal, and habitat modification. In *Biodiversity and ecosystem function*, ed. E.-D. Schulze & H. A. Mooney, pp. 327–44. Springer-Verlag, Berlin.

Tilman, D. (1994). Competition and biodiversity in spatially structured habitats. *Ecology* **75**: 2–16.

Tilman, D. & Downing, J. A. (1994). Biodiversity and stability in grasslands. *Nature* **367**: 363–5.

Timmins, S. M. & Williams, P. A. (1991). Weed numbers in New Zealand's forest and scrub reserves. *New Zealand Journal of Ecology* **15**: 153–62.

Towill, L. E. (1985). Low temperature and freeze-/vacuum-drying preservation of pollen. In *Cryopreservation of plant cells and organs*, ed. K. K. Kartha, pp. 171–97. CRC Press, Boca Raton, Florida.

Towill, L. E. (1989). Biotechnology and germplasm preservation. In *The National Plant Germplasm System of the United States*, ed. J. Janick, *Plant Breeding Reviews* **7**: 159–82. Timber Press, Portland, Oregon.

Turesson, G. (1922).The genotypical response of the plant species to the habitat. *Hereditas* **3**: 211–350.

Turkensteen, L. J. (1993). Durable resistance of potatoes against *Phytophthora infestans*. In *Durability of disease resistance*, ed. Th. Jacobs & J. E. Parlevliet, pp. 115–24. Kluwer Academic Publishers, Dordrecht.

Usher, M. B. (1988). Biological invasions of nature reserves: a search for generalisations. *Biological Conservation* **44**: 119–35.

Usher, M. B. & Edwards, M. (1986). The selection of conservation areas using the arthropod fauna of Antartic islands. *Environmental Conservation* **13**: 115–22.

Van der Plank, J. E. (1975). *Principles of plant infection*. Academic Press, New York.

van der Putten, W. H. & Troelstra, S. R. (1990). Harmful soil organisms in coastal foredunes involved in degeneration of *Ammophila arenaria* and *Calammophila baltica*. *Canadian Journal of Botany* **68**: 1560–8.

van der Putten, W. H., van Dijk, C. & Troelstra, S. R. (1988). Biotic soil factors affecting the growth and development of *Ammophila arenaria*. *Oecologia* **76**: 313–20.

Van Treuren, R., Bijlsma, R., Van Delden, W. & Ouborg, N. J. (1991). The significance of genetic erosion in the process of extinction. I. Genetic differentiation in *Salvia pratensis* and *Scabiosa columbaria* in relation to population size. *Heredity* **66**: 181–9.

Vane-Wright, R. I., Humphries, C. J. & Williams, P. H. (1991). What to protect? Systematics and the agony of choice. *Biological Conservation* **55**: 235–53.

Vavilov, N. I. (1926). Studies on the origin of cultivated plants. *Bulletin of Applied Botany, Genetics and Plant Breeding* **16**: 1–248.

Vavilov, N. I. (1949–50). The origin, variation, immunity, and breeding of cultivated plants. *Chronica Botanica* **13**: 1–366.

Vavrek, M. C., McGraw, J. B. & Bennington, C. C. (1991). Ecological genetic variation in seed banks. III. Phenotypic and genetic differences between young and old seed populations of *Carex bigelowii*. *Journal of Ecology* **79**: 645–62.

Vergara, B. S. (1988). Variability in rice and varietal developments for temperature-limiting tropical areas. In *Life support plant species. Diversity and conservation*, ed. R. S. Paroda, P. Kapoor, R. K. Arora & B. Mal, pp. 51–4. National Bureau of Plant Genetic Resources, New Delhi.

Vitousek, P. M. (1990). Biological invasions and ecosystem processes: towards an integration of population biology and ecosystem studies. *Oikos* **57**: 7–13.

Vitousek, P. M., Walker, L. R., Whiteaker, L. D., Mueller-Dombois, D. & Matson, P. A. (1987). Biological invasion by *Myrica faya* alters ecosystem development in Hawaii. *Science* **238**: 802–4.

Vogel, O. A., Allan, R. E. & Peterson, C. J. (1963). Plant and performance characteristics of semidwarf winter wheats producing most efficiently in Eastern Washington. *Agronomy Journal* **55**: 397–8.

Vuylsteke, D. R. (1989). *Shoot-tip culture for the propagation, conservation and exchange of* Musa *germplasm*. IBPGR, Rome.

Wade, M. J. & McCauley, D. E. (1988). Extinction and recolonization: their effects on genetic differentiation on local populations. *Evolution* **42**: 995–1005.

Warcup, J. H. (1981). Effect of fire on the soil microflora and other non-vascular plants. In *Fire and the Australian biota*, ed. A. M. Gill, R. H. Groves & I. R. Noble, pp. 203–14. Australian Academy of Science, Canberra.

Wark, D. C. (1963). *Nicotiana* species as sources of resistance to blue mould (*Peronospora tabacina* Adam) for cultivated tobacco. In *Proceedings of the 3rd World Tobacco Scientific Congress*, pp. 252–9. Salisbury.

Wark, D. C. (1970). Development of flue-cured tobacco cultivars resistant to a common strain of blue mould. *Tobacco Science* **171**: 147–50.

Waterhouse, D. F. (1994). *Biological control of weeds: Southeast Asian prospects.* Australian Centre for International Agriculture Research, Canberra.

Watson, I. A. (1970a). The utilization of wild species in the breeding of cultivated crops resistant to plant pathogens. In *Plant genetic resources – their exploration and conservation*, ed. O. H. Frankel & E. Bennett, *IBP Handbook No. 11*, pp. 441–57. Blackwell Scientific Publications, Oxford.

Watson, I. A. (1970b). Changes in virulence and population shifts in plant pathogens. *Annual Review of Phytopathology* **8**: 209–30.

Watt, A. S. (1957). The effect of excluding rabbits from grassland (Mesobrometum) in Breckland. *Journal of Ecology* **45**: 861–78.

Weir, B. S. (1990). *Genetic data analysis.* Sinauer Associates, Sunderland, Mass.

Weir, B. S. & Cockerham, C. C. (1973). Mixed self and random mating at two loci. *Genetical Research, Cambridge* **21**: 247–62.

Weir, B. S. & Cockerham, C. C. (1989). Complete characterization of disequilibrium at two loci. In *Mathematical evolutionary theory*, ed. M. E. Feldman, pp. 86–110. Princeton University Press, Princeton, NJ.

Wells, M. J., Engelbrecht, V. M., Balsinhas, A. A. & Stirton, C. H. (1983). Weed flora of South Africa 3: more power shifts in the veld. *Bothalia* **14**: 967–70.

Weltzien, E. & Fischbeck, G. (1990). Performance and variability of local barley landraces in Near-Eastern environments. *Plant Breeding* **104**: 58–67.

Weste, G. (1981) Changes in vegetation of sclerophyll shrubby woodland associated with invasion by *Phytophthora cinnamomi. Australian Journal of Botany* **29**: 261–76.

White, G. A., Willingham, B. C., Skrdla, W. H., Massey, J. H., Higgins, J. J., Calhoun W., Davis A. M., Dolan D. D. & Earle, F. R. (1971). Agronomic evaluation of prospective new crop species. *Economic Botany* **25**: 22–43.

White, J. C. (1989). Ethnoecological observations on wild and cultivated rice and yams in northeastern Thailand. In *Foraging and farming: the evolution of plant exploitation*, ed. D. R. Harris, & G. C. Hillman, pp. 152–8. Unwin Hyman, London.

Whyte, R. O. & Julén, G. (1963). *Proceedings of a Technical Meeting on Plant Exploration and Introduction. Genetica Agraria* **17**: 573 pp.

Wilcox, B. A. & Murphy, D. D. (1985). Conservation strategy: the effects of fragmentation on extinction. *American Naturalist* **125**: 879–87.

Wilkes, G. (1989). Maize: domestication, racial evolution, and spread. In *Foraging and farming: the evolution of plant exploitation*, ed. D. R. Harris, & G. C. Hillman, pp. 440–55. Unwin Hyman, London.

Williams, R. J., Reid, R., Schultze-Kraft, R., Sousa Costa, N. M. & Thomas B. D. (1984). Natural distribution of *Stylosanthes*. In *The biology and agronomy of* Stylosanthes, ed. H. M. Stace & L. A. Edye, pp. 73–101. Academic Press, Sydney.

Wills, R. T. (1993). The ecological impact of *Phytophthora cinnamomi* in the Stirling Range National Park, Western Australia. *Australian Journal of Ecology* **18**: 145–59.

Wilson, E. O. (ed.) (1988). *Biodiversity*. National Academy Press, Washington DC.

Wilson, E. O. (1992). *The diversity of life*. Allan Lane, Penguin Press, London.

Withers, L. A. (1987). Long-term preservation of plant cells, tissues and organs. *Oxford Surveys of Plant Molecular and Cell Biology* **4**: 221–72.

Withers, L. A. (1989). *In vitro* conservation and germplasm utilisation. In *The use of plant genetic resources*, ed. A. H. D. Brown, O. H. Frankel, D. R. Marshall & J. T. Williams, pp. 309–34. Cambridge University Press, Cambridge.

Withers, L. A. (1990). Tissue culture in the conservation of plant genetic resources. In *Conservation of plant genetic resources through* in vitro *methods*, ed. A. H. Zakri, M. N. Normah, A. G. Abdul Karim & M. T. Senawi, pp. 1–18. Forest Research Institute, Malaysia/National Committee on Plant Genetic Resources, Kuala Lumpur.

Withers, L. A., Wheelans, S. K. & Williams, J. T. (1990). *In vitro* conservation of crop germplasm and the IBPGR databases. *Euphytica* **45**: 9–22.

Wolfe, M. S. (1985). The current status and prospects of multiline cultivars and variety mixtures for disease resistance. *Annual Review of Phytopathology* **23**: 251–73.

Wolff, K. (1988). Genetic analysis of ecological relevant morphological variability in *Plantago lanceolata* L. 3. Natural selection in an F_2 population. *Theoretical and Applied Genetics* **75**: 772–8.

Wolff, K., Rogstad, S. H. & Schaal, B. A. (1994). Population and species variation of minisatellite DNA in *Plantago*. *Theoretical and Applied Genetics* **87**: 733–40.

Woods, F. W. & Shanks, R. E. (1959). Natural replacement of chestnut by other species in the Great Smoky Mountains National Park. *Ecology* **40**: 349–61.

Woodward, F. I. (1987). *Climate and plant distribution*. Cambridge University Press, Cambridge.

Worthington, E. B. (ed.) (1975). *The evolution of IBP. International Biological Programme 1*. Cambridge University Press, Cambridge.

Wright, S. (1946). Isolation by distance under diverse systems of mating. *Genetics* **31**: 39–59.

Wright, S. J. & Hubbell, S. P. (1983). Stochastic extinction and reserve size: a focal species approach. *Oikos* **41**: 466–76.

Wu, K.-S. & Tanksley, S. D. (1993). Abundance, polymorphism and genetic mapping of microsatellites in rice. *Molecular and General Genetics* **241**: 225–35.

WWF/IUCN/BGCS (1989). *The botanic gardens conservation strategy*. WWF and IUCN, Gland, Switzerland.

Yen, D. E. (1974). *The sweet potato and Oceania*. Bishop Museum Press, Honolulu, Hawaii.

Yen, D. E. (1985). Wild plants and domestication in Pacific islands. In *Recent advances in Indo-Pacific prehistory*, ed. V. N. Misra & Peter Bellwood, pp. 315–16. Oxford and IBH Publishing Co., New Delhi.

Zakri, A. H. & Ghani, F. D. (1988). Identification and characterisation of life support species in Malaysia. In *Life support plant species. Diversity and conservation*, ed. R. S. Paroda, P. Kapoor, R. K. Arora & B. Mal, pp. 108–15. National Bureau of Plant Genetic Resources, New Delhi.

Zeven, A. C. & de Wet, J. M. J. (1982). *Dictionary of cultivated plants and their regions of diversity: excluding most ornamentals, forest trees and lower plants*. 2nd edn. Pudoc, Wageningen.

Zhang, Q., Saghai Maroof, M. A. & Kleinhofs, A. (1993). Comparative diversity analysis of RFLPs and isozymes within and among populations of *Hordum vulgare* ssp. *spontaneum*. *Genetics* **134**: 909–16.

Zohary, D. (1989). Domestication of the Southwest Asian Neolitic crop assemblage of cereals, pulses, and flax: the evidence from the living plants. In *Foraging and farming: the evolution of plant exploitation*, ed. D. R. Harris, & G. C. Hillman, pp. 358–72. Unwin Hyman, London.

Zohary, D. & Feldman, M. (1962). Hybridization between amphidiploids and the evolution of polyploids in the wheat (*Aegilops–Triticum*) group. *Evolution* **16**: 44–61.

Zohary, D., Harlan, J. R. & Vardi, J. (1969). The wild diploid progenitors of wheat and their breeding value. *Euphytica* **18**: 58–65.

Zohary, D. & Hopf, M. (1988). *Domestication of plants in the old world*. Clarendon Press, Oxford.

Zohary, D. & Spiegel-Roy, P. (1975). Beginnings of fruit growing in the Old World. *Science* **187**: 319–27.

Glossary

BGCS	Botanic Gardens Conservation Secretariat
CGIAR	Consultative Group on International Agricultural Research
CIAT	International Center of Tropical Agriculture
CIMMYT	International Center for Maize and Wheat Improvement
CSIRO	Commonwealth Scientific and Industrial Research Organization
FAO	Food and Agriculture Organization
GRC	Genetic Resource Centres
GRMU	Genetic Resource Management Unit
HYV	High Yielding Varieties
IARC	International Agricultural Research Centres
IARI	Indian Agricultural Research Institute
IBP	International Biological Program
IBPGR	International Board for Plant Genetic Resources
ICARDA	International Center for Agricultural Research in the Dry Areas
ICM	Integrated Conservation Model
ICRISAT	International Crops Research Institute for the Semi-Arid Tropics
ICUC	International Centre for Underutilized Crops
IPGRI	International Plant Genetic Resources Institute
IRRI	International Rice Research Institute
IUCN	International Union for the Conservation of Nature and Natural Resources
MAB	Man and the Biosphere
MGM	Major Gene Mutation
MVP	Minimum Viable Population

NRC	National Research Council
NSSL	National Seed Storage Laboratory
ORSTOM	Institut Français de Recherche Scientifique pour la Developpement en Cooperation
PCR	Polymerase Chain Reaction
PVA	Population Viability Analysis
RAPD	Randomly Amplified Polymorphic DNA
RFLP	Restriction Fragment Length Polymorphism
SLOSS	Single Large Or Several Small [reserves]
SPNR	Society for the Promotion of Nature Reserves
TPC	Threatened Plants Committee
TPU	Threatened Plants Unit
UNESCO	United Nations Educational, Scientific and Cultural Organization
USDA	United States Department of Agriculture
WCMC	World Conservation Monitoring Centre
WHO	World Health Organization
WWF	World Wide Fund for Nature

Index

The emphasis in this book is on biological principles rather than particular plant species. Consistent with this emphasis and in the interest of simplicity and utility, the names of plants, scientific and/or common, are in the main excluded from this index.